Topics in Calculus

To Carol and Linda

Preface

This text was especially designed for a one-semester course in calculus for nonscience majors. If it can be said to have a philosophy or strategy, it is the following: First, we try to quickly give the student the tools and techniques he needs to solve interesting problems. Second, we do not worry about rigor, but rather attempt to develop the student's understanding of the concepts and ideas of calculus from a geometric or intuitive point of view. The text is self-contained, except for a few facts from high school geometry and algebra. Starred sections can (and probably should) be omitted in a one-semester course. Starred problems are more difficult than their unstarred brethren, but ample hints are supplied with many. Some proofs and material of unusual technical difficulty appear in appendices that are placed at the end of each chapter. Many sections contain more material than we would attempt to cover in a one-semester course. However, the student can profitably read over parts of the remainder on his own. Chapter 1 is mainly an introduction. Chapter 2 discusses limits from an intuitive standpoint. Chapters 3, 4, and 5 (the derivative, max–min, graphing, the integral, area) are the core of the book.

The remaining Chapters 6–9 are independent and can be covered in any order. Chapter 6 and 7 contain miscellaneous topics in differential and integral calculus that can be selected according to taste. Chapter 8 covers first order differential equations and some applications to the social and biological sciences. Many people may prefer to skip Chapter 6 and 7 to get to this more substantial fare. Chapter 9 on multivariant calculus represents a topic that is inherently more difficult than the preceding material. Most instructors will find it is not possible to include this in a one-semester course. At Indiana University, the authors have been able to cover Chapters 1–5 and Chapter 8, omitting all starred sections, in one semester. The entire text could easily be covered in a two-semester course.

Rather than burden the student with too many formulas, we have relegated the trigonometric functions to the starred sections. The functions still at our disposal; polynomials, logs, exponentials, etc., are sufficient to attack a wide and interesting range of problems.

We are grateful to George Springer for suggesting we write such a book, and in aiding us in many ways after we had started. Our colleagues have been a source of valuable suggestions and criticism, and we single out John Chadam, Peter Fillmore, and Glenn Schober in particular. We are indebted to Mrs. Jeanne Baird for her valuable assistance in preparing the manuscript for publication. It is impossible to list all those who typed a portion of the text. However, special recognition is accorded Mary Lynn Cook, who retyped the entire manuscript. Several classes at Indiana had to live with the preliminary (or phone book) edition, and did so cheerfully, for which we salute them. We also acknowledge thanks to the editors of Xerox College Publishing for their help and cooperation.

Finally, we wish to thank our families for their patience, understanding, and encouragement in this venture.

M.L.

J.G.S.

Bloomington, Indiana
December, 1969

Skeleton Syllabus

Chapter	Sections
1	1.1–1.6
2	2.1–2.2 (read 2.3–2.5)
3	3.1–3.7
4	4.1–4.4, 4.6 (read appendix)
5	5.1–5.4, 5.6

6.1, 6.3, 6.4   7.2, 7.3   8.1–8.4

This is the minimum backbone for a one-semester course. There is time to add additional topics.

Contents

Topics in Calculus

1

Introduction

1.1 Real Numbers

We shall begin our study of calculus with a discussion of real numbers. The real numbers are basic to many branches of mathematics. In particular, they are of primary importance to our further study of calculus. The answer to the question "What is a real number?" is not easy. It would be nice if we could simply state a one-sentence definition and let it go at that. However, it is not possible. We will assume that you already know something about numbers—namely addition, subtraction, multiplication, and division (by numbers other than 0).

The numbers most familiar to you, namely, 1, 2, 3, ... are called *positive integers*. If we add or multiply two positive integers, we again obtain a positive integer. What happens if we subtract two positive integers? First, observe that if we subtract 3 from 5, we again obtain a positive integer 2. However, if we subtract 5 from 3 (this is certainly possible!) then we do not obtain a positive integer. Clearly if you owe the bookstore $5 and only pay $3, you still owe them $2. Thus, we see the need for considering numbers of the form $-1, -2, -3, \ldots$. These numbers are called *negative integers*. The collection of positive integers, negative integers, and the number 0 is called the set of integers. It should also be clear that we need to consider fractions as part of our number system. We define a *rational number* to be any number that can be expressed as the quotient of two integers, where it is understood that the integer in the denominator cannot be zero. Thus, $\frac{1}{3}, \frac{17}{2}, -\frac{2}{3}, \frac{4}{2}, \frac{6}{8}$ are all rational numbers. Note that the integers themselves form part of the rational numbers. (An integer can be considered as a rational number with the number 1 in the denominator.) Now, one might think that we have characterized all of the real numbers—namely the collection of all integers and rational numbers. This is

1

certainly not true. There are other numbers that should also belong to any collection of real numbers. Suppose we are given a right isosceles triangle whose legs are of unit length. (See Figure 1.1.1.) The problem is to determine the length of the hypoteneuse.

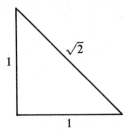

FIGURE 1.1.1

Recalling the Pythagorean theorem, it is easy to see that the length is given by the number $\sqrt{2}$. Such numbers as $\sqrt{2}, \sqrt{3}, \pi$, etc. must also have a place in our number system. These numbers cannot, however, be represented as the quotient of two integers. We call these numbers *irrational* numbers. The collection of all integers, rational numbers, and irrational numbers composes the *real number* system.

A geometrical interpretation of the number system may be provided by the following construction. On a straight line, called the "coordinate line," we mark off a segment 0 to 1 as in Figure 1.1.2. This establishes our unit of length. By the way, this unit of length

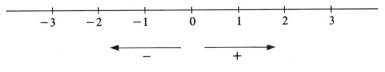

FIGURE 1.1.2

may be chosen at will. The positive and negative integers are then represented as a set of points equally spaced on the coordinate line, with positive integers to the right of 0 and negative integers to the left of 0. To represent a fraction with denominator n we simply divide each of the unit segments into n equal parts and represent the fractions with denominator n as the points of the subdivision. How do we put in the irrational numbers? This is accomplished by an approximation procedure. An irrational number can be represented by an infinite decimal expansion, that is, $\sqrt{2} = 1.41214\ldots$. Thus we can say that $\sqrt{2}$ must be somewhere between the numbers $1\frac{2}{5}$ and $1\frac{1}{2}$. In this way, we can determine approximately how each of the irrational numbers fits into our coordinate line. Sometimes, the coordinate line is referred to as the real line or the real axis. We shall assume that you know how to add, multiply, and divide real numbers. Remember that division by 0 is not permitted. Here is one reason. Suppose $a \neq 0$ and we assume that a can be divided by 0. This would mean that $a/0 = b$, where b is some number. However, all of us know that $a/0 = b$ means that $a = b \cdot 0$—but this would imply that $a = 0$, contrary to our statement $a \neq 0$. Observe that since $0 \cdot b = 0$ for all real numbers b, there is no number $1/0$ satisfying $0 \cdot 1/0 = 1$. One of the properties of real numbers we expect is that if a is any real number, then there exists a number $1/a$ such that $a \cdot (1/a) = 1$. Thus, any attempt to define $0/0$ will be meaningless.

Let us make one more remark concerning the coordinate axis. Observe that if you are given any real number you can determine a point on the line corresponding to this number. Conversely, given any point on the line we can determine a number to assign to this point. (This fact should be evident from our geometrical construction.) Thus we see that corresponding to every point on the line there exists one and only one real number we can assign to it and conversely, given any real number, there exists one and only one point on the line representing this number. This correspondence (that is, one number for one point and conversely) is called a *one-to-one* correspondence.

It should now be clear that we need some ordering properties for our number system. That is, we must know what we mean by saying that a number n is greater than or less than a number m. Geometrically, this simply means that any number to the right of a given number on the line is greater than the given number and any number to the left of the given number is less than that number. It would be of use to us to make these ideas more precise and to introduce some symbols to describe them.

Note, we already have the positive numbers placed to the right of 0 on our real axis. We may therefore say a number a is greater than 0 if and only if a is positive. In symbols, we write this as

$$a > 0,$$

where the symbol $>$ is read as "is greater than." We may now define precisely the term "a is greater than b," written $a > b$.

DEFINITION. *Let a and b be any two numbers. We say that $a > b$ if $a - b$ is positive, that is, $a - b > 0$.*

Thus, $5 > 2$ means that $5 - 2 > 0$, which is certainly obvious.

We can now define the concept "less than" also in terms of our new symbol $>$. A number n is said to be less than 0 or negative if $-n > 0$. Introducing a new symbol $<$, read as "less than," we say that $m < 0$ if $-m > 0$. Thus we say that $m < n$, where m and n are two numbers if $n - m = -(m - n) > 0$, or using our $<$ symbol, $m - n < 0$. All of these statements are equivalent. Thus $2 < 5$ means that $2 - 5 < 0$ or $-(2 - 5) > 0$. Note that $a < b$ if and only if $b > a$.

Let us now state some rules concerning the symbols $<$ and $>$. First, we observe that:

RULE 1.   *If $a > 0$ and $b > 0$ then $ab > 0$ and $a + b > 0$.*

RULE 2.   *If a is any real number then either $a > 0$ or $a = 0$ or $a < 0$ and these three possibilities are mutually exclusive.*

RULE 3.   *If $a < 0$ and $b < 0$ then $ab > 0$. (You recall that the product of two negative numbers is positive.)*

RULE 4.   *If $a < 0$ and $b > 0$ then $ab < 0$. (That is, the product of a positive number and a negative number is indeed negative.)*

Rules 1–4 are useful and should be remembered. Using Rules 1 and 2 it is possible to deduce Rules 3 and 4. If you are ambitious try this. Otherwise, for our purposes we will accept these rules as given. As an example, Rule 3 tells us that $(-2)\,(-5) > 0$—certainly $+10 > 0$. In like manner, Rule 4 tells us that $(-2)\,(5) < 0$—certainly $-10 < 0$ since $-(-10) > 0$! Try some examples to convince yourself that Rules 1–4 are plausible.

From Rules 1–4 we can establish many other results concerning $<$ and $>$. We state the most useful ones here. If a, b and c are any three real numbers then

RULE 5.   *If $a > b$, then $a + c > b + c$ and if $a > b$ and $c > 0$, then $ac > bc$.*

RULE 6.   *If $a > b$, then $(-a) < (-b)$. In particular if $c < 0$, then $ac < bc$.*

RULE 7.   *If $a > b > 0$, or $0 > a > b$, then $1/a < 1/b$.*

Example.   If $5 > 2$ then $-5 < -2$. Just look at the coordinate line to establish this. In like manner if $5 > 2$, then $(-2)\,5 < (-2)\,2$, which implies that $-10 < -4$! Rule 7 tells us that if $5 > 2$ then $\frac{1}{5} < \frac{1}{2}$.

Rules 1 through 7 are simply computational rules, which you will use over and over again.

A relation involving the symbols $<$ and $>$ is called an *inequality*. All of our discussions so far have involved inequalities. We may often write one symbol $m \geq n$ when we mean the number m is greater than or equal to the number n. In like manner we may use the symbol $\leq$ to designate less than or equal.

In order to define the concept of distance between points on the real axis we need the idea of the *absolute value* of a number.

DEFINITION.   *Let m be any real number. We define the absolute value of m, written $|m|$, as follows:*

$$|m| = \begin{cases} m & \text{if } m > 0 \\ 0 & \text{if } m = 0 \\ -m & \text{if } m < 0. \end{cases}$$

For example we observe that $|2| = 2$ while $|-2| = -(-2) = 2$, since $-2 < 0$. It is a simple matter to show that if n is any real number, then $|n| = |-n|$. If n is positive, then $|n| = n$ and $|-n| = -(-n) = n$. Hence, for $n > 0$, $|n| = |-n|$. Now, if $n < 0$, $|n| = -n$, while $|-n| = -n$ since $-n > 0$. Thus, for $n < 0$, $|n| = |-n|$. Clearly $|0| = |-0| = 0$. Hence, for all cases $|n| = |-n|$. It is clear from the definition that $|n| \geq 0$, and $|n| = 0$ if and only if $n = 0$. The proof given above illustrates something very useful when dealing with absolute values. You must always consider the separate cases $n \geq 0$, $n < 0$ when verifying statements concerning $|n|$. Another example might better illustrate this:

Problem.   For what real numbers x is $|x + 1| = 3$?

Solution.   We must solve the above equation for x. First, consider the case $x + 1 > 0$; then $|x + 1| = x + 1$ and $x + 1 = 3$ which implies $x = 2$. Next, consider the case $x + 1 < 0$, and $|x + 1| = -(x + 1)$; then $-(x + 1) = 3$ and hence $x + 1 = -3$, which yields $x = -4$. It is clear then that the above equation is satisfied for $x = -4$ and $x = 2$. (Note that $(x + 1) \neq 0$ otherwise; the equation could not be satisfied.)

In the next section we will see that $|x|$ simply represents the distance from the number x to the number 0 on the coordinate axis and $|x - a|$, where a is any real number, denotes the distance from x to the real number a. Thus, $|x - 3|$ is the distance from x to 3.

We will now state some properties of absolute values that will be useful for our further work. All of the results can be verified from the definition, and we leave it to the interested reader to do this.

PROPERTY 1. *If x is any real number then $|x|^2 = x^2$ and $\sqrt{x^2} = |x|$.*

PROPERTY 2. *If x, y are any real numbers then $|xy| = |x|\,|y|$.*

PROPERTY 3. *If x, y are any real numbers then $|x + y| \leq |x| + |y|$.*

Property 3 is called the *triangle inequality*. Let us illustrate the above properties with some examples.

Example 1. $|5|^2 = 5^2 = 25$, $\sqrt{5^2} = |5| = 5$; $\sqrt{(-4)^2} = |-4| = -(-4) = 4$.

Example 2. $|5(-2)| = |5|\,|-2| = (5)\,(-(-2)) = 5 \cdot 2 = 10$.

Example 3. Let $x = 3$, $y = -4$ then $|3 + (-4)| = |3 - 4| \leq |3| + |-4| = 3 + 4 = 7$. Since $|3 - 4| = |-1| = 1$, it is clear that in this case $|3 - 4| < 7$. If we let $x = 3$ and $y = 4$ then $|3 + 4| = |3| + |4| = 7$.

You should construct other numerical examples to help yourself justify Properties 1–3.

Let us pose the following question: Find all those real numbers x such that $|x| < 4$? In order to answer the question, we must first return to the definition of absolute value. First, if $x > 0$ we recall the $|x| = x$, and hence those numbers for which $x < 4$ and $x \geq 0$ certainly satisfy the inequality. Now we must see what happens if $x < 0$. In this case $|x| = -x$ and $-x < 4$ means from Rule 6 on inequalities that $x > -4$. Combining these two results we see that all those numbers x such that $x < 4$ *and* $x > -4$ satisfy $|x| < 4$. We write this in a combined form as follows: Those numbers x such that $-4 < x < 4$ satisfy the inequality $|x| < 4$. Let us see how this can be interpreted geometrically on the real axis. (See Figure 1.1.3.)

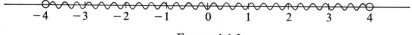

$$-4 \quad -3 \quad -2 \quad -1 \quad 0 \quad 1 \quad 2 \quad 3 \quad 4$$

FIGURE 1.1.3

You can easily convince yourself that all those numbers x represented on the line between -4 and 4 (the $\sim$ line) are those numbers for which $|x| < 4$. The $\circ$ at -4 and 4 indicates that these two numbers are *not* included. If the numbers -4 and 4 are meant to be included, the inequality must be of the form $|x| \leq 4$.

We can consider inequalities such as the above in a more general way. The verification of the results proceeds in much the same way as the preceding example. We then add the following properties of inequalities, which will be very useful for future work:

PROPERTY 4. *Let m be any given real number. The inequality $|x| \leq m$ holds if and only if $-m \leq x \leq m$. (Note: Here verification involves first showing that $|x| \leq m$ implies $-m \leq x \leq m$ and then showing the converse, $-m \leq x \leq m$ implies $|x| \leq m$.)*

Again, we leave the verification to the interested reader. Use exactly the same procedure indicated in the preceding example.

PROPERTY 5.   *Let m be any given real number. The inequality $|x| \geq m$ holds if and only if $x \leq -m$ or $x \geq m$. (Note: If $m \leq 0$, then $|x| \geq m$ for all real numbers x.)*

The Figure 1.1.4 indicates geometrically the collection of all those numbers x such that $|x| \leq m$ and $|x| \geq m$. The ⌇⌇⌇ region consists of all those numbers x where $|x| \geq m$.

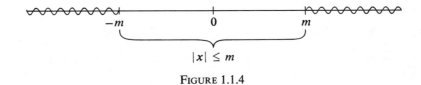

$$|x| \leq m$$

FIGURE 1.1.4

Before we proceed any further, let us solve some inequalities, that is, find some numbers x satisfying certain inequalities.

Example 1.   Find all those real numbers x such that $|x + 2| \leq 4$.

Solution.   Using Property 4, we see that the inequality $|x + 2| \leq 4$ is satisfied if and only if $-4 \leq x + 2 \leq 4$; that is, $-4 \leq x + 2$ and $x + 2 \leq 4$. From our Rule 5 on inequalities, it is clear that $x + 2 \leq 4$ implies $x \leq 2$ and $-4 \leq x + 2$ implies $-6 \leq x$. Therefore, those numbers x satisfying $|x + 2| \leq 4$ are $-6 \leq x \leq 2$.

Example 2.   Find all those numbers x such that $|x - 2| \geq 3$.

Solution.   Using Property 5 we see that the inequality $|x - 2| \geq 3$ is satisfied if and only if $x - 2 \geq 3$ or $x - 2 \leq -3$. Now, it is clear (Why?) that $x - 2 \geq 3$ implies that $x \geq 5$ and $x - 2 \leq -3$ implies that $x \leq -1$. Thus those numbers x satisfying $|x - 2| \geq 3$ are $x \geq 5$ or $x \leq -1$. (See Figure 1.1.5.)

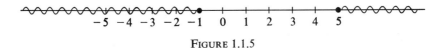

FIGURE 1.1.5

The ⌇⌇⌇ region in Figure 1.1.5 is the collection of those numbers x such that $x \leq -1$ or $x \geq 5$.

As our third example we shall solve a slightly different type of inequality.

Example 3.   Find all those numbers x such that $x(x + 5) \geq 0$.

Solution.   First we must ask when the product of two numbers is positive, that is, when $a \cdot b \geq 0$. Clearly $a \cdot b \geq 0$ if and only if both a and b are positive or both a and b are negative. Thus $x(x + 5) \geq 0$ if and only if $x \geq 0$ and $(x + 5) \geq 0$ or $x \leq 0$ and $(x + 5) \leq 0$. This implies that $x \geq 0$ and $x \geq -5$ or $x \leq 0$ and $x \leq -5$ and thus, $x \geq 0$ or $x \leq -5$. Hence, $x(x + 5) \geq 0$ for those x's such that $x \leq -5$ or $x \geq 0$. (See Figure 1.1.6.)

FIGURE 1.1.6

We conclude this section with a brief discussion of *sets*. You have probably observed that until now we have referred to a *collection* of numbers or *all* numbers satisfying certain conditions. This was repeated over and over again. It would be very convenient to have a notation for this. For this purpose it is helpful to use the notation and terminology from set theory. In mathematics, we define a set simply as a collection of objects. For example, we speak of the set of real numbers, meaning the collection of all real numbers. The members of the set are called *elements* of the set and are said to *belong to* or be *contained in* the set. We shall mainly be interested in sets of numbers in this text.

Notation 1. We shall designate sets by capital letters A, B, C, ... and elements of a set by small letters a, b, c, ...

Notation 2. We write "$x \in A$" to mean x is an element of the set A. (That is, the symbol $\in$ means "is an element of.")

We shall say that two sets A and B are equal and write $A = B$ if and only if they have the same elements.

Sets may be defined either (1) by listing the elements in the set, or (2) by stating the specific properties possessed by each element of the set. For example, suppose A is the set of positive integers less than or equal to 5. We may either list the elements of A and write

$$A = \{1, 2, 3, 4, 5\}$$

where the brackets { } will indicate the set, or we may write

$$A = \{x \mid x \text{ is a positive integer and } x \le 5\},$$

which is read: A is the set of all those numbers x such that x is a positive integer and $x \le 5$.

In general it is convenient to use the second method of defining a set—that is, where we use | to indicate the words "such that."

Notation. $A = \{x \mid x \text{ has property } P\}$. This notation means that A is the set of all those elements x such that x has property P. It is clear then that if we write $a \in A$, then a has property P. We determine whether or not an element is a member of a specific set "by checking to see" if the element possesses the property P.

Example 1. Write, using the second method of defining sets, the notation for the set of numbers whose absolute value is less than 5.

Solution. Let A designate the set. Then

$$A = \{x \mid |x| < 5\}.$$

In this example any element of A must have the property that its absolute value is less than 5. Note that 6 is not an element of A. We indicate this by writing $6 \notin A$.

Example 2.   Does the number $\frac{1}{2}$ belong to the set A where

$$A = \{x|\, |x + 2| > 3\}?$$

Solution.   Clearly $\frac{1}{2} \notin A$ since $|\frac{1}{2} + 2| < 3$. That is, $\frac{1}{2}$ does not possess the property of the elements of A.

Sometimes we will want to speak of only a particular collection of elements of a set, rather than the set itself. For example, if $A = \{x\,|\,x$ is a real number$\}$ we may only want to discuss the set $B = \{x\,|\,x$ is a positive number$\}$. Note, every element of B is an element of A, but A consists of many elements not included in B. In this case we say that B is a *subset* of A and write $B \subset A$—read as "the set B is contained in the set A." We use the word "proper" to indicate that A contains elements that are not in B. We can therefore make the following definition.

DEFINITION.   *We say that a set B is a subset of A, written $B \subset A$, if every element of B is an element of A.*

(*Note:* If $B \subset A$ every element of A is also an element of B, then $A = B$.)

Certain subsets of the real numbers occur so often in our work, that we give them special names. Below we first define these subsets and then illustrate them on the real axis. These subsets are called *intervals*.

DEFINITION. 1   *A closed interval on the real axis, written* $[a, b]$, *where a and b are any real numbers is defined by*

$$[a, b] = \{x\,|\,a \le x \le b\}.$$

Geometrically, $[a, b]$ *is illustrated by:*

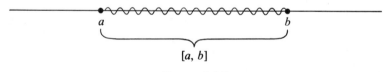

$$[a, b]$$

FIGURE 1.1.7

DEFINITION 2.   *An open interval, written* (a, b), *where a, b are any real numbers, is defined by*

$$(a, b) = \{x\,|\,a < x < b\},$$

that is, the numbers a and b are excluded from our set. On the real axis,

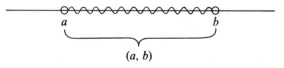

$$(a, b)$$

FIGURE 1.1.8

In obvious ways we can define:

Definition 3. $[a, b) = \{x \mid a \leq x < b\}$.

Definition 4. $(a, b] = \{x \mid a < x \leq b\}$.

These sets are called half-open intervals.

We could now go back to our inequalities and interpret our solutions as subsets of the real numbers. These subsets are often referred to as *solution sets*. For example, recall from the example on page 6 those numbers x satisfying the inequality $|x + 2| \leq 4$ where $-6 \leq x \leq 2$. We could now say that the solution set to this inequality is the closed interval $[-6, 2]$. In the exercises following this section, we shall give many other examples.

There are other properties and operations with sets that could be discussed. These concepts will be introduced as they are needed.

Exercises

1. Suppose x and y are two numbers such that $x > y$. Which of the following inequalities do you know to be true? (Simply refer to one of Properties 1–6 to justify your result.)

   (a) $\dfrac{x}{5} > \dfrac{y}{5}$

   (b) $x + 5 > y + 5$

   (c) $-x > -y$

   (d) $-x + 5 < -y + 5$

   (e) $\dfrac{5}{x} < \dfrac{5}{y}$

   (f) $\dfrac{1}{x} - 5 < \dfrac{1}{y} - 5$

   (g) $-5x < -5y$

2. For what real numbers x is:

   (a) $|x| = 3$

   (b) $|x - 2| = 3$

   (c) $|5x| = 2$

   (d) $|x - 2| = 2 - x$

   (e) $\left|\dfrac{x}{2}\right| \leq 3$

   (f) $\left|\dfrac{x}{3} - 9\right| \leq 7$

3. Find all those real numbers x satisfying the following inequalities and indicate the appropriate intervals on the real line.

   (a) $|2x| \leq 1$

   (b) $|x + 3| \leq 5$

   (c) $|x - 2| > 4$

   (d) $|x - 2| \leq 2 - x$

   (e) $|4 - 2x| \leq 1$

   (f) $|3 + 2x| \leq 7$

4. Find all those real numbers x satisfying the following inequalities and indicate the appropriate intervals on the real line.

   (a) $x(x + 3) \geq 0$

   (b) $x(3 - x) \geq 0$

   (c) $x(x + 3) \leq 0$

   (d) $x(3 - x) \leq 0$

   (e) $x(2x + 1) \leq 0$

   (f) $(x - 1)(2x + 3) \geq 0$

   (g) $\dfrac{x - 1}{x} \leq 0$

   (h) $x^2 + x \leq 0$

   (i) $x^2 + 3x + 2 \geq 0$

   [HINT: Write
   $x^2 + 3x + 2 = (x + 1)(x + 2)$.]

5. Explain why the following statements are true.
    (a) $|x|$ is the larger of the numbers x and $-x$.
    (b) If $-2 < x < 2$ then $|x| < 2$.

6. Assume that a used car dealer budgets \$400 for radio advertising and that a local radio station charges \$80 for a one-minute commercial during the night call show. How many of these commercials can the dealer afford? [HINT: Let x denote the unknown number of commercials. Clearly, their cost is $80x$. Hence, you must solve the inequality $80x \leq 400$.]

7. An automatic elevator designed to carry a maximum load of 8000 pounds is being used to lift pianos weighing 500 pounds each. What is the greatest number of pianos the elevator can lift?

8. Suppose that Mr. Fink, who has \$10,000 in his savings account, wants to use some of his money to buy a certain kind of stock, which sells for \$80 a share, but he does not want the balance of his savings account to go below \$4000. What is the greatest number of shares of stock that he can buy? [HINT: Let x denote the unknown number of shares. Clearly, $10,000 - 80x \geq 4000$.]

9. Let A and B be any two sets such that $A = B$. Show that this is true if and only if $A \subset B$ and $B \subset A$. [HINT: You must first show that if $A = B$ then $A \subset B$ and $B \subset A$; and then show if $B \subset A$ and $A \subset B$, $A = B$. Use the definitions to do this.]

10. Express your solutions to Exercises 3 and 4 as intervals.

11. (a) Show that if $c > 0$, the closed interval $[-c, c]$ consists of all those real numbers x such that $|x| \leq c$.
    (b) Show that if $c > 0$, the closed interval $[a - c, a + c]$ consists of all those real numbers x such that $|x - a| < c$.

12. Show by examples that the following are *not* correct statements.
    (a) If $a < b$ and $c < d$ then $ac < bd$.
    (b) If $1/a > 1/b$ and $c > d$ then $c/a > d/b$.

13. (a) Let x be any real number. Show that $-|x| \leq x \leq |x|$.
    (b) Using part (a) and the rules for working with absolute values, show that the triangle inequality $|x + y| \leq |x| + |y|$ is valid.

1.2 Coordinate Systems

Before proceeding any further, we need a precise way of measuring the distance between two points, say x_1 and x_2, on the real line. We define this concept as follows:

DEFINITION. *Given two points x_1, x_2 on the real line, the distance from x_1 to x_2 is* $|x_1 - x_2|$.

Thus if $x_1 = 1$ and $x_2 = 5$ then there are 4 units separating them. In this case $|x_1 - x_2| = |1 - 5| = |-4| = 4$, which agrees with the more intuitive interpretation given above. Note the following facts.

(1)   The distance from x_1 to x_2 is positive or zero.
(2)   If the distance from x_1 to x_2 is zero then $x_1 = x_2$.
(3)   Since $|(x_1 - x_2)| = |-(x_2 - x_1)| = |x_2 - x_1|$ the distance from x_1 to x_2 is the same as the distance from x_2 to x_1.
(4)   If $x_1 \neq x_2$ then the distance from x_1 to x_2 is greater than zero.

We have previously observed that once a unit of length is chosen, we can represent numbers as points on a line and conversely. Let us now extend this procedure to the plane and to pairs of real numbers. Choose a point in the plane and call this point the origin, 0. Draw a horizontal line through 0 and construct a line perpendicular to it through 0. The horizontal line is called the x axis and the vertical line is called the y axis. Both of these axes are real lines with 0 at the origin.

The order of the numbers on the x and y axes can most easily be described by the Figure 1.2.1. Observe that the two axes divide the plane into four quadrants. If we let

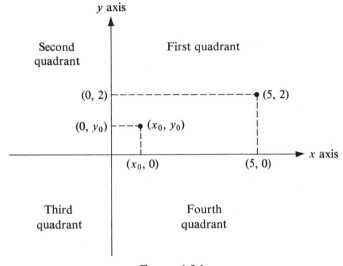

FIGURE 1.2.1

P be any point in the plane, and draw perpendicular lines to the horizontal and vertical axes, we obtain two numbers x_0 and y_0 as indicated in the above figure; x_0 is called the x coordinate of P, and y_0 is called the y coordinate of P. We shall use the notation (x_0, y_0) to describe the point in the plane whose x coordinate is x_0 and whose y coordinate is y_0. For example, to find the point $(5, 2)$ count 5 units along the x axis and then up 2 units. For a point in the first quadrant both x and y coordinates are positive. What is the situation for the other quadrants? A little thought reveals the following:

(1)   Every pair (a, b) of real numbers corresponds to a point in the plane. (Just measure a units along the x axis—to the right if a is positive and to the left if a is negative—and b units up to determine P if b is positive. If b is negative, measure b units down.)

(2)   To every point P in the plane we can associate a pair (a, b). We simply drop perpendicular lines L_1 and L_2 to the x and y axis and determine the coordinates of P by noting that L_1 intersects the x axis at $(a, 0)$ and that L_2 intersects the y axis at $(0, b)$.

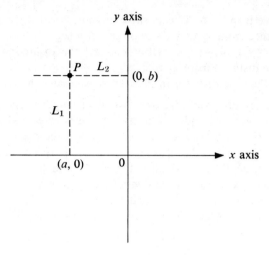

FIGURE 1.2.2

Thus the coordinates of P must be (a, b). (See Figure 1.2.2.) (Statements (1) and (2) say that there exists a 1–1 correspondence between points in the plane and pairs of real numbers.)

(3)   The order of the pair is *important*; thus $(-1, 5)$ is not the same point as $(5, -1)$. (See Figure 1.2.3.)

For this reason the pair (x_0, y_0) is referred to as an ordered pair.

(4) (a_1, b_1) and (a_2, b_2) represent the same point if and only if $a_1 = a_2$ and $b_1 = b_2$.

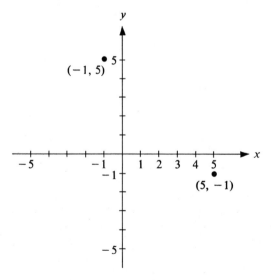

FIGURE 1.2.3

We now extend the concept of distance to points in the plane with the following definition.

DEFINITION.   *The distance from* $P_1 = (x_1, y_1)$ *to* $P_2 = (x_2, y_2)$ *(written as* $d[P_1, P_2]$*) is equal to*

$$\sqrt{(x_1 - x_2)^2 + (y_1 - y_2)^2}.$$

(*Note:* This distance is just the length of the line segment $P_1 P_2$ in Figure 1.2.4.)

If we consider the figure below it becomes clear that this definition follows easily from the Pythagorean theorem.   Recall that $|a|^2 = a^2$.

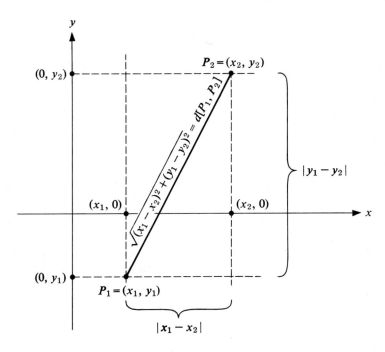

FIGURE 1.2.4

Example.   Find the distance from $P_1 = (1, 3)$ to $P_2 = (4, 6)$.

   Solution. $d[P_1, P_2] = \sqrt{(1 - 4)^2 + (3 - 6)^2} = \sqrt{18}$.

Note the following:

(1) $d[P_1, P_2] \geq 0$.

(2) $d[P_1, P_2] = 0$ if P_1 is the same point as P_2.

(3) $d[P_1, P_2] \neq 0$ if $P_1 \neq P_2$.

(4) $d[P_1, P_2] = \sqrt{(x_1 - x_2)^2 + (y_1 - y_2)^2}$
$\qquad\qquad = \sqrt{(x_2 - x_1)^2 + (y_2 - y_1)^2} = d[P_2, P_1]$.

Actually it does not matter in which order you write the x's and the y's as long as you group the x's together (and the y's together).

(5)   If P_1 and P_2 are points on the x axis then our two definitions of distance agree. Thus if $P_1 = (x_1, 0)$ and $P_2 = (x_2, 0)$ then

$$d[P_1, P_2] = \sqrt{(x_1 - x_2)^2 + (0 - 0)^2} = \sqrt{(x_1 - x_2)^2} = |x_1 - x_2|.$$

The same comment is valid for points on the y axis.

Exercises

1.   Graph the following points.

(a) $(-1, 9)$ (e) $(0, 0)$ (h) $(-\frac{2}{3}, -\frac{1}{2})$

(b) $(5, 2)$ (f) $(2, -1)$ (i) $(-1, 0)$

(c) $(-1, -2)$ (g) $(\frac{1}{2}, -\frac{2}{3})$ (j) $(0, -1)$

(d) $(-11, 6)$

2.   Find the distance from:

(a) $(1, 3)$ to $(5, 9)$ (d) $(-1, 0)$ to $(0, -1)$

(b) $(-1, 0)$ to $(6, -2)$ (e) $(-\frac{1}{2}, -\frac{1}{2})$ to $(\frac{1}{2}, \frac{1}{2})$

(c) $(19, -7)$ to $(0, 0)$

3.   Let L be the line segment joining (x_1, y_1) to (x_2, y_2). By means of the distance formula show that $[(x_1 + x_2)/2, (y_1 + y_2)/2]$ is the midpoint of L.

4.   An IBM 704 computer is to be shipped from Bloomington to Richmond, Indiana. It can be shipped directly by train at a cost of $5 per mile or by truck via Indianapolis at a cost of $3 per mile. Assuming that Bloomington, Richmond, and Indianapolis are located as shown below, which way is cheaper? (See Figure 1.2.5.)

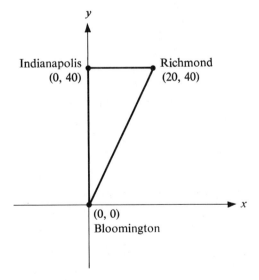

FIGURE 1.2.5

5. A vacuum cleaner salesman living in town *A* of Figure 1.2.6 regularly must visit customers in towns *B* and *C*. If all units are in miles, what is the total length of one round-trip?

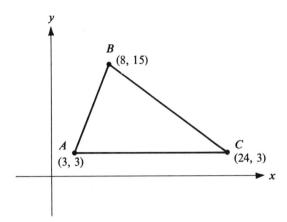

FIGURE 1.2.6

6. Let $C = \{(x, y) \mid$ distance from the point (x, y) to the origin is equal to $R\}$.
   (a) Show that $x^2 + y^2 = R^2$. (Note that if $x_0{}^2 + y_0{}^2 = R^2$, then the distance from the point (x_0, y_0) to the origin is R; this equation represents the equation of a circle that has its center at the origin and a radius R.) Draw a picture of this circle on a proper coordinate system.
   (b) Find the equation of the circle whose center is at the origin and whose radius is 5. Check by substitution whether or not the following points lie on this circle: $(2, 5)$, $(-1, 4)$, $(3, -4)$, $(-5, 0)$; $(0, -5)$.

1.3 Straight Lines

Next to points, lines are the simplest and probably the most useful geometric objects we encounter. Consider the equation $y = 2x + 5$. If we plot a few points satisfied by the equation, we notice that they lie on a line. (See Figure 1.3.1.) This is not an accident but an example of a general situation. The equation $y = mx + b$ describes a straight line for any pair of real numbers m and b. (Think of m and b as being fixed real numbers even though we do not know their values.)

There is one property of a line that should be obvious to anyone—its slope or steepness. What we need is a way of describing this concept of slope by a number. We do this as follows.

DEFINITION. *If* (x_1, y_1) *and* (x_2, y_2) *are two distinct points on the nonvertical line L then the slope of L is*

$$\frac{y_2 - y_1}{x_2 - x_1}.$$

An alert reader may inquire whether a different choice of points on L might not give a different number for the slope. Consider Figure 1.3.2; a few elementary computations

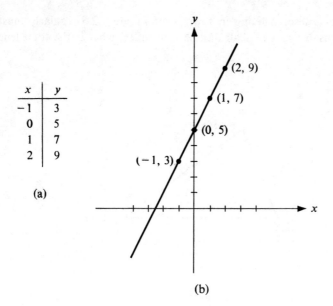

x	y
−1	3
0	5
1	7
2	9

(a)

(b)

FIGURE 1.3.1

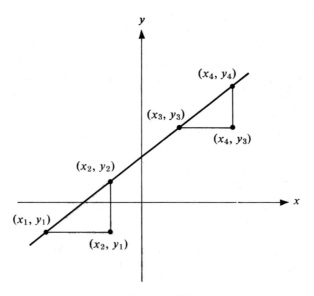

FIGURE 1.3.2

with similar triangles will show that the slope does not depend on the points chosen. One need only observe that

$$\frac{y_4 - y_3}{x_4 - x_3} = \frac{y_2 - y_1}{x_2 - x_1}.$$

If $x_1 = x_2$ (that is, if L is parallel to the y axis) we do not define the slope of L. Some people may feel the slope should be defined as ∞ in this case. There is nothing wrong with this but it is not very useful.

Observe that if the slope is positive the line rises to the right and if it is negative it "sinks" to the right. (See Figures 1.3.3 and 1.3.4.) It is not hard to see why. If

$$\frac{y_2 - y_1}{x_2 - x_1} = m > 0 \qquad \text{then } (y_2 - y_1) \geq m(x_2 - x_1) \geq 0 \qquad \text{for } x_2 > x_1;$$

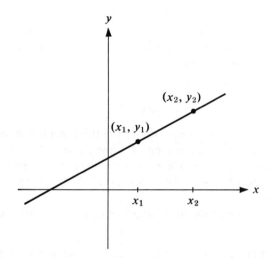

FIGURE 1.3.3

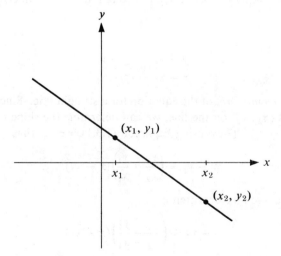

FIGURE 1.3.4

that is, $y_2 \geq y_1$ so the line is rising. A similar argument shows that if $(y_2 - y_1)/(x_2 - x_1)$ $= m < 0$ then the line sinks to the right.

We discuss the formulas for the equations of various lines by considering the three problems stated below. It is important to note that the different forms for the equation of a line are really all the same.

Problem 1. If a line L has slope m and passes through the point (x_1, y_1), what is the equation of L?

Solution.

$$(y - y_1) = m (x - x_1). \tag{1.3.1}$$

This is called the *point slope form* of the equation of a line. It is a simple matter to derive Equation (1.3.1). Think of (x, y) as being a fixed point on L. Then $m = (y - y_1)/(x - x_1)$ by the definition of slope and hence $(y - y_1) = m(x - x_1)$.

This equation may be reduced to the form

$$y = mx + b, \tag{1.3.2}$$

where $b = -mx_1 + y_1$. It is often convenient to use this form for the equation of a line. Suppose we wish to determine the equation of a line passing through $(1, 0)$ and having slope 1. Clearly, $m = 1$ so that $y = x + b$. We can determine b by using the fact that $y = 0$ when $x = 1$ and thus $b = -1$. The equation of the line is therefore $y = x - 1$.

Example. Find the equation of the line with slope equal to $-\frac{1}{2}$ that passes through the point $(2, 1)$.

Solution. By Equation (1.3.1), $(y - 1) = -\frac{1}{2}(x - 2)$ or $y = -\frac{1}{2}x + 2$, which is the result we would obtain by using Equation (1.3.2). $y = -\frac{1}{2}x + (2+1)$

Problem 2. Find the equation of the line L through the two (distinct) points (x_1, y_1) and (x_2, y_2).

Solution.

$$\frac{y - y_1}{x - x_1} = \frac{y_2 - y_1}{x_2 - x_1}. \tag{1.3.3}$$

We call this the *two point form* of the equation for a straight line. Since we are given two points (x_1, y_1) and (x_2, y_2) on the line, we can determine the slope of the line. Indeed, $m = (y_2 - y_1)/(x_2 - x_1)$. Thus, using Equation (1.3.1), we see that

$$y - y_1 = \left(\frac{y_2 - y_1}{x_2 - x_1}\right)(x - x_1).$$

We could equally as well have written

$$y - y_2 = \left(\frac{y_2 - y_1}{x_2 - x_1}\right)(x - x_2)$$

They represent the same line.

Example.   Find the equation of the line through $(1, 2)$ and $(-3, -4)$.

Solution.   By Equation (1.3.3), we have $(y - 2)/(x - 1) = (-4 - 2)/(-3 - 1)$ or $(y - 2)/(x - 1) = \frac{3}{2}$ or $(y - 2) = \frac{3}{2}(x - 1)$ or $y = \frac{3}{2}x + \frac{1}{2}$. The graph of this line is shown in Figure 1.3.5. Observe that two points are sufficient to graph a straight line.

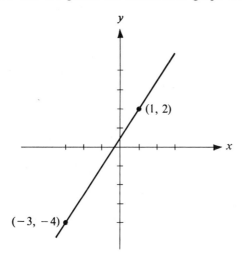

FIGURE 1.3.5

The different forms for the equations of a line are really all the same. That is, one can convert from one to the other by simple algebraic manipulation. The above example illustrates this fact.

Lines that are parallel to the x or y axes are special cases and the formulas do not always work out in these cases. However, a little thought should resolve any difficulties in such situations.

Example 1.   The line parallel to the y axis and passing through the point (c, d) has as its equation $x = c$. This line has no slope. (Why?)

Example 2.   The line parallel to the x axis and passing through the point (a, b) has as its equation $y = b$. The slope of this line is zero. Note that using the point slope form also yields the equation $y = b$.

An equation of the form $ax + by + c = 0$ also represents a line for any set a, b, c of real numbers. Conversely, *any* straight line has an equation of this form. We say that the equation $ax + by + c = 0$ is the most general form of the equation of a straight line. It is easy to verify that Equations (1.3.1), (1.3.2), and (1.3.3) are all of this form. If $b \neq 0$, then $y = (-a/b)x - c/b$ and the line has slope $(-a/b)$. If $b = 0$ then the line is parallel to the y axis and the equation reduces to $x = -c/a$.

Example.   Linear depreciation is one of several methods approved by the Internal Revenue Service for depreciating business property. If the original cost of the property is d dollars and it is depreciated linearly over N years, its value (undepreciated balance) y

Linear depreciation ♪

at the end of x years is given by

$$y = d - \left(\frac{d}{N}\right)x$$

or equivalently

$$y = d\left(1 - \frac{x}{N}\right).$$

Note that this is the equation of a straight line whose slope is given by $-d/N$. For instance, if football equipment worth \$2000 is depreciated over 5 years, the undepreciated balance after x years is given by

$$y = 2000\left(1 - \frac{x}{5}\right) = 2000 - 400x.$$

Note that after 5 years, we obtain $y = 2000 - 2000 = 0$, which means that the equipment is completely depreciated.

Problem 3.   Given two lines $L_1 : a_1 x + b_1 y + c_1 = 0$ and $L_2 : a_2 x + b_2 y + c_2 = 0$, find their point of intersection.

Note first that the lines need not intersect, for example, they might be parallel. Secondly, if the two lines are the same then every point is a point of intersection. Except for these special (and generally uninteresting) cases two lines will intersect in exactly one point.
  Suppose the point $P = (p, q)$ lies on L_1 and L_2; then $a_1 p + b_1 q + c_1 = 0$ since P lies on L_1 and $a_2 p + b_2 q + c_2 = 0$ since P lies on L_2. Solving these equations for (p, q) yields the desired point P.

Example.   Find the point of intersection of the following lines:

$$2x + y + 1 = 0 \quad \text{and} \quad x + 2y + 1 = 0.$$

Solution.   Solving the above system of equations we obtain: $3x + 1 = 0$ and $3y + 1 = 0$ and thus $x = -\frac{1}{3}, y = -\frac{1}{3}$. The graphs of the two equations and their point of intersection are shown in Figure 1.3.6.

  The following observations about lines are very useful. Let the lines L_1 and L_2 have slopes m_1 and m_2, respectively. Then
  (1) L_1 and L_2 are parallel if and only if $m_1 = m_2$.
  (2) L_1 and L_2 are perpendicular if and only if $m_1 m_2 = -1$.
In Exercises 14 and 15, we outline the method of deriving these formulas.

Example 1.   Find the slope of all lines parallel to $2y = -3x + 2$.

Solution.   Since this equation is equivalent to $y = -\frac{3}{2}x + 1$, the slope of all lines parallel to it must be $-\frac{3}{2}$.

Example 2.   Find the equation of the line perpendicular to the line $y = 3x - 4$ and passing through $(-14, 9)$.

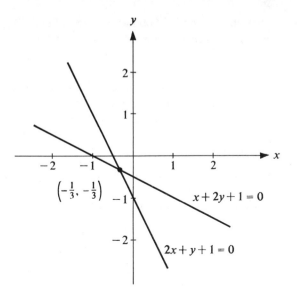

FIGURE 1.3.6

Solution. Let m be the slope of the line we are looking for. Then $3m =$ = $-\frac{1}{3}$. Thus the desired line has equation $y - 9 = -\frac{1}{3}(x + 14)$, from the formula, (1.3.1).

Exercises

1. Find the slope of the line through the following pairs of points.

   (a) $(1, -1); (2, 6)$

   (b) $(0, 0); (\frac{1}{9}, -1)$

   (c) $(1, \frac{2}{5}); (4, -7)$

   (d) $(3, -1); (3, 4)$

   (e) $(2\frac{1}{3}, 6); (-1, 6)$

   (f) $(\frac{19}{2}, -3); (0, 7)$

   (g) $(0, 2); (1, 0)$

   (h) $(-6, 3); (7, -\frac{2}{3})$

2. Find the equations of the lines passing through the pairs of points in Exercise 1.

3. Find an equation for the line with slope m and through the indicated point.

   (a) $5; (-1, 7)$

   (b) $-8; (3, 2)$

   (c) $-\frac{1}{2}; (7, -2)$

   (d) $0; (0, 0)$

   (e) $0; (-1, 7)$

   (f) $\frac{2}{3}; (9, 1)$

   (g) $7; (11, -\frac{1}{3})$

   (h) $-2; (6, 5)$

4. An apartment building then worth \$300,000 was built in 1950. What is its value (for tax purposes) in 1969 if it is being depreciated linearly over 50 years?

5. In 1960 the Hoosier Cab Co. purchased \$80,000 worth of new cabs. What is the value of these cabs in 1969, if they are being depreciated linearly over a period of 13 years?

6. A machine purchased new for $10,000 has a scrap value of $1500 after 10 years. If its value is depreciated linearly (from $10,000 to $1500), find the equation of the line that enables us to determine its value after it has been in use any given number of years. What is its value after 3 years? Generalize the technique used in this problem so that it applies to a machine purchased at an original cost d, has the scrap value s, and has been depreciated linearly (from d to s) over a period of N years; that is, find a formula giving the value of the machine after x years.

7. Find an equation of the line through $(1, 2)$ and perpendicular to the line $y = 5x + 3$.

8. Find the equation of the line through $(7, 11)$ and parallel to the line $y = -2x + 7$.

9. Find the slope of the lines perpendicular to the lines:

   (a) $y = 3x - 4$ (e) $4x + 9y + 36 = 0$
   (b) $4y + 5x - 2 = 0$ (f) $19x - 3y - 7 = 0$
   (c) $x = 2y + 7$ (g) $\frac{2}{3}x + \frac{3}{2}y - 1 = 0$
   (d) $2x = 3y - 9$

10. Find the intersection of the following pairs of lines.

    (a) $x - y = 1$ (d) $6x + 9y = 4$
        $2x + 4y = 3$ $5x - 4y = 3$

    (b) $x + \dfrac{y}{2} - 9 = 0$ (e) $x - y = 1$

        $y - x = 12$ $\dfrac{x}{3} + \dfrac{2y}{5} = 0$

    (c) $9x + 3y = 7$
        $3x + y = 6$

11. Find the equation of the line through the intersection of the two lines $x - y = 1$; $2x - 3y = 7$ and perpendicular to the line through $(1, 2)$ and $(-3, 1)$.

12. Find the point or points on the line $y = 2x + 1$ whose distance from $(1, 0)$ is 7.

13. Find the shortest distance from the point $(1, 2)$ to the line $L: y - 2x = -3$. [HINT: Find the equation of the line perpendicular to L through $(1, 2)$. Find its intersection with L. A little thought should tell you how to finish the problem.]

14. Show that two lines are parallel if and only if their slopes are equal. [HINT: Let the equation of L_1 be $y = m_1 x + b_1$ and that of L_2 be $y = m_2 x + b_2$. Assume that two lines are parallel if and only if they coincide or have no points in common. If $m_1 = m_2$, show that one of the two possibilities must occur. If $m_1 \neq m_2$ show that there is precisely one point of intersection. Combining all this finishes the problem.]

15. Show that two lines are perpendicular if and only if the product of their slopes is -1. (See Figure 1.3.7.)

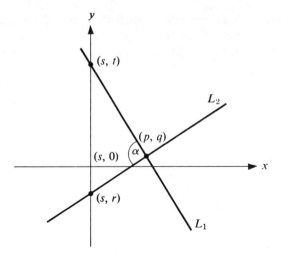

FIGURE 1.3.7

[HINT: If $L_1: y = m_1 x + b_1$ and $L_2: y = m_2 x + b_2$, find their intersection (p, q). Solve for t and r. This is easy since $t = m_1 s + b_1$ and $r = m_2 s + b_2$. (Why?) Now use the Pythagorean theorem to check given m_1 and m_2, whether α is a right angle, or to solve for m_1 and m_2 given α.

1.4 Functions. An Introduction

In our discussion of equations that correspond to straight lines we observed that once the real numbers m and b were fixed, we could determine y, where $y = mx + b$, by selecting the x's from the set of real numbers. For example, if we consider $y = 2x + 1$, then it is clear that for each real number x we obtain one and only one real number y corresponding to that x. Notice that the entire set of y's we find in this way is exactly the set of real numbers. Thus, we may consider the equation $y = 2x + 1$ to be a rule that assigns to every real number x one and only one real number y. More precisely, this rule says: take twice x and add 1. This leads us to the idea of a function, one of the very basic concepts in all of mathematics.

We can define a function in abstract terms as follows.

DEFINITION.   *Let A and B be two given sets. A function is a rule that assigns to each element, $a \in A$, one and only one element $b \in B$.*

The set A is called the *domain* of the function. We call the set of elements of B that correspond in this way to elements of A, the *range* of the function. For most of our discussions, both A and B will be sets of real numbers. Although the above definition is really not a formal one, it will suffice for our purposes. Before proceeding further we give some examples.

Example 1.   It is clear that a function was defined in the example in the first paragraph. The domain is clearly the set of real numbers.

Example 2.   Let A be the set of tests of 250 students in this calculus course. To each test we assign a number between 0 and 100. Note that we have now defined a function. The domain is the set of 250 tests, rather than a set of numbers. The range consists of those numbers between 0 and 100 that are actual test scores. It is clear that to each test there corresponds one and only one grade. However, *many* tests may have the same grade. (That is, many elements of A may be assigned the *same* element of B—this is still legitimate!)

Example 3.   Now we give an example that does *not* yield a function. Let A be a set of 10 specified books and let B be a set including all the authors of the books in A. Consider the association with each book in A of its author or authors in B. Note we have not defined a function since a book in A may have two or more authors.

Before we proceed further in our discussion of functions we need some notation. Since we will mainly be dealing with sets of real numbers, we will concentrate on notation for this case. We denote functions by small letters, $f, g, h, \ldots$. It will be clear from the context when a letter denotes an element of a set and when it denotes a function. If we say that f is a function, we mean that f is the rule that assigns numbers to elements in its domain. If f is a function, then the number that f assigns to a number x in its domain is denoted by $f(x)$—read as f of x" or "f evaluated at x." Sometimes it is called the value of f at x. Be careful to distinguish between f and $f(x)$! Observe that the symbol $f(x)$ makes sense only if $x \in$ domain of f; for other numbers x, the symbol $f(x)$ is *not* defined. The common procedure adopted for defining a function f is to indicate what $f(x)$ is for every number x in the domain of f. We illustrate this with some examples.

Example 1.   The function defined in the example given in the first paragraph is defined by

$$f(x) = 2x + 1.$$

Example 2.   Let

$$f(x) = \begin{cases} x & \text{if } 0 \le x \le 1 \\ 2x & \text{if } 1 < x \le 2. \end{cases}$$

Note that the domain of f is $[0, 2]$. The rule is: To each $x \in [0, 1]$ assign the number x; to $x \in (1, 2]$ assign the number $2x$. It is clear that to each $x \in [0, 2]$, we have assigned one and only one real number. What is $f(\frac{1}{2})$? Since $\frac{1}{2} \in [0, 1]$, we see that $f(\frac{1}{2}) = \frac{1}{2}$. What is $f(\frac{3}{2})$? Since $\frac{3}{2} \in (1, 2], f(\frac{3}{2}) = 3$. Observe that the range of f is $[0, 1] \cup (2, 4]$. Check this fact!

Example 3.   Let

$$f(x) = \sqrt{(1 - x^2)} \qquad \text{for} \quad -1 \le x \le 1.$$

We have clearly specified the domain to be the interval $[-1, 1]$. The rule is obvious—for each $x \in [-1, 1]$ assign the number $\sqrt{(1 - x^2)}$. Note that the rule does not make sense for $x \notin [-1, 1]$. Indeed if we take $x = 2$, then $\sqrt{1 - 2^2} = \sqrt{-3}$ is not a real number.

The above three examples illustrate that a function need *not* be defined only by one formula. (See Example 2.) Notice in Example 2 we carefully specified the domain of *f* while in Example 1 we did not include this in our definition. We adopt the following convention. *Unless the domain is explicitly stated, it is understood to consist of all numbers for which the definition makes any sense.* Thus, in Example 1, we understand that the domain is the set of all real numbers. In Example 3, if (without explicitly stating the domain) we considered the function written as $f(x) = \sqrt{(1 - x^2)}$, then our convention tells us that the domain must be a subset of $[-1, 1]$ since the definition would not make sense for other real numbers.

There are many useful ways of visualizing functions. Two of these in particular will be of great aid to us—(1) machine diagrams and (2) graphs.

A function can be thought of as an input and output machine. (See Figure 1.4.1.) If

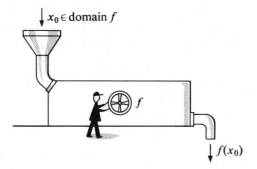

$x_0 \in$ domain f

$f(x_0)$

FIGURE 1.4.1

f is a function, then we can interpret the domain of *f* as the set of all those numbers that are fed into our machine. (The machine accepts as input only these numbers!) Once the machine is set to apply the rule *f*, it produces the number $f(x)$ as its output. Our only restriction on this machine is that if the same number is put in on two different occasions then the outputs must be the same. The range of the machine is the set of all outputs of the machine. As an example, we set up a machine for Example 2 above. (See Figure 1.4.2.) In this case the machine must be made to differentiate between the elements of $[0, 1]$, and the elements of $(1, 2]$.

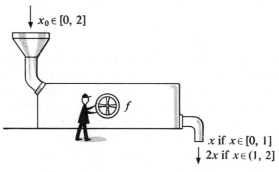

$x_0 \in [0, 2]$

x if $x \in [0, 1]$
$2x$ if $x \in (1, 2]$

FIGURE 1.4.2

The graph of a function plays a particularly important role in our applications. We define the graph of a function f as follows:

$$\text{graph of } f = \{(x, f(x)) \mid x \in \text{domain of } f\}.$$

Sometimes it is convenient when using the rectangular coordinate system to rewrite the definition as:

$$\text{graph of } f = \{(x, y) \mid x \in \text{domain of } f \text{ and } y = f(x)\}.$$

We consider first the graph of the function $f(x) = 2x + 1$. The definition tells us that:

$$\text{graph of } f = \{(x, y) \mid x \text{ is a real number and } y = 2x + 1\}.$$

Now, we draw this in a rectangular coordinate system. The graph of this function is clearly a straight line with slope 2 that goes through the points $(0, 1)$ and $(-\frac{1}{2}, 0)$. (See Figure 1.4.3.) Since it is very cumbersome to use the set notation each time, we simply state that the graph of f is the straight line whose equation is $y = 2x + 1$.

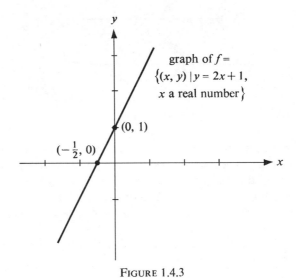

FIGURE 1.4.3

As a second example, we consider the graph of the function $f(x) = \sqrt{(1 - x^2)}$. Recall that the domain of $f = [-1, 1]$. By definition, graph of $f = \{(x, y) \mid x \in [-1, 1]$ and $y = \sqrt{(1 - x^2)}\}$. Hence, we set $y = \sqrt{(1 - x^2)}$ and observe that $y \geq 0$ (since $\sqrt{a}$ will *always* denote the positive root of a) and squaring both sides $x^2 + y^2 = 1$. All points on the graph must be in the first two quadrants. It is not hard to see that $x^2 + y^2 = 1$ represents a circle of radius 1 with center at the origin. Let $P = (x, y)$ be a point in the plane whose distance from the origin is 1. Then the distance from P to 0 is given by $\sqrt{(x - 0)^2 + (y - 0)^2}$. Thus $\sqrt{(x - 0)^2 + (y - 0)^2} = 1$, and after a simplification we obtain $x^2 + y^2 = 1$. Hence the graph of our functions is the upper semicircle illustrated in Figure 1.4.4.

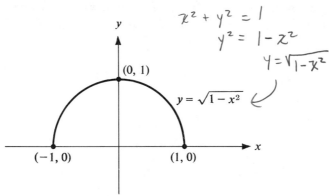

FIGURE 1.4.4

As a third example, we consider the function defined in Example 2 above:

$$f(x) = \begin{cases} x & \text{if } 0 \leq x \leq 1 \\ 2x & \text{if } 1 < x \leq 2. \end{cases}$$

First set $y = f(x)$; that is, for $x \in [0, 1]$, $y = x$; and for $x \in (1, 2]$, $y = 2x$. We observe that the graph is a straight line with slope 1 for $x \in [0, 1]$ and a straight line with slope 2 for $x \in (1, 2]$. (See Figure 1.4.5.) The circle at $(1, 2)$ means that this point is not included in our graph. Observe the gap in the graph at $x = 1$. In the next chapter, we have more to say about functions with gaps in their graphs.

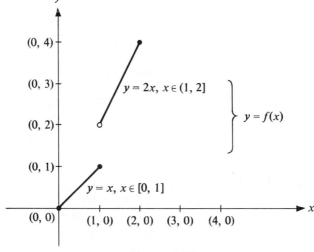

FIGURE 1.4.5

For our last example, consider the function defined by $f(x) = |x|$. Note that the domain of this function is the set of all real numbers. If $x \geq 0$, then $f(x) = x$ and if $x < 0$, $f(x) = -x$ by our definition of absolute value. Hence, we can see that the range of the function must be the set of all nonnegative real numbers. The graph of f is shown in Figure 1.4.6. Observe that for $x > 0$, graph of $f = \{(x, y) \mid y = x, x > 0\}$, and for $x < 0$, graph of $f = \{(x, y) \mid y = -x, x < 0\}$.

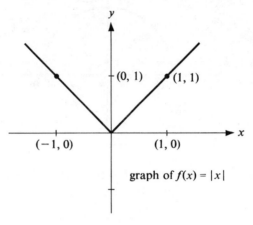

graph of $f(x) = |x|$

FIGURE 1.4.6

Before proceeding further, we stop to make some simple observations concerning functions and their graphs.

Observe that in the notation for a function $f(x)$, the x plays the role of a "dummy" in the sense that the function $f(x)$ is the same as $f(a), f(b)$, etc. For example if $f(x) = x^2 + 1$, then $f(a) = a^2 + 1, f(b) = b^2 + 1$, etc. The function is the *same* in all cases. Also, observe that if f is defined by the above formula, then $f(x + 2) = (x + 2)^2 + 1$, and $f(x - 2) = (x - 2)^2 + 1$. That is, no matter what we put in for x we always come out with that thing squared plus one. Even if we put in a "pink elephant" we get ("pink elephant")$^2 + 1$. This of course, assumes "pink elephant" is in the domain of f! It is essential to thoroughly comprehend this. Of course, it is also tacitly assumed that whatever is substituted for x still lies in the domain of f. Let us look at some examples.

Example 1. Let $f(x) = (x + 3)^3 + 1$. Find (i) $f(n + h)$, (ii) $f(c)$, and (iii), $f(x/4)$.

  Solution. (i) $f(n + h) = [(n + h + 3)]^3 + 1$,

  (ii) $f(c) = (c + 3)^3 + 1$,

  (iii) $f\left(\dfrac{x}{4}\right) = \left(\dfrac{x}{4} + 3\right)^3 + 1$.

Example 2. In discussing the graph of f, notice that the domain of f is always found on the x axis while the range is the set of values given on the y axis. For example, let

$$f(x) = \begin{cases} x, & x \in [0, 1) \\ 2, & x = 1 \\ 3x, & x \in (1, 2]. \end{cases}$$

The graph of $f = \{(x, y) \mid y = f(x), x \in [0, 2]\}$ is shown below, with the domain and range of f indicated in Figure 1.4.7.

In Chapter 4, we carefully study the properties of graphs of functions and give simple procedures for accurately drawing the graphs. However, it would be useful for us to have at our disposal certain elementary functions and their graphs. We list these functions and their graphs in the next section.

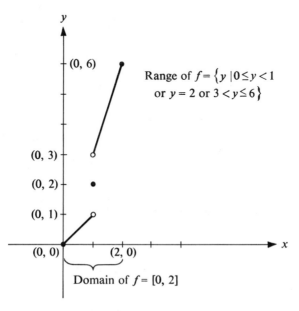

FIGURE 1.4.7

Exercises

1. Determine whether the situations described below represent a function. If so, state the domains and ranges of the functions.
   (a) Let A be a set of 20 authors and B the set of 30 books containing those written by the authors in A. Consider the rule which assigns to each author in A the book or books he has written.
   (b) Let A and B be the set of all real numbers. Consider the rule that assigns to each number n in A, a number n^5 in B.
   (c) Consider the rule that assigns to each mother her oldest child. Next, consider the rule that assigns to each mother her children.
   (d) The rule that associates to each person his social security number.
   (e) The rule that associates to each person the first letter of his first name.
   (f) The rule that associates with each letter of the alphabet all those people whose last name begins with that letter.
2. Consider the function

$$f(x) = \begin{cases} \dfrac{x+1}{x-1}, & \text{if } x \neq 1 \\ 1, & \text{if } x = 1. \end{cases}$$

(a) What is the domain of f?

(b) Determine each of the following:

(i) $f(2)$ (iii) $f(a + 2)$

(ii) $f(-10)$ (iv) $f(a^2)$

(v) $f\left(\dfrac{1}{a}\right)$

3. Graph the function $g(x) = (1 - x)/(1 - x)$. Is this the same function as $h(x) = 1$?
    Why not?

4. Let $f(x) = |x| - x$. What is $f(-1); f(1); f(-20)$? If $x \geq 0$, write another formula
    for $f(x)$. Do the same if $x < 0$.

5. Sketch a graph of the following functions and indicate at least three points on each
    graph.

(a) $f(x) = 5x + \frac{1}{3}$

(b) $f(x) = \dfrac{x}{2} + 1$

(c) $f(x) = |x| + x$

(d) $f(x) = \begin{cases} x & \text{if } x < 0 \\ 2 & \text{if } x = 0 \\ x & \text{if } x > 0 \end{cases}$

(e) $f(x) = \dfrac{|x|}{x}$

6. (a) Graph the following function and indicate its domain and range on the graph.

$$f(x) = \begin{cases} 2x & \text{if } x \in [0, 1) \\ 3 & \text{if } x = 1 \\ 5 - 3x & \text{if } x \in (1, \frac{5}{3}] \end{cases}$$

(b) If $0 \leq x \leq \frac{1}{2}$ what is

(i) $f(x + 0.1)$ (iii) $f(x + 1)$

(ii) $f(x + 0.1) - f(x)$ (iv) $f(x + \frac{1}{2})$?

7. Can a circle be the graph of a function? Why? How about a vertical line?

8. Sketch the function whose graph goes through the point $(1, 1)$ and is (a) perpendicu-
    lar to the graph of the function $f(x) = 2x + 1$ and (b) parallel to the graph of the
    function $f(x) = 2x + 1$.

9. A book publisher agrees to pay an author royalties according to the following
    scheme: \$1.00 for the nth copy sold if $1 \leq n \leq 10{,}000$; \$1.50 for the nth copy if
    $10{,}000 < n \leq 20{,}000$; \$2.00 for the nth copy if $20{,}000 < n$. Let f be the function
    that assigns to the nth copy sold, the author's income for the first n books sold.
    Find formulas for $f(n)$ and graph f. Find $f(15{,}000), f(20{,}000), f(25{,}000)$, and
    $f(60{,}000)$.

10. The mathematics department's "magnetic tape Selectric typewriter" is supposed to
    be checked once a month. If this is not done (but the machine is checked at least
    once a year), the expected cost of repairs is \$10 plus ten times the square of the
    number of months the machine has gone without being checked. Express by
    means of a formula this relationship between the number of months the machine
    has gone without being checked and the expected cost of repairs, and use the
    formula to calculate the expected cost of repairs when the machine has not been
    checked for (a) 2 months, (b) 4 months, (c) 6 months, (d) 8 months, and (e) 12
    months.

11. The manager of a chain of bookstores has the following data on the supply s of certain paperbacks and their price p in dollars per cartons of books.

Price (dollars)	p	4	6	8	10	17
Supply (hundred boxes)	$s(p)$	0	12	21	25	27

(a) Display this information graphically.

(b) Check visually whether the function given by

$$s(p) = 30 - \frac{60}{p - 2}$$

can be used to approximate the above data by calculating $s(p)$ for each of the given values of p, plotting the corresponding points on the diagram constructed in part (a) and joining them by means of a smooth curve.

(c) Use the formula of part (b) to calculate $s(62)$ and $s(122)$, and discuss the practical significance of the results.

1.5 Some Useful Functions

Before proceeding further, it would be advantageous for us to have at our disposal certain elementary functions and their graphs. We list these functions and their graphs below.

(1) *The constant functions*

Any function of the form, $f(x) = c$, where c is any constant, is called a constant function. Its graph is shown in Figure 1.5.1.

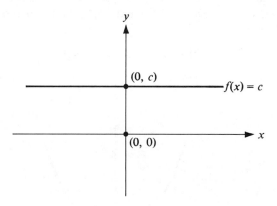

Graph of $f(x) = c$, where $c > 0$

FIGURE 1.5.1

(2) *Linear functions*

Any function of the form, $f(x) = ax + b$, where a and b are constants, is called a linear function. The word linear is used because the graph of this function (see Section 1.3) is a straight line with slope a. Its graph is shown in Figure 1.5.2.

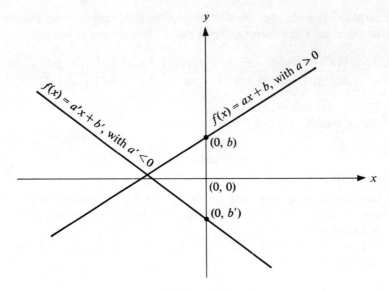

FIGURE 1.5.2

(3) *Power functions*

$f(x) = x^n$, where n is a positive integer. We observe that if $n = 1$, then the graph of f is a straight line with slope 1. For $n \geq 2$, the general shape of the graph depends upon whether n is an even or an odd number. If n is even, the shape is similar to the graph of $f(x) = x^2$, shown below. (The graph of this function is called a parabola.) If n is odd and > 3, then the graph is similar to $f(x) = x^3$—but flatter near zero and steeper outside of $[-1, 1]$. (See Figures 1.5.3 and 1.5.4.) Notice that if $n = 2$, and we replace x by $-x$ we obtain

$$f(-x) = (-x)^2 = x^2 = f(x).$$

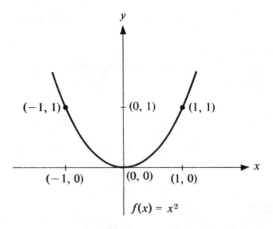

FIGURE 1.5.3

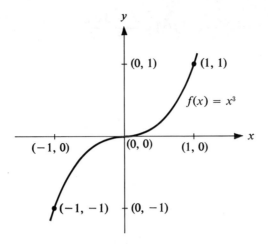

FIGURE 1.5.4

This geometrically means that the graph of f is *symmetric* about the y axis—that is, if (x, y) is on the graph, then so is $(-x, y)$. In fact, observe that this is true for any even $n \geq 2$. A function having the property that $f(-x) = f(x)$ is called an *even* function.

Also note that if $n = 3$ and we replace x by $-x$, we obtain $f(-x) = (-x)^3 = -x^3 = -f(x)$. Geometrically, the graph of f is symmetric about the origin—that is, if (x, y) is on the graph then so is $(-x, -y)$. Again, this is true for n odd and ≥ 3. A function possessing the property that $f(-x) = -f(x)$ is called an *odd* function. The notion of odd and even functions is introduced as a labor-saving device. If a function turns out to be odd or even, once we graph half of it the other half is obvious.

(4) *Polynomial functions*

A *polynomial function* is a function, defined for all real numbers x, of the form

$$f(x) = c_0 + c_1 x + c_2 x^2 + c_3 x^3 + \cdots + c_n x^n,$$

where $c_0, c_1, \ldots, c_n$ are given real numbers. Observe that the preceding three functions given above are special cases of this function. For example, if all of the c's except c_n are zero and $c_n = 1$, we obtain the power function. We show methods of quickly constructing graphs of polynomials in Chapter 4.

(5) *Rational functions*

A *rational function* is defined to be a function of the form

$$f(x) = \frac{g(x)}{h(x)},$$

where g and h are polynomial functions. Remember that we must exclude all those real numbers x for which $h(x) = 0$ from our domain. Some examples of rational functions are:

Example 1.
$$f(x) = \frac{x^2 + 3x + 1}{x^2 - 4}.$$

Thus, $\pm 2 \notin$ domain of f.

Example 2.

$$f(x) = \frac{x^3 + 2x^2 + x + 1}{x^2 + 1}.$$

Note that the domain of f includes all real numbers since $x^2 + 1 \neq 0$ for any real number x.

Example 3.   A very important rational function is

$$f(x) = \frac{1}{x}$$

whose graph we draw in Figure 1.5.5. We note that $0 \notin$ domain of f. Since $f(-x) = -f(x)$ (Why?), we need only draw the graph in the first quadrant and by symmetry obtain the part of the graph in the third quadrant. It should be intuitively clear that as we choose larger and larger values of x, the values $f(x)$ become smaller and smaller—that is, become closer to zero.

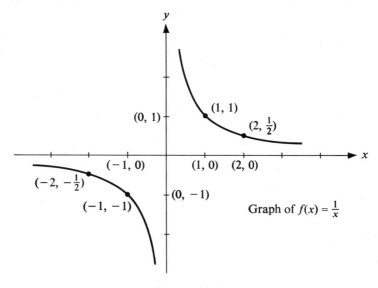

FIGURE 1.5.5

(6) *Root functions*

If we look at the graph of $f(x) = x^2$ in Figure 1.5.3 we are led to believe that there exists *exactly* one nonnegative number p such that $p^2 = w$, where $w \geq 0$. This is easy to see as follows: Suppose there were two numbers p and q such that $p^2 = w$ and

$q^2 = w$. Then $p^2 = q^2$ and hence $p^2 - q^2 = 0$, which implies that $(p - q)(p + q) = 0$ and thus $p = q$ or $p = -q$—but since p is not negative, $p = q$. Thus, there is one and only one nonnegative number p such that $p^2 = w$. We designate this number p by $w^{1/2}$. In fact, it is possible (though we shall not do it here) to show that if n is even there is exactly one nonnegative number p such that $p^n = w$. We then define this number to be $p = w^{1/n}$. (This simply says that $w^{1/n}$ is that number that raised to the nth power yields w.) We say that f is a *root function* if

$$f(x) = x^{1/n}.$$

Sometimes we use the notation $x^{1/n} = \sqrt[n]{x}$. The domain of f is the set of all nonnegative real numbers. A similar discussion for n odd shows that there is exactly one real number p such that $p^n = w$ for every real number w. This is easy to see if we take $n = 3$. We again call this unique number $w^{1/n}$ and define the nth root function by $f(x) = x^{1/n}$. In this case, the domain of f is the set of *all* real numbers.

The graphs of the resulting functions are shown below in Figures 1.5.6 and 1.5.7. Notice that for n even, the graph is obtained by reflecting the right-hand branch of x^n through the line $y = x$.

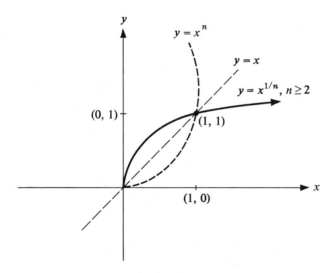

FIGURE 1.5.6

(7) *Exponential functions*

Let a be a real number, where $a > 0$. We shall proceed to define the function $f(x) = a^x$, where x is any real number. This will take several steps. If you believe you can eliminate some of the steps, stop and think.

DEFINITION. *If a is any positive number, then a^x is defined as follows:*
   (i) *if x is a positive integer, say $x = n$, $a^x = a^n = \underbrace{(a \cdot a \cdot a \cdots a)}_{(n\,\text{times})}$ (that is, there are n factors of a).*
   (ii) *if $x = 0$, then $a^0 = 1$.*

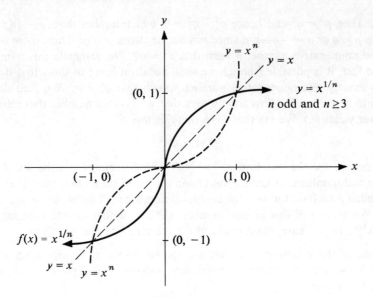

FIGURE 1.5.7

(iii) *if x is a negative integer, say x = −n, then $a^x = a^{-n} = 1/a^n$.*
(iv) *if x is a rational number, say x = p/q, then $a^x = a^{p/q} = \sqrt[q]{a^p}$ (the qth root of a to the pth power).*
(v) *After we have defined a^x for x rational, the graph of the function $f(x) = a^x$ for x rational would appear as in Figure 1.5.8.*

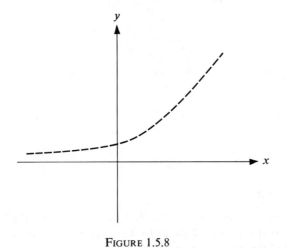

FIGURE 1.5.8

To define a^x on the irrational numbers we simply fill in the gaps (points) in the obvious way (that is, we follow the obvious pattern).

It therefore follows that the function $f(x) = a^x$ has for its domain the set of real numbers.

The graph of this function for $a > 1$ is shown in Figure 1.5.9 below. Observe that as x becomes very large (positively) $f(x)$ also becomes large. We observe that in like manner if x becomes large negatively then $f(x)$ becomes close to zero. We also see that a^x is always *positive*.

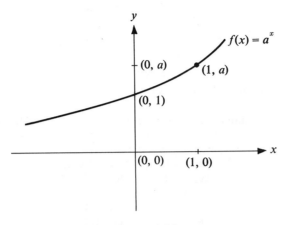

FIGURE 1.5.9

In our future discussions we find that the irrational number $a = e = 2.71828\ldots$ appears very often. The function $f(x) = e^x$ is called the exponential function and is sometimes written $f(x) = \exp(x)$.

From the definition of a^x, we can easily show that the following properties hold:

PROPERTY 1. $a^x \cdot a^y = a^{x+y}$. [*In function notation this states that $f(x) = a^x$ satisfies the equation $f(x)f(y) = f(x + y)$.*]

PROPERTY 2. $(a^x)^y = a^{xy}$.

PROPERTY 3. $\dfrac{a^x}{a^y} = a^x \cdot a^{-y} = a^{x-y}$.

PROPERTY 4. *If $a \neq 1$, then $a^x = a^y$ if and only if $x = y$.*

These properties of exponential functions are very useful and should be remembered. Also observe the difference between a power and an exponential function—for a power function the *base* is variable and the exponent is fixed, and for an exponential function base a is fixed and the exponent variable!

(8) *The logarithm function*

Let $a > 0$. We define $\log_a x = y$ if $a^y = x$. In words, the log to the base a of x is y if $a^y = x$. This is really simpler than it sounds. Consider the following table.

$$2^0 = 1 \qquad 2^6 = 64$$
$$2^1 = 2 \qquad 2^7 = 128$$
$$2^2 = 4 \qquad 2^8 = 256$$
$$2^3 = 8 \qquad 2^9 = 512$$
$$2^4 = 16 \qquad 2^{10} = 1024$$
$$2^5 = 32$$

What is $\log_2 64$ (that is, to what powers do you have to raise 2 to get 64)? By glancing at the chart it is easy to see that $2^6 = 64$ or $\log_2 64 = 6$.
What is $\log_2 512$? *Answer:* 9.

Using the information in the table we can construct a partial table of logs to the base 2.

$$\log_2 1 = 0 \qquad \log_2 64 = ?$$
$$\log_2 2 = 1 \qquad \log_2 128 = ?$$
$$\log_2 4 = 2 \qquad \log_2 256 = ?$$
$$\log_2 8 = 3 \qquad \log_2 512 = ?$$
$$\log_2 16 = 4 \qquad \log_2 1024 = ?$$
$$\log_2 32 = ?$$

Of course, this table will not help us find $\log_2 17$.

DEFINITION.   *Let $a > 0$ be fixed. Given $x > 0$ there is exactly one real number y such that $a^y = x$. This unique number y is called the logarithm of x to the base a, and we write $y = \log_a x$.*

We now consider the function $f(x) = \log_a x$, where $a > 0$. The domain of f is all positive real numbers, that is, all x such that $x > 0$. We are mainly interested in the function $f(x) = \log_a x$ when $a > 1$. Its graph appears in Figure 1.5.10. Note that the graph of $\log_a x$ is obtained by reflecting the graph of $y = a^x$ about the line $y = x$. Thus, if you can remember the graph of one of them, you can easily obtain the other.

Again, from the standpoint of calculus, a very convenient base is $a = e = 2.71828\ldots$. We call the logarithm to the base e, the *natural* logarithm and write

$$f(x) = \log_e x = \log x = \ln x.$$

All of the above notations are frequently used for natural logarithms. Our convention will be that we always mean the natural logarithms whenever we omit the base in our notation.

Properties similar to those for exponential functions hold for logarithmic functions. In fact, their derivation follows directly from the analogous ones of exponential functions.

PROPERTY 1. $\log_a(xy) = \log_a x + \log_a y$.

PROPERTY 2.   *If r is any real number, then $\log_a x^r = r \log_a x$. In particular,*

$$\log_a a^x = x \log_a a = x.$$

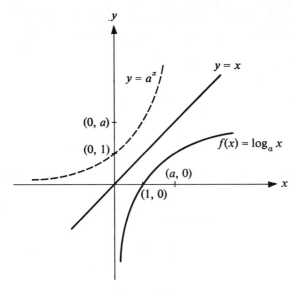

FIGURE 1.5.10

PROPERTY 3. $\log_a \dfrac{x}{y} = \log_a x - \log_a y$.

PROPERTY 4. $\log_a x = \log_a y$ *if and only if* $x = y$.

We outline a method of obtaining Properties 1–4 for logarithmic functions in the exercises which follow.

We illustrate the usefulness of the preceding properties via the following example.

Example. Find $16 \cdot 32$.

Solution. This is not terribly difficult, but it is even less odious if done by logarithms. Now, $\log_2(16 \cdot 32) = \log_2 16 + \log_2 32$ by Property 1, and from the table, $\log_2 16 + \log_2 32 = 4 + 5 = 9$. Thus, $\log_2(16 \cdot 32) = 9$. Hence from the table and Property 4 we conclude that $16 \cdot 32 = 2^9 = 512$.

Exercises

1. (a) Sketch a graph of the following functions.

  (i) $f(x) = \dfrac{x^2}{2}$ (iii) $f(x) = \dfrac{x^3}{3}$ (v) $f(x) = x^5$

  (ii) $f(x) = 2x^2$ (iv) $f(x) = 3x^3$

  (b) In (a), which of the functions are odd functions? even functions?

2. Sketch a graph of the following functions.

  (a) $f(x) = \dfrac{1}{2x}$

(b) $f(x) = 2x^{1/2}$ [HINT: First graph $g(x) = 4x^2$.]

(c) $f(x) = 3x^{1/3}$ [HINT: First graph $g(x) = 27x^3$.]

3. A psychology student believes that the score of a student on a given test worth 100 points can be expressed by the function

$$S(t) = \frac{I^2(50 + 5t - t^2)}{2000}$$

where I is the IQ of the given student and t is the number of hours spent studying for the test. Thus the score for a given individual is a function of the time spent studying. Find $S(t)$ for the following students with the given IQ's when $t = 0, 2, 10$.

Student	IQ
A	90
B	100
C	120
D	150

Do you see anything wrong with this formula?

4. Taxis charge 25 cents for the first quarter mile of transportation and 5 cents for each additional quarter mile or part thereof. Graph this function for $0 \leq x \leq 4$, where x represents number of miles travelled.

5. A travel agency advertises all-expenses-paid trips to the Cotton Bowl for special groups. Transportation is by train. The agency charters one car on the train seating 50 passengers, and the charge is $100 plus an additional $5 for each empty seat. (Thus, if there are 4 empty seats each person pays $120, if there are 5 empty seats each person pays $125, etc.) If there are x empty seats, how many passengers are there on the train, how much does each passenger have to pay, and what are the travel agency's total receipts? Plot the graph of the function that relates the travel agency's total receipts to number of empty seats and judge under what conditions its receipts will be greatest.

6. In connection with Exercise 5, how many passengers would have to go on the trip to give the travel agency total receipts of $3000?

7. (a) $2^2 \cdot 2^3 = 2^?$ (g) Does $3^{(2^4)} = (3^2)^4$?

(b) $2^3 \cdot 3^3 = ?$

(c) $5^3 \cdot 5^{-4} = 5^?$ (h) $\dfrac{5^4}{5^3} = 5^?$

(d) $7^4 \cdot 7^{1/3} = 7^?$ (i) $(8)^{1/3} = ?$

(e) $(2^2)^5 = 2^?$ (j) $(4^{1/2})^3 = (4^3)^{1/2} = ?$

(f) $6^2 = 2^? \cdot 3^?$

8. Sketch a graph of $f(x) = 2^x$ and $g(x) = x^2$ on the same axes, noting the differences.

9. A simple example of an exponential function arises in the study of *compound interest*. This is interest added to the principal at regular intervals of time and thereafter the interest itself earns interest. The interval of time between successive calculations of interest is called the conversion period. If the conversion period is one year, we say that the interest is compounded annually and if we borrow (or invest) P dollars compounded annually at the interest rate i, then the amount we owe (or have coming) at the end of n years is given by

$$A = P(1 + i)^n.$$

   (a) If \$10,000 is invested at 10% compounded annually, find the value of this investment after one year, 5 years, and 25 years. [HINT: For 5 years, $A = 10,000(1 + 0.10)^5 = 10,000(1.10)^5$. Note that to find $(1.10)^5$ we can use logarithms. Namely, $\log(1.1)^5 = 5 \log(1.1) \cong 5(0.095) = 0.475$ (from Table A.2) and $(1.1)^5 = {}^{0.475} \cong 1.6$ from Table A.1. ($\cong$ means approximately).]
   (b) If someone borrows \$3000 at 6% compounded annually, what is the value of this investment after 3 years, 5 years, and after 20 years?
   (c) Suppose we invest \$5000 at 4% compounded quarterly. What amount do we have coming after 1 year? 3 years? 20 years? [HINT: Now $i = 0.04/4$; n must now represent conversion period in terms of months. Thus, $n = 4$. (There are 4 conversion periods in a given year.)]

10. (a) $\log_3 9 = ?$ (d) $\log_3 1 = ?$ (g) $\log_{13} 169 = ?$
     (b) $\log_3 27 = ?$ (e) $\log_4 256 = ?$
     (c) $\log_3 243 = ?$ (f) $\log_4 16 = ?$

11. (a) Consider 3^{3^3}. This is ambiguous without parenthesis. Which is larger, $(3^3)^3$ or $3^{(3^3)}$?
     (b) What is the largest number you can write with four 2's? Make a quick guess as to the relative size of 22^{22}, 2^{222}, $2^{[2^{(22)}]}$, 2222, $2^{(222)}$. (*Note:* The last number is larger than 1 followed by one million zeros.)

12. Show how one can obtain the graph of the function $f(x) = \log_2 x$ from the graph of $g(x) = 2^x$. Sketch both graphs on the same axes. [HINT: Look at Figure 1.5.10.]

13. Using Properties 1–4 for exponential functions, show that
     (i) if $f(x) = a^x$, then $f(x + h) - f(x) = a^x(a^h - 1)$.
     (ii) if $f(x) = a^x$, then $f(x - 1)f(1) = f(x)$.
     (iii) if $f(x) = a^x$, then $f(x + h)/f(h) = f(x)$.

14. Using Properties 1–3, establish the corresponding Properties 1–4 for logarithmic functions. [HINT: To prove Property 1, let $w = \log_a x$ and $v = \log_a y$; then by definition, $x = a^w$ and $y = a^v$. Hence, $xy = a^w \cdot a^v = a^{w+v}$ (by Property 1 for exponential functions). Thus, $w + v = \log_a xy$ and hence $\log_a(xy) = \log_a x + \log_a y$. The other three properties are similarly derived.)

15. Using logarithmic Properties 1–4, show that
     (a) if $f(x) = \log_a x$, then $f(x + h) - f(x) = \log_a(1 + h/x)$.
     (b) if $f(x) = \log_a x$, then $f(x) - f(x^2) = -f(x)$.
     (c) if $f(x) = \log_a x$, then $f(pq) - f(p/q) = 2f(q)$.
     (d) if $f(x) = \log_a x$, then $a^{f(x^2)} = x^2$.

1.6 Operations on Functions

It is possible to combine functions in several ways to arrive at new functions. To simplify matters we assume that f and g are two functions having the real line as their domain. This way we can forget about domains and concentrate on more basic issues. First, it is possible to add f and g together. We could designate this new function simply as $f + g$. This does not make explicit what number $f + g$ associates with 14, for example. So instead we write it another way, namely

$$[f + g](x) = f(x) + g(x).$$

The reason for putting square brackets around $f + g$, that is, writing $[f + g]$, is to indicate that this is now a function in its own right. The domain of the new function $f + g$ is simply the set of numbers common to both the domain of f and the domain of g.

Example. If $f(x) = x + 1$ and $g(x) = |x|$ then graph $[f + g](x)$.

  Solution.   First we draw the graphs of f and g as in Figure 1.6.1 (a) and (b). The graph

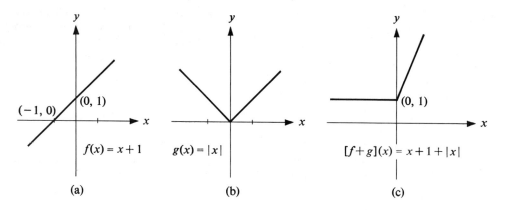

FIGURE 1.6.1

of $f + g$ is drawn by inspecting the graphs of f and g. This method works in very simple cases. One could have done the example by simplifying the expression for $[f + g]$. More precisely, $[f + g](x) = x + |x| + 1$. If $x \geq 0$, $x + |x| = 2x$ while if $x < 0$, $x + |x| = 0$. Hence,

$$[f + g](x) = \begin{cases} 2x, & \text{if } x \geq 0 \\ 1, & \text{if } x < 0. \end{cases}$$

This is precisely the graph shown in Figure 1.6.1 (c). Note that the domain of f is the set of all real numbers and the domain of g is the same. Thus, the domain of $f + g$ is also the set of all real numbers.

  The function obtained by subtracting g from f, that is, $f - g$ or $[f - g](x) = f(x) - g(x)$, is similar to the addition case and we illustrate it with an example and go on.

Example. Let $f(x) = 2x + 1$. Let $g(x) = 2^x$. Graph $[f - g](x)$.

Solution. Note that $[f - g](x) = f(x) - g(x) = 2x + 1 - 2^x$.

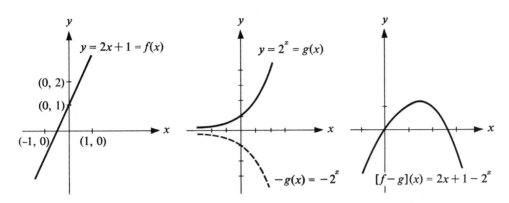

FIGURE 1.6.2

Since the domain of both f and g is the set of all real numbers, so is the domain of $f - g$. (Those numbers common to both the domain of f and of g.) (See Figure 1.6.2.)

Observe that a polynomial can be considered to be the sum of several power functions.

Next, we consider the function obtained by multiplying the functions f and g. This will be designated by $f \cdot g$ or $[f \cdot g](x) = f(x) \cdot g(x)$. Again, the domain of the new function $f \cdot g$ is the set of real numbers common to the domain of f and of g. In many cases it is easy to see what $f \cdot g$ looks like by manipulating the product in terms of x. That is, if $f(x) = (1 - x^4)$ and $g(x) = x^2/(1 + x^2)$ then

$$[f \cdot g](x) = (1 - x^4) \cdot x^2/(1 + x^2)$$
$$= (1 - x^2)(1 + x^2) \cdot x^2/(1 + x^2)$$
$$= (1 - x^2)x^2 = x^2 - x^4.$$

The graph of $[f \cdot g](x)$ is shown in Figure 1.6.3. The domain of f and g is the set of all real numbers. Hence, so is the domain of $f \cdot g$. (Cancelling the term $(1 + x^2)$ could lead to difficulties but since the term is never zero it is permissible. In this text we will be casual about such cavalier treatment of rational functions but it should be mentioned from time to time.)

On the other hand, in some cases one can see fairly quickly what $f \cdot g$ is by graphing f and g.

Example. Let

$$f(x) = \begin{cases} 2 & \text{for } x \le 1 \\ -2 & \text{for } x > 1 \end{cases} \quad \text{and} \quad g(x) = 2^x.$$

Graph the function $[f \cdot g](x)$.

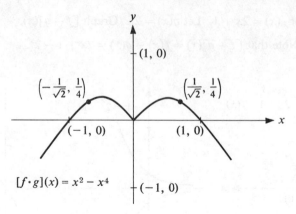

FIGURE 1.6.3

Solution.   We first note that

$$[f \cdot g](x) = f(x)g(x) = \begin{cases} 2^{x+1} & \text{for } x \le 1 \\ -2^{x+1} & \text{for } x > 1. \end{cases}$$

Again, the domain of $f \cdot g$ is the set of all real numbers. The graphs of f, g, and $f \cdot g$ are shown in Figure 1.6.4.

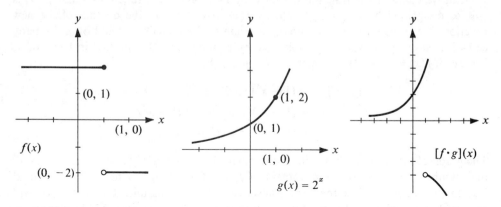

FIGURE 1.6.4

Consider, now, the function obtained by dividing two functions. This is obviously very similar to multiplication. We designate it by f/g or $[f/g](x) = f(x)/g(x)$, provided that $g(x) \ne 0$ for real numbers x. In this case, the domain of f/g is the set of real numbers common to the domain of f, the domain of g, and the set of x's such that $g(x) \ne 0$.

Example.   Let $f(x) = x^2 + 1$ and $g(x) = x^2 - 1$. (See Figure 1.6.5.) Graph $[f/g](x)$.

Solution.   Before we graph $[f/g](x) = (x^2 + 1)/(x^2 - 1)$ let us make a few observations. First we will rewrite it as $[f/g](x) = (x^2 + 1)/[(x - 1)(x + 1)]$. Note that the

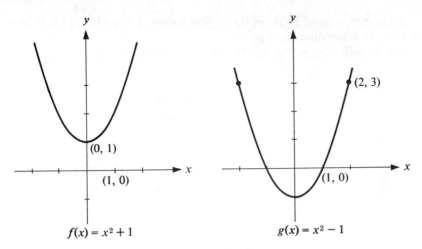

FIGURE 1.6.5

denominator vanishes for $x = \pm 1$. Moreover the expression

$$x^2 - 1 \quad \text{is} \quad \begin{cases} \text{positive for } x < -1 \\ \text{negative for } -1 < x < 1 \\ \text{positive for } x > 1. \end{cases}$$

Now it is not too hard to see what f/g look like. (See Figure 1.6.6.) The domain is the set of all real numbers except $x = \pm 1$.

We can also get a new function h from f and g by setting $h(x) = g(f(x))$. In this case h is a function of a function and we say that h is the composition of g and f and write

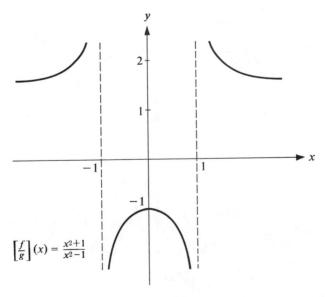

FIGURE 1.6.6

$h = [g \circ f]$ or $h(x) = [g \circ f](x) = g[f(x)]$. The domain of $g \circ f$ is $\{x \mid x$ is in the domain of x and $f(x)$ is in the domain of $g\}$.

Consider the following picture of the rule for h (Figure 1.6.7).

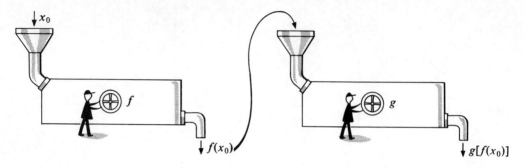

FIGURE 1.6.7

Thus, if x_0 is fed into the first machine, $f(x_0)$ is the output. Then, $f(x_0)$ is fed into the second machine whose output is $g(f(x_0))$.

Let us consider several examples.

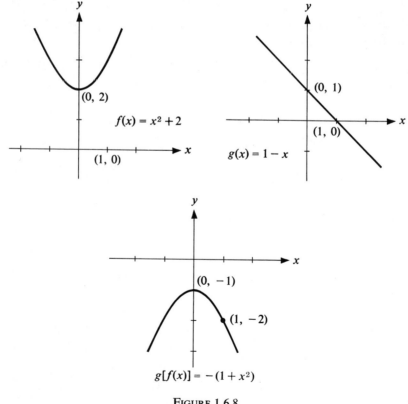

FIGURE 1.6.8

Example 1.   Let $f(x) = x^2 + 2$ and $g(x) = 1 - x$. Then $[g \circ f](x) = g[f(x)] = 1 - f(x) = 1 - (x^2 + 2) = -(1 + x^2)$ or $g[f(x)] = g(x^2 + 2) = 1 - (x^2 + 2) = -(1 + x^2)$. The domain of $g \circ f$ is clearly the set of all real numbers. The graphs of f, g, and $g \circ f$ are shown in Figure 1.6.8.

Example 2.   Let

$$f(x) = \begin{cases} x \text{ if } 0 \le x \le 1 \\ 2 \text{ if } x > 1 \end{cases} \quad \text{and} \quad g(x) = x^2.$$

Then

$$[g \circ f](x) = g[f(x)] = \begin{cases} g(x) \text{ if } 0 \le x \le 1 \\ g(2) \text{ if } x > 1 \end{cases}$$

therefore

$$[g \circ f](x) = \begin{cases} x^2 \text{ if } 0 \le x \le 1 \\ 4 \ \ \text{ if } x > 1. \end{cases}$$

The domain of $g \circ f$ is $\{x \mid x \ge 0\}$. The graphs of f, g, and $g \circ f$ are shown in Figure 1.6.9.

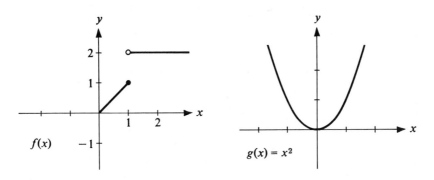

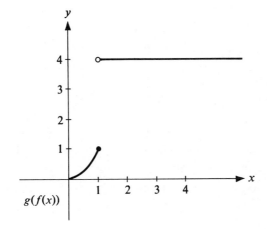

FIGURE 1.6.9

Note that

$$f[g(x)] = \begin{cases} g(x) & \text{if } 0 \le g(x) \le 1 \\ 2 & \text{if } g(x) > 1 \end{cases} = \begin{cases} x^2, & x^2 \le 1 \\ 2, & x^2 > 1. \end{cases}$$

Thus,

$$f[g(x)] = \begin{cases} x^2 & -1 \le x \le 1 \\ 2 & \text{for all other } x. \end{cases}$$

The domain of $f \circ g$ is the set of all real numbers. (See Figure 1.6.10.) Hence, we see that, in general, $g[f(x)] \ne f[g(x)]$.

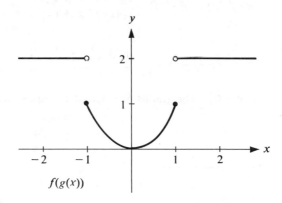

$f(g(x))$

FIGURE 1.6.10

Exercises

1. Let $f(x) = 4x$ and $g(x) = x + 1$. Graph the following.
   (a) $[f + g](x)$ (b) $[f \cdot g](x)$ (c) $[f \circ g](x)$
   What are the domains of the functions in (a), (b), and (c)?

2. Let $f(x) = 1$ and $g(x) = x$. Graph f/g and g/f.

3. Let f and g be given functions.
   (a) Is $[f + g] = [g + f]$? Why?
   (b) Is $[f \cdot g] = [g \cdot f]$? Why?
   (c) Is $[f/g] = [g/f]$? Why? (Consider Exercise 2.)

4. Let $f(x) = 3^x$ and $g(x) = 1 + x$. Graph $(g \circ f)(x)$. Is $[g \circ f](x) = [f \circ g](x)$?

5. Let $f(x) = \log x$ and $g(x) = e^x$. Graph $[g \circ f](x)$ and $[f \circ g](x)$. (*Note of caution:* $g[f(x)]$ takes on *no* negative values.)

6. Write the following functions as the composition of two functions.

   Example. $h(x) = e^{(x^2 + 9)}$.

Solution. $h(x) = g[f(x)]$ when $f(x) = x^2 + 9$ and $g(x) = e^x$.

(a) $h(x) = \log(x^2 + 7)$ (f) $h(x) = e^{|x|}$

(b) $h(x) = \sqrt{\dfrac{x+1}{x-1}}$ (g) $h(x) = 2^{(3x)}$

(c) $h(x) = \log(x + e^{-x} + 1)$ (h) $h(x) = 3^{(x+4)^3}$

(d) $h(x) = e^{(1+x+x^2+x^3)}$ (i) $h(x) = \log(\log x)$ for $x > 1$

(e) $h(x) = \left[\dfrac{\log(x^2+1)}{x^4+1}\right]^{1/3}$ (j) $h(x) = (\log|x+1|)^2$

                                     (k) $h(x) = e^{\sqrt{1+x}}$ for $x \geq 0$.

7. The domain of f consists of all numbers x such that $-2 \leq x \leq 2$ and

$$f(x) = \begin{cases} -2 & \text{if } x \text{ is an integer} \\ 2 & \text{if } x \text{ is not an integer.} \end{cases}$$

 (a) Draw the graph of f.
 (b) What is $f(2)$? $f(\frac{1}{2})$?
 (c) Let g be a function with the same domain as f, defined by $g(x) = [f \circ f](x)$. Draw the graph of g.

8. Let $f(x) = x^2$ and $g(x) = x^{1/2}$. Is $[f \circ g](x) = [g \circ f](x)$? Why?

9. Let f be a function whose domain is the set of real numbers.
 (a) Let $g(x) = 7 + f(x)$. Geometrically how is the graph of g obtained from the graph of f?
 (b) Let $h(x) = f(x - 3)$. How is the graph of h obtained from the graph of f? Can you get the graph of h by leaving f fixed and moving the coordinate axis? Write h as the composition of f and something.

10. Give an explicit example of a function (whose domain is the real numbers) such that
 (a) $f(x) < 2$ for $x < 0$,
 (b) $f(x) > 3$ for $x > 0$,
 (c) $|f(0)| < \frac{1}{2}$.
 (You may need a different "rule" for different sections of the x axis.)

11. Let $f(x) = |x| - 1$. Let $g(x) = |x|$.
 (a) Graph $g[f(x)]$. How many corners does the curve have? How could you get a curve with 5 corners?
 (b) Graph $f[g(x)]$. Note that $f[g(x)] = f(x)$. But $g[f(x)] \neq f(x)$ nor anything like it. This should reinforce the suspicion that $g(f) = f(g)$ is not very likely if f and g are chosen at random.

1.7* Trigonometric Functions

In our study of calculus, it is useful in order to define the trigonometric functions to measure angles in radians. We now define radians. Let us assume we have a circle of radius 1 and a fixed angle α. (See Figure 1.7.1.) We have a tape measure but no pro-

---

* Sections with asterisks may be omitted without loss of continuity.

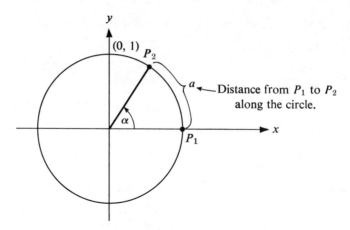

FIGURE 1.7.1

tractor. How can we measure the angle α? The obvious answer is to measure the distance from P_1 to P_2 *along the circle.* Call this distance a. Then the angle α has radian measure a. Since a circle is 360 degrees it is clear that 360 degrees $= 2\pi$ radians. If we started with a circle of radius r, then the angle α would have radian measure a/r. That is, the length of the subtended arc is simply αr. We now return to the trigonometric functions.

We define the *sine* and *cosine* functions as follows: Consider a circle of radius 1 with its center at the origin of the rectangular coordinate system. (See Figure 1.7.2.) For any *nonnegative* u, let $P_u = (a, b)$ be the point obtained by rotating the point $P = (1, 0)$ a

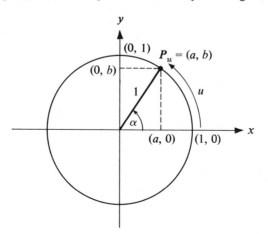

FIGURE 1.7.2

distance of u radians around the circle in a *counterclockwise* direction. Note that since the length of the circumference of the circle is 2π units, if $u > 2\pi$, then P will be rotated about the circle more than once. We define

$$\sin u = b \quad \text{and} \quad \cos u = a.$$

We immediately see, from the Pythagorean theorem, that $a^2 + b^2 = 1$ and hence for any u, $\sin^2 u + \cos^2 u = 1$.

We may define the sine and cosine functions for u negative by taking P_u to be the point obtained by rotating P, $|u|$ units in a *clockwise* direction. (See Figure 1.7.3.)

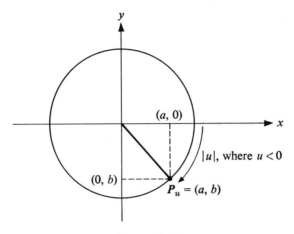

FIGURE 1.7.3

From Figure 1.7.3, we note that if P_u is in the fourth quadrant, then $b < 0$ and $a > 0$—that is, $\cos u > 0$ and $\sin u < 0$. In like manner if P_u is in the second quadrant, $\cos u < 0$, $\sin u > 0$. If P_u lies in the third quadrant, then both $\sin u$ and $\cos u$ are negative. Using the fact that $u = 2\pi$ measures the length of the circumference of the circle we observe that the points $(1, 0)$, $(0, 1)$, $(-1, 0)$, $(0, -1)$ correspond to $u = 0$, $\pi/2$, π, and $3\pi/2$, respectively. Thus, we can state the following:

PROPERTY 1. Sin $0 = 0$, sin $\pi/2 = 1$, sin $\pi = 0$, sin $3\pi/2 = -1$, cos $0 = 1$, cos $\pi/2 = $ cos $3\pi/2 = 0$, cos $\pi = -1$.

PROPERTY 2. If $0 \leq u \leq \pi/2$, then $\cos u \geq 0$ and $\sin u \geq 0$.

PROPERTY 3. If $\pi/2 \leq u \leq \pi$, then $\cos u \leq 0$ and $\sin u \geq 0$.

PROPERTY 4. If $\pi \leq u \leq 3\pi/2$, then $\cos u \leq 0$ and $\sin u \leq 0$.

PROPERTY 5. If $3\pi/2 \leq u \leq 2\pi$, then $\cos u \geq 0$ and $\sin u \leq 0$.

PROPERTY 6. The largest value of both $\cos u$ and $\sin u$ for $0 \leq u \leq 2\pi$ is $+1$ and the smallest value is -1. (That is, $|\cos u| \leq 1$ and $|\sin u| \leq 1$.)

PROPERTY 7. Also observe that $\sin (u + 2\pi) = \sin u$ and $\cos (u + 2\pi) = \cos u$. (This is easily deduced from our definition of the sine and cosine.)

Property 7 states that the sine and cosine functions are *periodic*, with period 2π. In general, we say that a function f is periodic with period p if $f(x + p) = f(x)$. (We assume p is such that $x + p \in$ domain of f.)

From the information given in Properties 1–7, we can easily construct the graphs of the functions, $f(x) = \sin x$ and $g(x) = \cos x$. The x may be thought of as the angle measured in *radians*. Their graphs are shown in Figures 1.7.4 and 1.7.5.

The domain of both the cosine and sine functions is the set of real numbers, while the range of these functions is $[-1, 1]$. From the definitions of the sine and cosine, it is

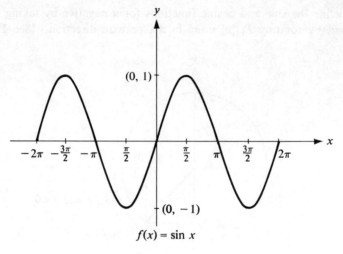

$$f(x) = \sin x$$

FIGURE 1.7.4

easy to see that the cosine function is an even function—[cos $(-x) = \cos x$] and the sine function is an odd function—[sin $(-x) = -\sin x$]. [Figures 1.7.2 and 1.7.3 show that sin $u = b$ and sin $(-u) = -b$, while cos $u = a$ and cos $(-u) = a$].

The remaining trigonometric functions can all be defined in terms of the cosine and sine functions.

DEFINITION 1.   *The tangent is defined by*

$$\tan x = \frac{\sin x}{\cos x}.$$

DEFINITION 2.   *The secant is defined by*

$$\sec x = \frac{1}{\cos x}.$$

DEFINITION 3.   *The cotangent is defined by*

$$\cot x = \frac{\cos x}{\sin x}.$$

DEFINITION 4.   *The cosecant is defined by*

$$\operatorname{cosec} x = \frac{1}{\sin x}.$$

Other properties of the trigonometric functions are discussed as we need them.

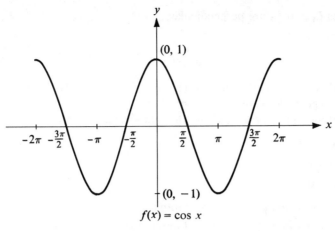

$$f(x) = \cos x$$

FIGURE 1.7.5

Exercises

1. Using Figures 1.7.4 and 1.7.5, sketch the graph of the following functions.

    (a) $f(x) = 2 \sin x$ (e) $f(x) = 3 \cos x$

    (b) $f(x) = -\sin x$ (f) $f(x) = -\cos x$

    (c) $f(x) = |\sin x|$ (g) $f(x) = |\cos x|$

    (d) $f(x) = \sin |x|$ (h) $f(x) = \cos |x|$

2. Using the definitions of the sine and cosine functions, derive the following identities.

    (a) $\sin (t + \pi/2) = \cos t$

    (b) $\cos (t + \pi/2) = -\sin t$

    [HINT: Let $0 < t < \pi/2$, then $y = \sin (t + \pi/2)$ in the figure below. Clearly, $y = \sin (t + \pi/2)$, $x = \cos (t + \pi/2)$, $a = \cos t$ and $b = \sin t$. We must show that $y = a$.

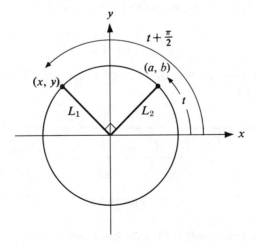

FIGURE 1.7.6

Since the lines L_1 and L_2 are perpendicular,

$$\frac{b}{a}\left(\frac{y}{x}\right) = -1.$$

(Why?) Hence $b^2y^2 = a^2x^2$, and so $(1 - a^2)y^2 = a^2(1 - y^2)$. Thus $y = a$. (Why?)]
A similar type of derivation yields (b).

3. Find all those numbers x satisfying the following equations.

    (a) $\sin x = \cos x$

    (b) $\cos x = 0$

    (c) $\sin x = 0$

4. Let $f(x) = \log x$ and $g(x) = \cos x$. Graph $f + g$.
5. Let $f(x) = \cos x$ and $g(x) = 1 + x$. Graph $g[f(x)]$. What is $[f \circ g](x)$?
6. If $f(x) = e^x + \sin x$, and $g(x) = \log x$, find $[f \circ g](x)$ and $[g \circ f](x)$.
7. If $f(x) = x^2$ and $g(x) = \sin x$, graph $g[f(x)]$. What is $[f \circ g](x)$?
8. Using the definition of $\tan \theta$ and the identity $\sin^2 \theta + \cos^2 \theta = 1$, derive the following identities.

    (a) $\cot \theta = \dfrac{\cos \theta}{\sin \theta}$ (c) $1 + \cot^2 \theta = \csc^2 \theta$

    (b) $1 + \tan^2 \theta = \sec^2 \theta$

1.8 Summary of Chapter 1

We summarize the important results of this chapter below.

(1) Let a and b be two real numbers. We say that $a \geq b$ if $a - b \geq 0$. In like manner, $a \leq b$ if $b - a \geq 0$.

(2) (a) If a is any real number, we define the absolute value of a, written $|a|$ as follows:

$$|a| = \begin{cases} a & \text{if } a \geq 0 \\ -a & \text{if } a < 0. \end{cases}$$

    (b) If b is any real number, then $|x| \leq b$ if and only if $x \geq -b$ and $x \leq b$. (See Figure 1.8.1.)

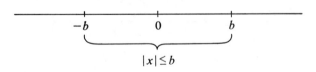

FIGURE 1.8.1

    (c) If b is any real number, then $|x| \geq b$ if and only if $x \leq -b$ and $x \geq b$. (See Figure 1.8.2.)

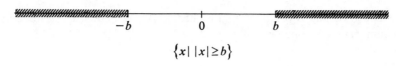

$$\{x|\ |x|\geq b\}$$

FIGURE 1.8.2

(3) Let a, b be any two real numbers. Then
   (a) $[a, b] = \{x|a \leq x \leq b\}$,
   (b) $(a, b) = \{x|a < x < b\}$,
   (c) $[a, b) = \{x|a \leq x < b\}$,
   (d) $(a, b] = \{x|a < x \leq b\}$.

(4) If $P_1 = (x_1, y_1)$ and $P_2 = (x_2, y_2)$ are two points in the coordinate plane then the distance from P_1 to P_2, denoted by $d[P_1, P_2]$, is

$$d[P_1, P_2] = \sqrt{(x_1 - x_2)^2 + (y_1 - y_2)^2}.$$

In particular, if $y_1 = y_2 = 0$, then

$$d[P_1, P_2] = |x_1 - x_2|.$$

(5) Let (x_1, y_1) and (x_2, y_2) be any two distinct points on a line L. The *slope* of the line L, m, is given by

$$m = \frac{y_2 - y_1}{x_2 - x_1}.$$

(6) Forms of the equation for a straight line:
   (a) *Point-slope formula:* If (x_1, y_1) is any point on a line L, with slope m, the equation of the line is given by

$$y - y_1 = m(x - x_1).$$

   (b) *Slope-intercept formula:* If L is a line with slope m and passes through the point $(0, b)$, then its equation is given by

$$y = mx + b.$$

   (c) *Two-point formula:* If (x_1, y_1) and (x_2, y_2) are any two points on a line L, its equation is given by

$$y - y_1 = \left[\frac{y_1 - y_2}{x_1 - x_2}\right](x - x_1).$$

   (d) If a, b, and c are any triple of real numbers, the equation

$$ax + by + c = 0 \qquad (1.8.1)$$

represents the most general form of an equation for a straight line. That is, the equation of any straight line may be written like (1.8.1), and Equation (1.8.1) always represents the equation of a line with slope $-a/b$.

(7) Let m_1 and m_2 be the respective slopes of lines L_1 and L_2. Then
   (a) L_1 is parallel to L_2 if and only if $m_1 = m_2$, and
   (b) L_1 is perpendicular to L_2 if and only if $m_1 m_2 = -1$.

(8) Let A and B be any two sets of real numbers. A *function* is a rule that assigns to each number, $x \in A$, one and only one number $y \in B$. The set A is called the domain of the function. If f denotes the function, we write $y = f(x)$. Unless the domain is explicitly given, it is understood to consist of all numbers for which the definition makes any sense. The graph of a function is

$$\{(x, y) \mid x \in \text{domain of } f \text{ and } y = f(x)\}.$$

(9) *Useful functions*

(a) *The constant function:* $f(x) = c$, where c is a given real number.

(b) *The linear function:* $f(x) = ax + b$, where a and b are given real numbers.

(c) *The power function:* $f(x) = x^n$, where n is a positive integer.

(d) *The root function:* $f(x) = x^{1/n}$, where n is a positive integer. Note that $y = x^{1/n}$ if and only if $x = y^n$.

(e) *The polynomial function:* $f(x) = a_0 + a_1 x + a_2 x^2 + \cdots + a_n x^n$, where the a_i's, $i = 1, 2, \ldots, n$, are given real numbers.

(f) *The rational function:* $f(x) = p(x)/q(x)$, where $p(x)$ and $q(x)$ are polynomials.

(g) *The exponential function:* $f(x) = a^x$, where $a > 0$. The function most useful in calculus is $f(x) = e^x$, where $e = 2.71828 \ldots$. The domain of the exponential function is the set of all real numbers.

(h) *The logarithm function:* $f(x) = \log_a x$, where $a > 0$. This is the "logarithm of x to the base a." Note that $y = \log_a x$ if and only if $x = a^y$. The domain of the logarithm function is the set of all those real numbers $x > 0$. If $a = e$, we say that $\log x$ is the natural logarithm of x and write $\log x = \log_e x = \ln x$.

The graphs of the logarithm and exponential functions are shown in Figure 1.8.3.

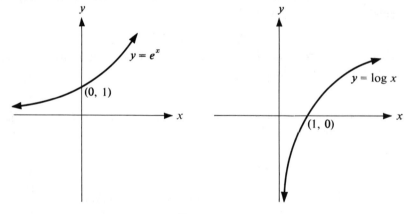

FIGURE 1.8.3

(10) *Operations on functions:*

(a) $[f + g](x) = f(x) + g(x)$

(b) $[f \cdot g](x) = f(x)g(x)$

(c) $[f/g](x) = \dfrac{f(x)}{g(x)}$, provided that x is such that $g(x) \neq 0$

(d) $[f \circ g](x) = f[g(x)]$ and $[g \circ f](x) = g[f(x)]$

REVIEW EXERCISES

1. (a) Find all those real numbers x satisfying the inequalities:

    (i) $|x + 3| \leq 4$ (ii) $|x + \sqrt{2}| > 1$ (iii) $|x - 5| \leq 2$.

    (b) Determine the real numbers x for which $x(x - 1)(x + 2) < 0$.
    In both (a) and (b), illustrate your results on the coordinate axes.

2. What does each statement imply about a, b, c in the line whose equation is

    $$ax + by + c = 0?$$

    (a) The slope is $\frac{3}{2}$.
    (b) The line goes through the points $(4, 0)$ and $(0, 3)$.
    (c) The line goes through the origin.
    (d) The line goes through $(1, 1)$.
    (e) The line is parallel to the x axis.
    (f) The line is perpendicular to the x axis.
    (g) The line is parallel to the line whose equation is $3y = 2x - 4$.
    (h) The line is perpendicular to the line whose equation is $2x - 5y = 7$.
    (i) The line is identical with $y = 3x - 4$.

3. Consider the two lines L_1 and L_2 whose equations are given by

    $$L_1 : x - y = 1 \qquad L_2 : 2x + y = 2.$$

    (a) Graph the two lines and find their point of intersection.
    (b) Find the equation of the line through the point of intersection of the above
    lines and perpendicular to the line

    $$-\tfrac{1}{2}x + 2y = \tfrac{1}{4}.$$

    (c) Sketch a graph of the line found in (b).

4. Suppose that the total cost of two different lawn mowers A and B is \$400. If it is
    known that the total cost of 3 of the A mowers and 7 of the B mowers is no more
    than \$3000, what is the minimum cost of mower A? [HINT: Let x be the cost of
    mower A. Then the cost of mower B is $400 - x$. Now, solve $3x + 7(400 - x) \leq
    3000$.]

5. Two milk trucks must make deliveries over a distance totalling 200 miles. Truck
    A gets 20 miles per gallon of fuel and Truck B gets 15 miles per gallon. If the total
    fuel consumption is to be no more than 10 gallons, what is the minimum number
    of miles Truck A must travel?

6. Let

    $$f(x) = \begin{cases} 2^x, & 0 \leq x < 1 \\ 1, & x = 1 \\ 3 - x, & 1 < x \leq 3. \end{cases}$$

    (a) Sketch a graph of f, clearly indicating the domain and range of the function.
    (b) If a is a number such that $\frac{1}{2} \leq a \leq 1$, what is $f(a + \frac{1}{2})$?

7. Let $f(x) = x^2$ and $g(x) = \sqrt{(1 - x^2)}$.
    (a) What is $[f + g](x)$? What is the domain of $f + g$?
    (b) What is $[f \cdot g](x)$? What is the domain of $f \cdot g$?
    (c) What is $[f \circ g](x)$? What is $[g \circ f](x)$? What is the domain of $f \circ g$? What is the domain of $g \circ f$? Sketch a graph of $f \circ g$?

8. (a) Show how one can obtain the graph of $f(x) = \log_3 x$ from the graph of $g(x) = 3^x$. Draw the graphs of both functions on the same set of axes.
    (b) What is the value of y if $y = 2^{\log_2 34}$?

9. (a) Suppose a and b are positive constants. Let the point $(0, a)$ represent an off-shore rock in the ocean. Let the x axis denote the shore line, with $(b, 0)$ a lifeguard station. The mile is the unit of distance. A man swims from the rock to the point $(x, 0)$ and then runs to the lifeguard station. If he swims s miles per hour and runs r miles per hour, and if T is the total time required for the trip, express T as a function of x. Consider only values of x such that $0 \le x \le b$.
    (b) Choose $a = \frac{3}{4}, b = 1, s = 2, r = 6$, and construct a rough graph of the function, using $x = 0, \frac{1}{4}, \frac{1}{2}, \frac{3}{4}, 1$.

10. A bank pays interest compounded quarterly at a 4% interest rate. What deposit should be made to yield \$2000 at the end of six years? (Refer to Exercise 9 in Section 1.5.)

11. The rate at which a certain chemical salt dissolves in a liquid obeys a law nearly equal to the law $y = 2^{-x}$, where y is the rate of solution in grams per second and x is the time in seconds. Make a sketch of the graph of the function defined by $y = 2^{-x}$ and give the domain and range of the function.

12. (a) If $f(x) = x^2 + x$ compute the value of

$$\frac{f(2 + h) - f(2)}{h}, \qquad \text{where } h > 0.$$

    (b) If $f(x) = |x|$, compute

$$\frac{f(1 + h) - f(1)}{h}, \qquad \text{where } h > 0.$$

    Compute $f(h)/h$, where $h > 0$ and $h < 0$.

13. Show that if $f(x) = (x + 3)/2x - 1)$, then $f \circ f$ gives the identity function g, defined by $g(x) = x$.

2

Limits
and Continuity

2.1 Introduction

After the concept of function, probably the most important concept and perhaps the most difficult one to comprehend is that of a limit. We shall first illustrate the use of the word limit and some related words as they occur in situations already known to the reader.

Suppose a rubber ball is dropped from the roof of a two-story house. It may, in its motion, be said to get closer and closer to the ground. However, there is a distinction between two ways in which the distance of the ball from the ground "approaches zero." If the ball reaches the ground and continues bouncing, its distance from the ground becomes smaller than $\frac{1}{2}$ in., smaller than 0.001 in. and in fact, smaller than any positive distance you like to name. The distance of the ball from the ground can then be said to approach *zero as its limit*. If, on the other hand, the ball gets caught in a tree or lands on the first-floor balcony, then its distance from the ground will remain greater than many distances that could be named. The phrase "as its limit" does not apply in this situation. Thus we see a relationship between the words "approach," "closer and closer," "smaller and smaller," and "limit."

Next, suppose we wish to measure the steepness or slope of a mountainside at some point. What could we do? We could get an *estimate* of the steepness by standing at the point of interest and sending a friend up the mountain a little way. Then, we could compare the difference in height with the horizontal distance between us, giving a ratio which would serve as an approximate slope. This would be the slope of the broken line in Figure 2.1.1. It is clear from Figure 2.1.1 that if we were to send our friend very far away, say over the top of the mountain then the approximate values we get for the slope would be meaningless. On the other hand, if we tell our friend to come closer and closer to us, the estimates would be closer and closer to what we would like to call the slope or

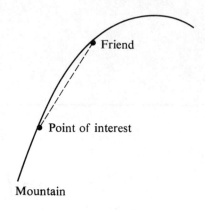

FIGURE 2.1.1

steepness of the mountainside at the point of interest. In fact many more estimates could be obtained by sending him down the mountain, and these would be better if he were not sent too far away. Thus, we would interpret the slope of the mountainside at the point of interest as the number *approached* by the approximations to the slopes of the lines joining us to our friend as our friend comes closer and closer to us down the mountain. We say that this number is the limiting value of the respective slopes. In the next chapter, we represent this situation analytically. Meanwhile, we see that this is a special instance of the following question: If f is a function and a is any given number, then what happens to $f(x)$ as x gets closer and closer to a? Alternatively we could ask, "What is the limit of $f(x)$ as x approaches a?" Thus, again we are in some sense equating the word "limit" with the words "closer and closer." In the next two sections, we attempt to give a satisfactory explanation that answers this question.

2.2 A Geometric View of Limits

Imagine a fly walking along a wall as in Figure 2.2.1. We watch the fly up to the time he reaches the heavy black line and then try to predict at what point he will cross it.

For example, if the fly took the path in Figure 2.2.2 or Figure 2.2.3, then we would predict that he would cross the heavy black line at the dot. However, consider the path of the fly in Figure 2.2.4. His path moves back and forth between the dotted lines infinitely

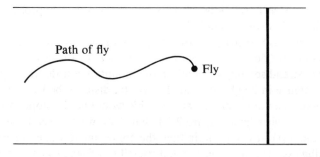

FIGURE 2.2.1

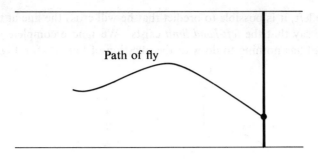

Path of fly

FIGURE 2.2.2

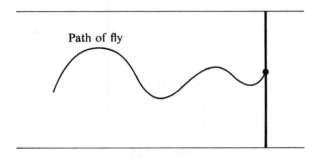

Path of fly

FIGURE 2.2.3

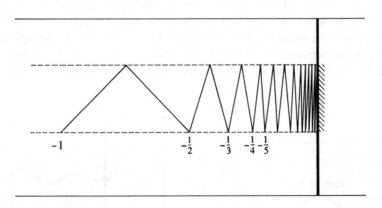

FIGURE 2.2.4

often. Here, it is not possible to predict just where he will cross the heavy black line. Indeed it appears he may cross it at any or every point in the crosshatched region.

At this point in the discussion it is convenient to replace the fly by his path. Thus, instead of the fly, we have a curve $y = f(x)$. Rather than ask where the fly crosses the line $x = a$ (*the heavy black line*), we ask what is the limit of $f(x)$ as x approaches a from the left. (Notice the fly was walking from left to right. He was always to the left of the heavy black line $x = a$.) We write $\lim_{x \to a^-} f(x) = L$ if as the fly approaches the line

$x = a$ from the left, it is possible to predict that he will cross the line at the point (a, L). In this case we say that the *left-hand limit* exists. We ignore completely the value of f at a, that is, $f(a)$ has nothing to do with the question of limits. (See Figure 2.2.5.)

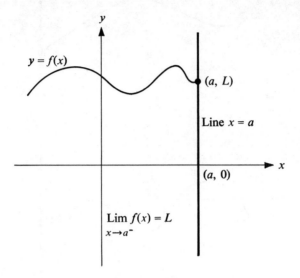

FIGURE 2.2.5

Let us now consider the analogous situation where the fly approaches from the right. Again we ask if it is possible to predict where the fly will cross the heavy black line. In Figure 2.2.6(a) he will clearly cross the heavy black line at the dot. Note that the fly is now approaching the black line from the right. In this case we say the right-hand limit exists. Again, we replace the path of the fly by a curve $y = f(x)$ and replace the heavy black line by the line $x = a$. If we can predict where the fly will cross the line $x = a$, we

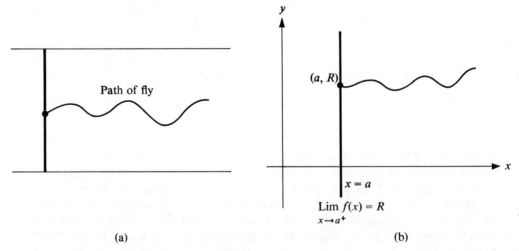

(a)   (b)

FIGURE 2.2.6

say $\lim_{x \to a^+} f(x)$ exists. If the fly will cross at the point (a, R), we say that $\lim_{x \to a^+} f(x) = R$. [See Figure 2.2.6(b).]

The following rule is extremely important.

HEURISTIC RULE.   *We are given a curve $y = f(x)$. We say $\lim_{x \to a} f(x) = L$, that is the limit of f as x approaches a exists and is equal to L, if the left-hand limit at a and the right-hand limit at a both exist and are equal to L.*

The statement that $\lim_{x \to a} f(x)$ exists can be assigned the following geometric interpretation. Given the curve $y = f(x)$ consider it to be the path of two flies; one fly approaching the line $x = a$ from the left and one from the right. If you can predict where each fly will cross the line $x = a$ and if both flies will cross at the same point (that is, hit head on), then $\lim_{x \to a} f(x)$ exists. Otherwise it does not.

We now present several examples to clarify this rule. To reinforce or comment on the unimportance of $f(a)$ in the examples, we will not bother to graph f at a.

Example 1.

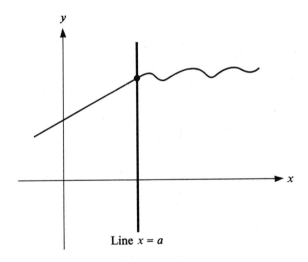

Line $x = a$

FIGURE 2.2.7

In Figure 2.2.7 it is clear that the fly approaching from the left will cross at the dot; the fly approaching from the right will also cross at the dot. Thus both the right- and left-hand limits of f at a exist. Moreover, the crossing point is the same in both cases. (The flies will coincide.) Thus, since the right- and left-hand limits exist and are equal, we conclude that $\lim_{x \to a} f(x)$ exists.

Example 2.

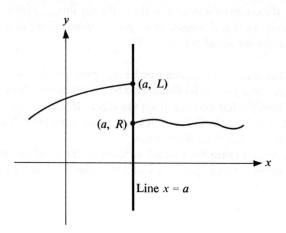

FIGURE 2.2.8

In Figure 2.2.8, both the right- and left-hand limits exist but they are different. Thus, $\lim_{x \to a} f(x)$ does *not* exist.

Example 3.

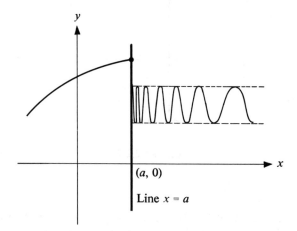

FIGURE 2.2.9

In Figure 2.2.9, the left-hand limit exists but the right-hand limit does not exist. Hence, $\lim_{x \to a} f(x)$ does *not* exist.

Example 4.

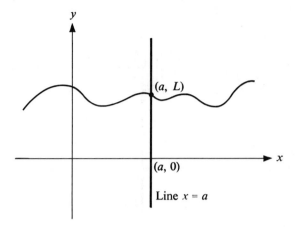

FIGURE 2.2.10

In Figure 2.2.10 both the right- and left-hand limits exist and they are equal. Thus, $\lim_{x \to a} f(x)$ *exists.*

Example 5.

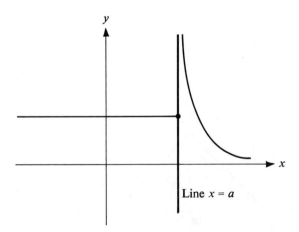

FIGURE 2.2.11

In Figure 2.2.11, the left-hand limit exists but the right-hand limit does not exist. Thus, $\lim_{x \to a} f(x)$ does *not* exist. One might argue that the fly approaching from the right will cross the line $x = a$ at $+\infty$. Geometrically this is correct. However, we do not want to work with infinities and so we will say that the right-hand limit (or left as the case may be) does not exist in such situations.

Example 6.

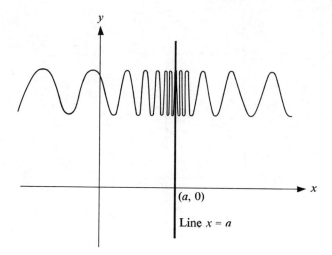

FIGURE 2.2.12

In Figure 2.2.12, neither the right-hand limit nor the left-hand limit exists. Thus, $\lim_{x \to a} f(x)$ does *not* exist.

Observe that we have ignored completely the value of the function at a in all of the above examples. The value $f(a)$ has no effect on the existence or nonexistence of a limit.

Example 7.

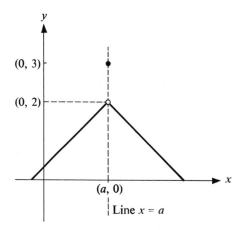

FIGURE 2.2.13

In Figure 2.2.13, we see that the left-hand limit and the right-hand limit exist and that they are equal. Thus, $\lim_{x \to a} f(x)$ exists. In fact, we see in the next section that we can assign the limit a numerical value, namely $\lim_{x \to a} f(x) = 2$. However, $f(a) = 3$. Again we see that the value of f at a has *nothing* to do with the existence of a limit.

In most of the examples above we have not included any numerical values. In simple examples such as these, one can decide whether or not the limit exists by simply examining the picture. Numbers are not necessary. In the next section we study limits by more algebraic techniques.

Exercises

1. Decide from Figure 2.2.14 whether the left-hand limit, the right-hand limit, and the $\lim_{x \to a} f(x)$ exist.

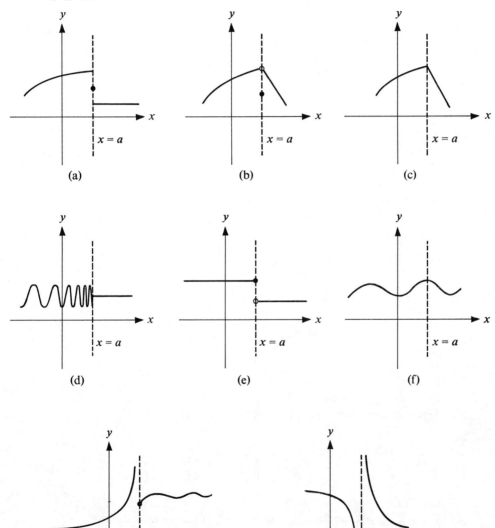

FIGURE 2.2.14

2. Graph the following functions and decide from the graph whether the right-hand limit, the left-hand limit, and the $\lim_{x \to a} f(x)$ exist for a as specified in each part.

(a) $f(x) = \begin{cases} 2x - 1, & x \le 3 \\ -x + 1, & x > 3 \end{cases}$ $a = 3$

(b) $f(x) = \begin{cases} 9 - x, & x < 4 \\ 2x - 3, & x \ge 4 \end{cases}$ $a = 4$

(c) $f(x) = \begin{cases} x + 1, & x < 3 \\ 2, & x = 3 \\ -2x + 10, & x > 3 \end{cases}$ $a = 3$

(d) $f(x) = \begin{cases} 3x - 4, & x < 1 \\ 0, & x = 1 \\ x + 1, & x > 1 \end{cases}$ $a = 1$

(e) $f(x) = \begin{cases} x^2, & x < 0 \\ 5, & x = 0 \\ x, & x > 0 \end{cases}$ $a = 0$

(f) $f(x) = \begin{cases} -3x + 4, & x < 2 \\ 1037, & x = 2 \\ x - 4, & x > 2 \end{cases}$ $a = 2$

(g) $f(x) = \begin{cases} x + 11, & x < -3 \\ 8, & x = -3 \\ 3 + 2x, & x > -3 \end{cases}$ $a = -3$

3. Suppose f is defined as follows:

$$f(x) = \begin{cases} |x|, & -1 \le x \le 0 \\ 2, & 0 < x \le 1. \end{cases}$$

(a) Draw the graph of f.
(b) Do $\lim_{x \to 0^-} f(x)$ and $\lim_{x \to 0^+} f(x)$ exist?
(c) Does $\lim_{x \to 0} f(x)$ exist? Why?
(d) Try to evaluate $\lim_{x \to -1^+} f(x)$ and $\lim_{x \to 1^-} f(x)$.

4. Let

$$g(x) = \begin{cases} x^2, & x < 0 \\ x, & x \ge 0. \end{cases}$$

(a) Graph the function g.
(b) Do $\lim_{x \to 0^+} g(x)$ and $\lim_{x \to 0^-} g(x)$ exist? If so, try to evaluate them.
(c) Does $\lim_{x \to 0} g(x)$ exist?
(d) Does (i) $\lim_{x \to 0^-} [g(x) - g(0)]/x$ exist?
    Does (ii) $\lim_{x \to 0^+} [g(x) - g(0)]/x$ exist?
[HINT: First construct the graph of $[g(x) - g(0)]/x$.]
(e) Does $\lim_{x \to 0} [g(x) - g(0)]/x$ exist? Why?

2.3   Limits of Functions

In the discussion that follows, we rely heavily on our intuition. This part is not intended to be precise and can not be used in a rigorous way. Let us begin by restricting our attention to the function defined by

$$f(x) = x^2, \qquad x \in [0, 3].$$

It is clear that $f(2) = 4$. However, we might ask what the value of f is when x takes on values slightly larger than 2 and slightly smaller than 2. That is, we are no longer concerned with the value of f at 2, but rather the values of f when x gets close to 2 through values larger than 2 and for values smaller than 2. Observe that $f(2.2) = 4.84$, $f(2.1) = 4.41$, $f(2.01) = 4.0401, f(2.001) = 4.004001$, etc. Thus, we see that as x gets closer to 2 through values of $x > 2$, $f(x)$ gets closer and closer to 4. Comparing this with our geometrical discussion in Section 2.2, we see that this is equivalent to stating that the right-hand limit, $\lim_{x \to 2+} f(x) = 4$. In like manner, we see that $f(1.8) = 3.24$, $f(1.9) = 3.61$, $f(1.99) = 3.9601, f(1.999) = 3.996001$, etc. Again, we observe that as x gets closer to 2 through values of $x < 2$, $f(x)$ gets closer and closer to 4. This statement simply says that the left-hand limit, $\lim_{x \to 2-} f(x) = 4$. Thus, as x tends to 2, $f(x)$ tends to 4. Note that the value of $f(x)$ for $x = 2$ plays *no* role. This is roughly what we mean by saying that the limit of $f(x)$ as x tends to 2 is 4. We abbreviate the preceding statement as follows:

$$\lim_{x \to 2} x^2 = 4.$$

You are reminded of the fact that in order to find $\lim_{x \to 2} x^2$, we had to observe values of x^2 for both $x < 2$ and $x > 2$ (that is, we look at $\lim_{x \to 2+} x^2$ and $\lim_{x \to 2-} x^2$)—not caring at all about $x = 2$. Let us investigate what this means geometrically. (See Figure 2.3.1.) In Figure 2.3.1, one clearly sees that $\lim_{x \to 2+} f(x) = \lim_{x \to 2-} f(x) = 4$.

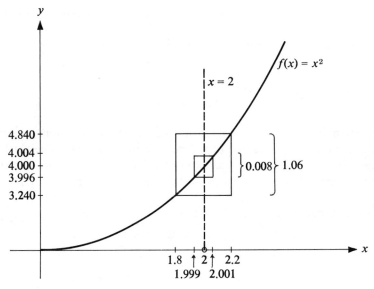

FIGURE 2.3.1

If in Figure 2.3.1 we restrict our attention to the interval (1.8, 2.2), the graph of *f* lies in a rectangle of height 1.06, while for $x \in$ (1.999, 2.001) the graph of *f* lies in a rectangle of height 0.008000. That is, the heights of the rectangles get closer and closer to 0. Note that all the rectangles contain the line $y = 4$ — and thus the heights as $x \to 2$ cluster about the line $y = 4$ — indicating geometrically that $\lim_{x \to 2} f(x) = 4$. This is certainly not a precise formulation of the concept of a limit, but should give us some feeling for the meaning of a limit. Note that the point (2, 0) is to be excluded from our intervals. [This is the reason for the circle about (2, 0)!]

Next we consider the following function defined by

$$f(x) = \frac{x - 2}{2\,|x - 2|}.$$

First, observe that $2 \notin$ domain of *f*. We wish to investigate this function for values of *x* near 2—in fact, we ask whether $\lim_{x \to 2} f(x)$ exists. First observe that if $x > 2$ then $|x - 2| = x - 2$ and hence $f(x) = (x - 2)/2(x - 2) = \frac{1}{2}$; while if $x < 2$, $|x - 2| = -(x - 2)$ and $f(x) = -(x - 2)/2(x - 2) = -\frac{1}{2}$. Thus, it is clear that as *x* approaches 2 through values of $x > 2$, $f(x)$ approaches $+\frac{1}{2}$, and as *x* approaches 2 through values of $x < 2$, $f(x)$ approaches $-\frac{1}{2}$. Hence, $\lim_{x \to 2^+} f(x) = \frac{1}{2}$ and $\lim_{x \to 2^-} f(x) = -\frac{1}{2}$. This is clearly illustrated in Figure 2.3.2. In this case, we must say that $\lim_{x \to 2} f(x)$ does not exist since the left- and the right-hand limits are not equal. (*Note:* The fact that $2 \notin$ domain of *f* did not enter into our discussion.) The graph of *f* is shown in Figure 2.3.2 below.

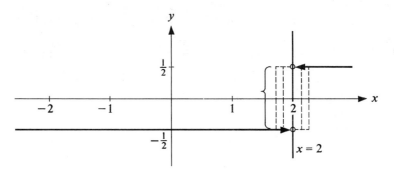

FIGURE 2.3.2

Observe that if in Figure 2.3.2 we restrict our attention to various intervals about (2, 0) (exluding this point), the graph of *f* always lies in a rectangle of height 1. (In fact, this rectangle has $y = \frac{1}{2}$ and $y = -\frac{1}{2}$ as two of its sides.) Thus, in this case the heights *never* approach zero. This is another geometrical argument that $\lim_{x \to 2} f(x)$ does not exist.

Let us now consider the function defined by

$$f(x) = \begin{cases} \dfrac{2x^2 - x - 3}{x + 1}, & x \neq -1 \\[2mm] 2, & x = -1. \end{cases}$$

We wish to investigate the behavior of $f(x)$ as x approaches -1. (That is, we are really asking if we can determine $\lim_{x \to -1} f(x)$ heuristically.) Again we first look at values of x close to -1 with $x > -1$. Clearly, $f(0) = -3$, $f(-0.9) = -4.8$, $f(-0.95) = -4.90$, $f(-0.99) = -4.98$, $f(-0.999) = -4.998$, etc. Thus, as x approaches -1 for $x > -1$, we see that $f(x)$ approaches -5. We can then say that $\lim_{x \to -1^+} f(x) = -5$. In like manner we observe that $f(-2) = -7$, $f(-1.5) = -6$, $f(-1.1) = -5.2$, $f(-1.01) = -5.02$, $f(-1.001) = -5.001$, etc. Hence, it appears that as x approaches -1 for $x < -1$, $f(x)$ approaches -5 and thus $\lim_{x \to -1^-} f(x) = -5$. Since the right- and left-hand limits are equal, we see that $\lim_{x \to -1} f(x) = -5$. Note that $f(-1) = 2$ did not enter into our discussion at all. We see in fact that $\lim_{x \to -1} f(x) = -5$ even though $f(-1) = 2$. The graph of f is shown in Figure 2.3.3.

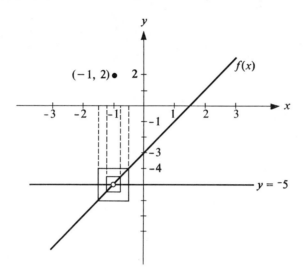

FIGURE 2.3.3

The circle indicates that the point $(-1, f(-1))$ is excluded from the straight line. If we choose intervals about $(-1, 0)$, excluding that point, we can see that the graph of f lies in rectangles whose heights get closer and closer to zero. In fact, they seem to cluster about the line $y = -5$. Once again, we have an indication that $\lim_{x \to -1} f(x) = -5$.

Our arguments for the existence of a limit are fairly awkward to go through each time. As you can see, we rely very heavily on our intuition. In the preceding example, a convenient use of algebra leads us to the conclusion much more quickly. We note that

$$f(x) = \frac{2x^2 - x - 3}{x + 1} = \frac{(2x - 3)(x + 1)}{(x + 1)}.$$

If $x \neq -1$, then

$$f(x) = 2x - 3$$

and one can easily see that as $x \to -1$ for $x > -1$, $f(x) \to -5$ (that is, $\lim_{x \to -1^+} f(x) = -5$) and as $x \to -1$, for $x < -1$, $f(x) \to -5$ (that is, $\lim_{x \to -1^-} (f(x) = -5)$). Hence, we conclude that $\lim_{x \to -1} f(x) = -5$. We emphasize once again the fact that the value of f at $x = -1$ did not play a role in our discussion.

Let us consider one more example. Suppose we define a function f as follows

$$f(x) = \begin{cases} 2x, & 0 \le x < 1 \\ 4, & x = 1 \\ 5 - 3x, & 1 < x \le 2. \end{cases}$$

We wish to determine whether or not $\lim_{x \to 1} f(x)$ exists. (Note that we have not really defined what is meant by "$\lim_{x \to 1} f(x)$ exists"—we rely solely on our previous intuitive discussion.) First, we investigate the behavior of $f(x)$ for values of x close to but greater than 1. Observe that $f(1.2) = 1.4, f(1.1) = 1.7, f(1.01) = 1.97, f(1.001) = 1.997$, etc. It appears that as $x \to 1$, $x > 1$, $f(x) \to 2$. Now we see that $f(0.8) = 1.6, f(0.9) = 1.8$, $f(0.99) = 1.98, f(0.999) = 1.998$, etc. We also note that as $x \to 1$, $x < 1$, $f(x) \to 2$. Thus, $\lim_{x \to 1^-} f(x) = \lim_{x \to 1^-} f(x) = 2$. Hence, we may conclude that $\lim_{x \to 1} f(x) = 2$. Again, note the fact that $f(1) = 4$ plays no role in the determination of $\lim_{x \to 1} f(x)$. The graph of f is shown in Figure 2.3.4.

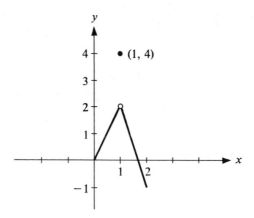

FIGURE 2.3.4

We could again give an intuitive geometrical argument, similar to those of the preceding examples to add further verification that $\lim_{x \to 1} f(x) = 2$. We leave that discussion to the reader. (It would be worthwhile for you to go through this.)

Before proceeding to a workable (but *not* formal) definition of a limit, we again illustrate a case where the limit does not exist. Consider the function

$$f(x) = \frac{1}{x}.$$

We wish to investigate $\lim_{x \to 0} f(x)$. First, we observe values of $f(x)$ for x near 0 and $x > 0$. Note that $f(0.1) = 10, f(0.01) = 100, f(0.001) = 1000, f(0.0001) = 10^4$, etc. That is, we observe that as $x \to 0$, $x > 0$, $f(x)$ becomes larger and larger and does not seem to approach any particular number. (We say that as $x \to 0$, $x > 0$, $f(x) \to \infty$.) Now, observe that $f(-0.1) = -10, f(-0.01) = -100, f(-0.001) = -1000, f(-0.0001) =$

-10^4, etc. It appears that as $x \to 0$, $x < 0$, $|f(x)|$ becomes larger and larger and does not appear to approach any particular number. (We say that as $x \to 0$, $x < 0$, $f(x) \to -\infty$.) Since, as $x \to 0$, $|f(x)|$ gets larger and larger and does not approach a fixed number, we conclude that $\lim_{x \to 0} f(x)$ does not exist. The graph of f is shown in Figure 2.3.5. It is

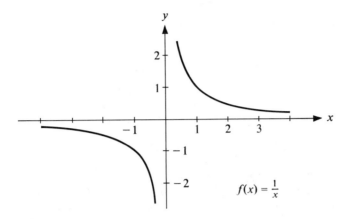

FIGURE 2.3.5

nearly impossible to give an intuitive geometrical argument as in the preceding examples.

We conclude this section with one more negative result for the existence of a limit. Consider the function f defined by

$$f(x) = \begin{cases} x, & 0 \le x \le 1 \\ 3 - x, & 1 < x \le 4. \end{cases}$$

We wish to determine whether or not $\lim_{x \to 1} f(x)$ exists. Looking at values of $f(x)$ for x close to 1 and $x < 1$, we see that $f(0.9) = 0.9$, $f(0.99) = 0.99$, $f(0.999) = 0.999$, etc.— thus we intuitively see that as $x \to 1$, $x < 1$, $f(x) \to 1$. Now for x close to 1 and $x > 1$, we note that $f(1.1) = 1.9$, $f(1.01) = 1.99$, $f(1.001) = 1.999$, etc.—thus as $x \to 1$, $x > 1$ we see that $f(x) \to 2$. Hence, $\lim_{x \to 1^-} f(x) = 1$ and $\lim_{x \to 1^+} f(x) = 2$. We say that $\lim_{x \to 1} f(x)$ does *not* exist since $f(x)$ approaches two different numbers as we let x approach 1. In order that a limit may exist, both numbers must be the same. The graph of f is shown in Figure 2.3.6. Again, a geometrical argument similar to that given in the second example this section of could be given to add further justification to our conclusion.

After discussing all of these examples, one might ask whether it is possible to give a working definition of a limit. The answer to this question is certainly affirmative. The definition we give is clearly *not* a formal one but only one that will aid us in determining the existence of a limit.

PROVISIONAL OR WORKING DEFINITION OF A LIMIT. *We say that* $\lim_{x \to a} f(x)$ *exists and is equal to a number L if the numbers* $f(x)$ *remain arbitrarily close to L when we take x sufficiently close to a, for values of x greater than a and values of x less than a, but* $x \ne a$. *If the numbers* $f(x)$ *remain arbitrarily close to L when* $x > a$ *is sufficiently close to a, we say that the right-hand limit exists and write* $\lim_{x \to a^+} f(x) = L$. *In like manner we define the left-hand limit,* $\lim_{x \to a^-} f(x) = L$.

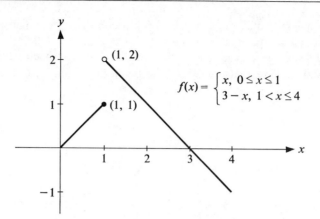

FIGURE 2.3.6

Observe certain facts about the definition:

(1) If $f(x)$ approaches two different numbers, say L_1 and L_2, depending on whether we take x close to a for values of $x > a$ or $x < a$, then we say that $\lim_{x \to a} f(x)$ does *not* exist. That is, if $\lim_{x \to a^+} f(x) = L_1$ and $\lim_{x \to a^-} f(x) = L_2$, then $\lim_{x \to a} f(x)$ does *not* exist.

(2) The fact that f is or is not defined at a does not play any role in the determination of the existence of a limit.

Before proeceeding we consider the following example.

Example.   Let $f(x) = x$ for $0 \le x \le 1$. (See Figure 2.3.7.)

Observe that $\lim_{x \to 1^-} f(x) = 1$. However, since $f(x)$ is not defined for $x > 1$, we cannot speak of $\lim_{x \to 1^+} f(x)$. Thus, we might say that $\lim_{x \to 1} f(x)$ does not exist. However, we use the convention that if 1 is the *right-hand end point* of the domain of f and if $\lim_{x \to 1^-} (x)$ exists, we say that $\lim_{x \to 1} f(x)$ exists. In like manner, $\lim_{x \to 0^+} f(x) = 0$, while we cannot discuss $\lim_{x \to 0^-} f(x)$ since $f(x)$ is not defined for $x < 0$. We use the convention that if 0 is a *left-hand end point* of the domain of f and if $\lim_{x \to 0^+} f(x)$ exists, then $\lim_{x \to 0} f(x)$ exists, and $\lim_{x \to 0} f(x) = \lim_{x \to 0^+} f(x)$.

In general, if $f(x)$ is defined for $x \in [a, b]$, we say that $\lim_{x \to a} f(x)$ exists if $\lim_{x \to a^+} f(x)$

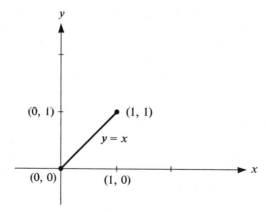

FIGURE 2.3.7

exists and $\lim_{x \to b} f(x)$ exists if $\lim_{x \to b^-} f(x)$ exists. In this case, we write $\lim_{x \to a^+} f(x) = \lim_{x \to a} f(x)$ and $\lim_{x \to b^-} f(x) = \lim_{x \to b} f(x)$.

We give a formal definition of a limit in an appendix to this chapter, but at this point it is desirable to add a historical note.

The problem of giving a precise meaning to $\lim_{x \to a} f(x) = L$ is not an easy one. The calculus reached a reasonable stage of development at the time of Newton and Leibnitz (about 1700) but a satisfactory definition of limit was not arrived at until after the year 1800. Nevertheless an immense amount of mathematics was produced in the eighteenth century. Part of the difficulty seems to be a psychological fluke. The natural (but wrong) approach to the problem is to let the x values get close to a and then claim the $f(x)$ values to be close to L. Slightly reworded, this definition takes the form "as x gets closer and closer to a, $f(x)$ gets closer and closer to L." Unfortunately this is not mathematically precise. The correct approach is to take an open interval about L and claim that $f(x)$ lies in this interval for the x's in some interval about a, but $x \neq a$. If this can be done for any open interval about L then we say $\lim_{x \to a} f(x) = L$.

Thus the trick, which took one hundred years to discover, is to start *not* with a small interval about a, but with a small interval about L. This procedure is examined in more detail in the appendix to this chapter.

Exercises

In all the following exercises, use arguments similar to those given in Section 2.3 to determine the existence of a limit. Construct a graph of each of the functions.

1. If $f(x) = x + 2$, does $\lim_{x \to -1} f(x)$ exist? If so, what is it?
2. If $f(x) = x^2 + 2$, does $\lim_{x \to 1} f(x)$ exist? If so, what is it?

3. Let
$$f(x) = \frac{2x - 2x^2}{x - 1}.$$

Does $\lim_{x \to 1} f(x)$ exist? If so, what is the limit?
[HINT: Use an algebraic manipulation similar to that used in the third example of Section 2.3.]

4. Let
$$f(x) = \frac{|x|}{2x},$$

Does $\lim_{x \to 0} f(x)$ exist? If so, what is the limit?

5. Consider the following graphs, representing functions $f(x)$, $g(x)$, $h(x)$, and $s(x)$, where $f(1) = 1$, $g(1) = 2$, $h(2) = 3$, and s is *not* defined at $x = \frac{1}{2}$. From the graphs in Figure 2.3.8 (using intuitive arguments) do you expect that the following limits exist?

(a) $\lim_{x \to 1} f(x)$ (c) $\lim_{x \to 2} h(x)$

(b) $\lim_{x \to 1} g(x)$ (d) $\lim_{x \to \frac{1}{2}} s(x)$

6. Let
$$f(x) = \begin{cases} x, & -1 \leq x < 0 \\ 1, & x = 0 \\ 2, & 0 < x \leq 1. \end{cases}$$

Does $\lim_{x \to 0} f(x)$ exist? If so, what is the limit?

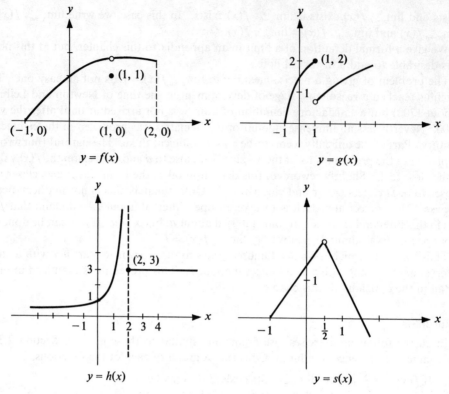

FIGURE 2.3.8

7. Let

$$f(x) = \begin{cases} x, & -1 \le x < 0 \\ 1, & x = 0 \\ 2x, & 0 < x \le 1. \end{cases}$$

Does $\lim_{x \to 0} f(x)$ exist? If so, what is the limit?

8. If $f(x) = 1/(x - 1)$, does $\lim_{x \to 1} f(x)$ exist? Graph this function.

9. Let $f(x) = x^2$.
    (a) Calculate
    $$\frac{f(x) - f(2)}{x - 2}.$$
    (b) Does $\lim_{x \to 2} [f(x) - f(2)]/(x - 2)$ exist? If so, what is it?

10. Let $f(x) = |x|$.
    (a) Find
    $$\lim_{x \to 0^+} \frac{f(x) - f(0)}{x}, \qquad \lim_{x \to 0^-} \frac{f(x) - f(0)}{x}.$$
    (b) Does $\lim_{x \to 0} [f(x) - f(0)]/x$ exist? Why?

2.4 Properties of Limits

The methods for evaluating limits in Section 2.3 often prove to be very cumbersome and awkward. It would be advantageous for us to develop some general properties of limits in order to make our calculations easier. We do not have at our disposal the necessary machinery to derive these properties rigorously. The formal definition given in the appendix to this chapter would allow us to rigorously derive the properties listed below. Although we do not use these properties very often in our future work, the reader should at least understand the statements of them.

PROPERTY 1. *If $f(x) = c$, where c is a constant, then $\lim_{x \to a} f(x) = c$ for any a.*

The graph of $f(x) = c$ is given in Figure 2.4.1.

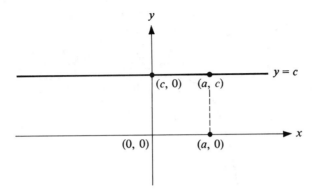

FIGURE 2.4.1

PROPERTY 2. *If $f(x) = x$, then $\lim_{x \to a} f(x) = a$.*

You can easily see the validity of this property from the graph shown in Figure 2.4.2.

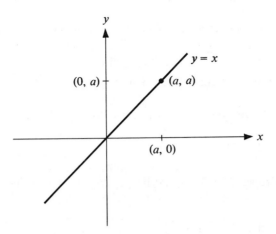

FIGURE 2.4.2

PROPERTY 3.   *Let $f(x) = x^n$ for **any** positive number n; then $\lim_{x \to a} f(x) = a^n$.*

For example, if $f(x) = x^3$, then $\lim_{x \to 2} f(x) = 2^3 = 8$; if $f(x) = x^{2/3}$, then $\lim_{x \to 3} f(x) = 3^{2/3}$; if $f(x) = x^{99}$, then $\lim_{x \to 10} f(x) = 10^{99}$. In order to verify this property for any n, we must resort to the formal definition of a limit. (Figure 2.4.3 shows $n = 3$ and $a = 2$.)

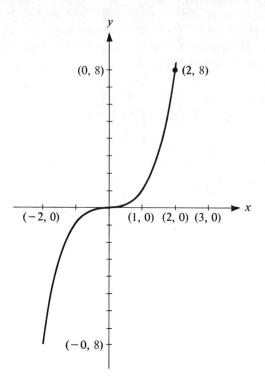

FIGURE 2.4.3

PROPERTY 4.   *Suppose that f and g are defined for x close to a number a, (a need **not** be in the domain of f and g!), and assume that $\lim_{x \to a} f(x)$ exists and $\lim_{x \to a} g(x)$ exists. Then $\lim_{x \to a} [f + g](x) = \lim_{x \to a} f(x) + \lim_{x \to a} g(x)$.*

In other words, Property 4 states that the limit of the sum of two functions is the sum of the limits.

Let us now look at some examples illustrating Properties 1–4.

Example 1.   Find $\lim_{x \to 1} (1 + x^2)$.

Solution.   By Property 4, $\lim_{x \to 1} (1 + x^2) = \lim_{x \to 1} 1 + \lim_{x \to 1} x^2$, and by Property 1, $\lim_{x \to 1} 1 = 1$, and Property 3 tells us that $\lim_{x \to 1} x^2 = 1^2 = 1$. Hence, $\lim_{x \to 1} (1 + x^2) = 1 + 1 = 2$.

Example 2.   Find $\lim_{x \to 4} [x^{1/2} + x^{3/2}]$.

Solution.   By Property 4,

$$\lim_{x \to 4} [x^{1/2} + x^{3/2}] = \lim_{x \to 4} x^{1/2} + \lim_{x \to 4} x^{3/2} = 4^{1/2} + 4^{3/2} = 10.$$

A similar property holds for the difference of two functions. That is, $\lim_{x \to a} [f - h](x) = \lim_{x \to a} f(x) - \lim_{x \to a} h(x)$ provided the appropriate limits exist. This result *follows* from Property 4, by letting $g(x) = -h(x)$!

PROPERTY 5. *Let f and g be two functions defined near a, such that* $\lim_{x \to a} f(x)$ *and* $\lim_{x \to a} g(x)$ *exist. Then* $\lim_{x \to a} [f(x)][g(x)] = [\lim_{x \to a} f(x)][\lim_{x \to a} g(x)]$.

In other words, the limit of the product is the product of the limits.

Consider the following example. Let us find $\lim_{x \to 3} 2x^3$. By Property 5, $\lim_{x \to 3} 2x^3 = (\lim_{x \to 3} 2)(\lim_{x \to 3} x^3) = 2 \cdot 3^3 = 54$, by Properties 1 and 3.

We might also note that Property 3 really follows from Properties 2 and 5. Namely,

$$\lim_{x \to a}[x \cdot x \cdot x \cdots \cdot x] = \underbrace{\left(\lim_{x \to a} x\right)\left(\lim_{x \to a} x\right) \cdots \left(\lim_{x \to a} x\right)}_{n \text{ times}} = \underbrace{a \cdot a \cdot a \cdots \cdot a}_{n \text{ times}} = a^n.$$

PROPERTY 6. *Let f and g be two functions defined near a such that* $\lim_{x \to a} f(x)$ *and* $\lim_{x \to a} g(x)$ *exist and such that* $\lim_{x \to a} g(x) \neq 0$. *Then* $\lim_{x \to a} f(x)/g(x)$ *exists and*

$$\lim_{x \to a} \frac{f(x)}{g(x)} = \frac{\lim_{x \to a} f(x)}{\lim_{x \to a} g(x)}.$$

In other words, the limit of the quotient of two functions is the quotient of the limits provided that the limit of the function in the denominator does not vanish.

Example 1. Let f be defined as follows

$$f(x) = \frac{x^2 + 1}{x + 3x + 2}.$$

We wish to see if $\lim_{x \to -1} f(x)$ exists. From Properties 3 and 4 we see that $\lim_{x \to -1} (x^2 + 1) = 2$ and $\lim_{x \to -1} (x + 3x + 2) = -2$. Hence, we may use Property 6 and obtain

$$\lim_{x \to -1} \frac{x^2 + 1}{x + 3x + 2} = \frac{\lim_{x \to -1}(x^2 + 1)}{\lim_{x \to -1}(x + 3x + 2)} = \frac{2}{-2} = -1.$$

Example 2. Suppose that

$$f(x) = \frac{(x^2 - 1)(x + 2)}{(x - 1)(x - 2)}.$$

We ask does $\lim_{x \to 1} f(x)$ exist. First, we observe that $\lim_{x \to 1} (x^2 - 1)(x + 2) = \lim_{x \to 1} (x^2 - 1) \lim_{x \to 1} (x + 2) = 0 \cdot 3 = 0$, by Properties 3, 4, 5. Using Properties 4 and 5 we find that $\lim_{x \to 1} (x - 1)(x - 2) = 0 \cdot (-1) = 0$. Hence we may *not* apply Property 6 since the limit of the function in the denominator vanishes. This does *not* mean that the limit does not exist—just that Property 6 does *not* apply. However, a convenient algebraic manipulation does yield a limit. First note that if $x \neq 1$, then

$$f(x) = \frac{(x - 1)(x + 1)(x + 2)}{(x - 1)(x - 2)} = \frac{(x + 1)(x + 2)}{(x - 2)}.$$

Thus, since we are only concerned with numbers near 1 and do not care about $x = 1$, we see that

$$\lim_{x \to 1} f(x) = \lim_{x \to 1} \frac{(x + 1)(x + 2)}{(x - 2)} = \frac{\lim_{x \to 1}(x + 1)(x + 2)}{\lim_{\to 1}(x - 2)}$$

$$= \frac{\lim_{x \to 1}(x + 1)\lim_{x \to 1}(x + 2)}{\lim_{x \to 1}(x - 2)} = \frac{2 \cdot 3}{-1} = -6.$$

Here we have combined Properties 3, 4, 5, and 6.

This second example is quite instructive. If it turns out that the denominator tends to zero in the limit, then an attempt should be made to see if there is an algebraic manipulation that "gets rid of" the troublesome expression as in the above example.

There are two other properties of limits involving inequalities that are extremely useful in the sequel. We state these properties here but their application appears in the next two chapters.

PROPERTY 7.   *Let f and g be defined near a but not necessarily at a. Suppose that for every x in the domain of f and g, $f(x) \le g(x)$ and $\lim_{x \to a} f(x)$ and $\lim_{x \to a} g(x)$ exist. Then, $\lim_{x \to a} f(x) \le \lim_{x \to a} g(x)$.*

Example.   Suppose $g(x) = x$ and $f(x) = x^2$, both having domain [0, 1]. Then $f(x) \le g(x)$, $x \in [0, 1]$. In particular, $\lim_{x \to \frac{1}{2}} f(x) = \frac{1}{4} \le \lim_{x \to \frac{1}{2}} g(x) = \frac{1}{2}$.

PROPERTY 8.   *Suppose the assumptions give in Property 7 hold and in addition that $\lim_{x \to a} f(x) = \lim_{x \to a} g(x)$. Let h be another function having the same domain as f and g and such that $f(x) \le h(x) \le g(x)$. Then $\lim_{x \to a} h(x)$ exists and $\lim_{x \to a} h(x) = \lim_{x \to a} g(x) = \lim_{x \to a} f(x)$.*

This property is often referred to as the *squeezing* or *pinching* process.

Example.   Suppose we let $h(x)$ be a function defined on [0, 2] such that $|h(x)| \le 2(x - 1)$ for $x \in [0, 2]$. We wish to determine $\lim_{x \to 1} h(x)$. Now, since $|h(x)| \le 2(x - 1)$, we know that $-2(x - 1) \le h(x) \le 2(x - 1)$. Observe that $\lim_{x \to 1} 2(x - 1) = \lim_{x \to 1} -2(x - 1) = 0$. If we let $f(x) = -2(x - 1)$ and $g(x) = 2(x - 1)$, then applying Property 8, $\lim_{x \to 1} -2(x - 1) \le \lim_{x \to 1} h(x) \le \lim_{x \to 1} 2(x - 1)$. Thus, $\lim_{x \to 1} h(x) = 0$.

At this point, we remark that all of the preceding eight properties hold for right- and left-hand limits. For example, if $\lim_{x \to a^+} f(x) = L_1$ and $\lim_{x \to a^+} g(x) = L_2$, then $\lim_{x \to a^+} [f(x) + g(x)] = L_1 + L_2$, etc.

Example 1.   Let $f(x) = 1/(2x + 1)$. Evaluate $\lim_{x \to 2} [f(x) - f(2)]/(x - 2)$.

Solution.   First observe that

$$f(x) - f(2) = \frac{1}{2x + 1} - \frac{1}{5} = \frac{5 - 2x - 1}{5(2x + 1)} = \frac{2(2 - x)}{5(2x + 1)}$$

and

$$\frac{f(x) - f(2)}{x - 2} = \frac{2(2-x)}{5(2x+1)(x-2)} = -\frac{2(x-2)}{5(2x+1)(x-2)} = -\frac{2}{5(2x+1)},$$

since $x \neq 2$. Thus,

$$\lim_{x \to 2} \frac{f(x) - f(2)}{x - 2} = \lim_{x \to 2} \left[-\frac{2}{5(2x+1)} \right] = -\frac{2}{25},$$

by Properties 4 and 6.

Example 2. Let $f(x) = 1/(2x+1)$. Evaluate $\lim_{x \to x_0} [f(x) - f(x_0)]/(x - x_0)$, where $x_0 \neq -\frac{1}{2}$.

Solution. First observe that

$$f(x) - f(x_0) = \frac{1}{2x+1} - \frac{1}{2x_0+1} = \frac{2x_0 + 1 - 2x - 1}{(2x+1)(2x_0+1)} = \frac{2(x_0 - x)}{(2x+1)(2x_0+1)}$$

Hence,

$$\frac{f(x) - f(x_0)}{x - x_0} = \frac{\dfrac{2(x_0 - x)}{(2x+1)(2x_0+1)}}{(x - x_0)} = \frac{2(x_0 - x)}{(2x+1)(2x_0+1)(x - x_0)}$$

$$= \frac{-2(x - x_0)}{(2x+1)(2x_0+1)(x - x_0)} = \frac{-2}{(2x+1)(2x_0+1)},$$

since $x \neq x_0$.

Thus,

$$\lim_{x \to x_0} \frac{f(x) - f(x_0)}{x - x_0} = \lim_{x \to x_0} \frac{-2}{(2x+1)(2x_0+1)} = \frac{-2}{(2x_0+1)^2},$$

by Properties 4 and 6.

Exercises

In the exercises below find the limits of the indicated functions, stating which of Properties 1–6 you use in each case.

1. Find:

    (a) $\lim_{x \to 1} 5x^2 - 9x + 3$ (c) $\lim_{x \to -2} (x+5)(x^2-9)$

    (b) $\lim_{x \to 3} x^3 - \dfrac{1}{x}$ (d) $\lim_{x \to 4} x^{1/2} - \dfrac{4}{x^2}$

(e) $\lim\limits_{x\to25} 3x^{1/2} + x - 4$

(i) $\lim\limits_{x\to1} \dfrac{x-1}{x+1}$

(f) $\lim\limits_{x\to-1} x^{35} - x^2 + 11$

(j) $\lim\limits_{x\to1} \dfrac{x^2-x}{x-1}$

(g) $\lim\limits_{x\to12} \dfrac{x+3}{x-7}$

(k) $\lim\limits_{x\to1} \dfrac{x^3-x}{x-1}$

(h) $\lim\limits_{x\to1} \dfrac{x^2+1}{2x^2-1}$

(l) $\lim\limits_{x\to1} \dfrac{(x^2-1)^2}{x-1}$

2. Find:

(a) $\lim\limits_{x\to2} \dfrac{x^2-x-6}{x-3}$

(c) $\lim\limits_{x\to1} \dfrac{x^2+2x-3}{x^2+x-2}$

(b) $\lim\limits_{x\to3} \dfrac{x^2-x-6}{x-3}$

(d) $\lim\limits_{x\to-2} \dfrac{3(x^2-4)}{5x(x+2)}$

3. Find:

(a) $\lim\limits_{x\to2} \dfrac{f(x)-f(2)}{x-2}$ if $f(x) = \dfrac{1}{x^2}$.

(b) $\lim\limits_{x\to x_0} \dfrac{f(x)-f(x_0)}{x-x_0}$ if $f(x) = \dfrac{1}{x^2}$ and $x_0 \in \operatorname{dom} f$.

4. (a)   Find $\lim\limits_{x\to1} \dfrac{|x-2|}{x-2}$.

(b)   Does $\lim_{x\to2} |x-2|/(x-2)$ exist? Why? (Recall the discussion in Section 2.3.) Sketch a graph of the function $f(x) = |x-2|/(x-2)$.

5.* Let n be a positive integer. Evaluate the following $\lim_{x\to a} (x^n - a^n)/(x-a)$. [HINT. From elementary algebra,

$$x^n - a^n = (x-a)(x^{n-1} + x^{n-2}a + x^{n-3}a^2 + \cdots + a^{n-1}).]$$

---

* Exercises marked with an asterisk are more difficult and/or deal with subjects that may be omitted without loss of continuity.

6.* Let

$$f(x) = \begin{cases} 1 & \text{if } x \text{ is an integer} \\ 0 & \text{if } x \text{ is } not \text{ an integer.} \end{cases}$$

Intuitively, do you think that $\lim_{x \to 2} f(x)$ exists? If so, what is its value?

7.* Evaluate the following limits

(a) $\lim\limits_{x \to 0} \dfrac{ax + b}{cx + d}$,   where a, b, c are given real numbers.

(b) $\lim\limits_{x \to \infty} \dfrac{ax + b}{cx + d}$,   where a, b, c are given real numbers.

[HINT. Use the fact that $\lim_{x \to \infty} 1/x = 0$ (that is, as x becomes *very large*, $1/x$ becomes very near zero); and write $(ax + b)/(cx + d) = (a + b/x)/(c + d/x)$.]

(c) $\lim\limits_{x \to 0} \dfrac{ax^2 + bx + c}{dx^2 + ex + f}$,   where a, b, c, c, d, e, f are given real numbers.

(d) $\lim\limits_{x \to \infty} \dfrac{ax^2 + bx + c}{dx^2 + ex + f}$,   where a, b, c, d, e, f are given real numbers.

8.* Suppose that f is a function defined on $[a, b]$ such that $|f(x)| \le M|x - c|$, where $c \in (a, b)$. Show that $\lim_{x \to c} f(x) = 0$.

2.5 Continuity

In Sections 2.2 and 2.3 we observed several examples of functions where $\lim_{x \to a} f(x)$ exists but $\lim_{x \to a} f(x) \ne f(a)$. We kept emphasizing that the limit is completely independent of the value of f at a—in fact, f need not even be defined at a! (See the third example in Section 2.3.) The various possibilities are shown in Figure 2.5.1.

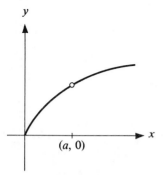

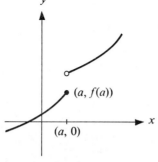

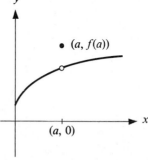

(a) Lim $_{x \to a}$ $f(x)$ exists but $f(a)$ is not defined.

(b) Lim $_{x \to a}$ $f(x)$ does not $f(a)$ is defined.

(c) Lim $_{x \to a}$ $f(x)$ exists, $f(a)$ defined, but Lim $x \to a$ $f(x) \ne f(a)$.

FIGURE 2.5.1

We like to think of the behavior of the functions in Figure 2.5.1 near *a* as *abnormal* and give a name to functions that do not exhibit such peculiarities. The term used is "continuous." Intuitively, we say that a function is continuous if its graph contains no jumps, breaks, or wild oscillations, that is, if we can draw its graph without removing our pencil from the paper. This description would enable you to judge, from looking at its graph, whether or not a function is continuous. Obviously there may be pitfalls to this approach and thus we need a more precise definition of continuity.

DEFINITION.   *A function f is continuous at a point* **a** *if* (i) *a* ∈ *domain of f and* (ii) $\lim_{x \to a} f(x)$ *exists and* $\lim_{x \to a} f(x) = f(a)$. *We say that f is continuous over an interval if f is continuous at every point in the interval.*

The definition of continuity involves $\lim_{x \to a} f(x)$, for which we have only given a provisional definition. In all the cases we will consider and for all "reasonable functions," the provisional definition is sufficiently discerning for us to decide whether the function in question is or is not continuous.

Let us now look again at the examples given in Section 2.2 to determine whether or not those functions are continuous.

Example 1.   Let $f(x) = x^2$. Is f continuous at $x = 2$?

Solution.   It is clear, from Section 2.2, that $2 \in \operatorname{dom} f$ and $\lim_{x \to 2} x^2 = 4 = f(2)$. Hence f is continuous at $x = 2$. We can observe from Figure 2.5.2 that f is continuous at every point in its domain.

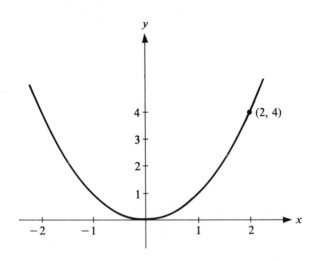

FIGURE 2.5.2

Example 2. Let $f(x) = (x - 2)/2|x - 2|$. Is f continuous at $x = 2$?

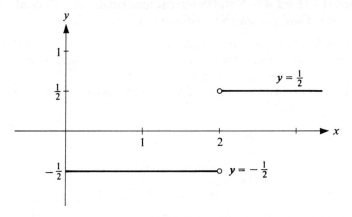

FIGURE 2.5.3

Solution. Since $2 \notin$ dom f, the first condition for continuity is violated and hence f is not continuous at $x = 2$. From the graph of f in Figure 2.5.3, we see that f is continuous at every other point in its domain. This is an example of a *jump* discontinuity—since at $x = 2$ we have a jump in the graph.

Example 3. Let

$$f(x) = \begin{cases} 2x - 3, & x \neq -1 \\ 2, & x = -1. \end{cases}$$

Is f continuous at $x = -1$?

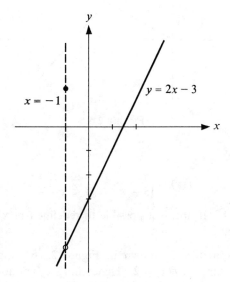

FIGURE 2.5.4

Solution. Note that $\lim_{x \to -1^-} f(x) = -5 = \lim_{x \to -1^+} f(x)$. Hence $\lim_{x \to -1} f(x) = -5$. However, $f(-1) = 2 \neq -5$, and thus the second condition is violated. ($-1 \in \text{dom } f$, so (i) is satisfied.) Thus, f is not continuous at $x = -1$.

A natural question to ask is: Can we redefine f at $x = -1$ so that f is continuous at that point? The answer is clearly yes! Define $f(-1) = -5$ rather than 2. Then $\lim_{x \to -1} f(x) = -5 = f(-1)$ and hence we have continuity.

Note that (as can be seen from Figure 2.5.4) f is continuous at every other point in its domain.

Example 4. Let

$$f(x) = \begin{cases} 2x, & 0 \le x < 1 \\ 4, & x = 1 \\ 5 - 3x, & 1 < x \le 2. \end{cases}$$

Is f continuous at $x = 1$? If not, can we redefine f at 1 in order to make f continuous at that point?

Solution. The graph of f is shown in Figure 2.5.5. Clearly $1 \in \text{dom } f$. We know also that $\lim_{x \to 1^-} f(x) = 2 = \lim_{x \to 1^+} f(x)$. Hence $\lim_{x \to 1} f(x) = 2$. However, $f(1) \neq 2$; hence f is not continuous at $x = 1$. If we define $f(1) = 2$, then f is continuous at $x = 1$. Again for every other $x \in [0, 2]$, it is clear that f is continuous.

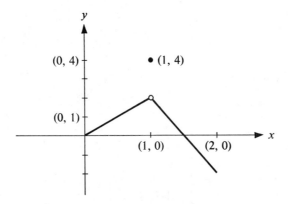

FIGURE 2.5.5

Example 5. Let

$$f(x) = \begin{cases} x, & 0 \le x \le 1 \\ 3 - x, & 1 < x \le 4. \end{cases}$$

If f continuous at $x = 1$? If not, is it possible to redefine f at $x = 1$ in order to make f continuous at that point?

Solution. The graph of f is shown in Figure 2.5.6. Clearly $1 \in \text{dom } f$. Now, $\lim_{x \to 1^-} f(x) = 1$ while $\lim_{x \to 1^+} f(x) = 2$. Hence $\lim_{x \to 1} f(x)$ does not exist. Thus, f is *not* continuous at that point. Since $\lim_{x \to 1^-} f(x) \neq \lim_{x \to 1^+} f(x)$, we *cannot* redefine f at 1 in order to make f continuous.

There are certain properties of continuous functions that follow immediately from Properties 1–6 in Section 2.3. The reader can easily verify this. We simply list those properties.

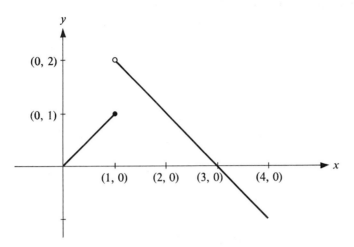

FIGURE 2.5.6

PROPERTY 1. *If $f(x) = c$, where c is a constant, then f is continuous at every point in its domain.*

PROPERTY 2. *If $f(x) = x^n$, for any positive number n, then f is continuous at every point in its domain.*

PROPERTY 3. *If f and g are continuous at any a, where $a \in \text{dom } f$ and $\text{dom } g$, then $f + g$ and $f - g$ are continuous at a. (That is, the sum and difference of any two continuous functions are again continuous.)*

PROPERTY 4. *If f and g are continuous at any point $a \in \text{dom } f$ and $\text{dom } g$, then $f \cdot g$ is continuous at a.*

PROPERTY 5. *If f and g are continuous at a and $g(a) \neq 0$, then f/g is continuous at a.*

We shall now give a supply of continuous functions that will be useful for our future work.

Example 1. (*Polynomials*) A function $f(x) = a_0 + a_1 x + a_2 x^2 + \cdots + a_n x^n$, where $a_0, a_1, \ldots, a_n$ are constants, is called a polynomial of degree n. We assert that if x_0 is any point in the domain of f, then f is continuous at x_0. This result follows from Properties 2, 3, and 4 above. It is necessary to verify that $\lim_{x \to x_0} f(x) = f(x_0)$. However, $\lim_{x \to x_0} f(x) = \lim_{x \to x_0} (a_0 + a_1 x + \cdots + a_n x^n) = a_0 + a_1 x_0 + \cdots + a_n x_0^n = f(x_0)$.

Example 2. (*Rational functions*) Let $f(x) = p(x)/q(x)$, where $p(x)$ and $q(x)$ are polynomials. If $x_0 \in \text{dom } f$ such that $q(x_0) \neq 0$, then f is continuous at x_0. (This is just a direct consequence of Property 5 above.)

Example 3. (*Exponential function*)   We recall that in Chapter 1 we discussed the function given by $f(x) = e^x$. The graph of this function is shown in Figure 2.5.7. It is intuitively clear from Figure 2.5.7 that f is continuous at every point in its domain (otherwise there would be gaps in the graph of f). The continuity of e^x could be rigorously derived by proving $\lim_{x \to x_0} e^x = e^{x_0}$—but we omit the details and rely only on our intuition.

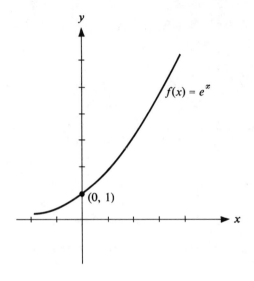

FIGURE 2.5.7

Example 4. (*The logarithmic function*)   Let $f(x) = \log x$ (where we understand the base to be e). The graph of f is shown in Figure 2.5.8. It should be intuitively clear from Figure 2.5.8 that f is continuous at every nonnegative point except $x = 0$. However, 0 is not in the domain of f—hence the first condition for continuity is violated. Thus, if $x_0 > 0$, $\lim_{x \to x_0} \log x = \log x_0$. Again, we rely only on our intuition for this function. A rigorous proof is beyond the scope of this course.

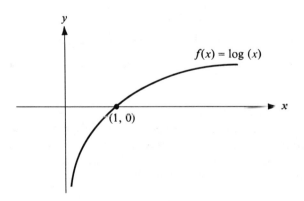

FIGURE 2.5.8

Exercises

1. Define f as follows:

$$f(x) = \begin{cases} x, & -1 \leq x < 0 \\ 1, & x = 0 \\ 2x, & 0 < x \leq 1 \end{cases}$$

   (a) Draw a graph of f.
   (b) Is f continuous at $x = -\frac{1}{2}$? Why? What about $x = +\frac{1}{2}$?
   (c) Is f continuous at $x = 0$? Why? If not, can you redefine f at 0 in order to make f continuous at that point?

2. Define f as follows:

$$f(x) = \begin{cases} -2, & -2 \leq x < -1 \\ -1, & -1 \leq x < 0 \\ 0, & 0 \leq x < 1 \\ 1, & 1 \leq x \leq 2. \end{cases}$$

   (a) Draw a graph of f.
   (b) At what points in its domain is f discontinuous? Why? Is it possible to redefine f at these points in order to make it continuous there?

3. The domain of definition of each of the following functions is to be $[-2, +2]$. Specify a function and draw its graph if:
   (a) it has a limit at each point and is continuous at each point.
   (b) it has a limit at each point but is not continuous at $x = 0$ and $x = 1$.
   (c) it has no limit when $x = 0$ and is not continuous at $x = 0, 1, 2$.

4. Let f be defined by

$$f(x) = \begin{cases} \dfrac{1}{x}, & x \neq 0 \\ 0, & x = 0. \end{cases}$$

Is f continuous at $x = 0$? Why?

5. Draw a graph of the function f given by $f(x) = e^{3x}$. From its graph, what can you say about the continuity of f? Do the same for $f(x) = e^{-3x}$.

6. Draw a graph of the function $f(x) = \log(x - 1)$, for $x > 1$. What can you infer about the continuity of f? If $f(x) = \log(x - x_0)$, where $x > x_0$ (x_0 a constant), where would you expect f to be continuous?

7. A discount in freight rates is often offered on a large shipment. Consider the cost function

$$c(x) = \begin{cases} 0.50x, & 0 < x \leq 100 \\ 0.45x, & 100 < x \leq 500 \\ 0.42x, & 500 < x, \end{cases}$$

where x is the number of pounds shipped and $c(x)$ is the cost.
   (a) Find $\lim_{x \to 50} c(x)$, $\lim_{x \to 100} c(x)$, and $\lim_{x \to 500} c(x)$.
   (b) Find those points where c is *not* continuous.

8. The current postage rate for airmail letters is 10 cents per ounce (and each fractional part of an ounce) up to 7 ounces; from then on it is 80 cents up to one pound. Sketch a graph showing the airmail postage of a letter weighing any amount up to one pound. At what points is this graph discontinuous?

9. Let

$$f(x) = \begin{cases} 1 & \text{if } x < -1 \\ |x| & \text{if } -1 < x < 1 \\ \frac{1}{2} & \text{if } 1 \le x \le 2 \\ \frac{1}{4}x & \text{if } 2 < x. \end{cases}$$

(a) Graph the function carefully.

(b) Give all points at which f is *not* continuous and reasons why.

2.6 Summary of Chapter 2

(1) *Heuristic definition of limit*

(a) We say that the limit as x approachs a from the right exists and is L_1 if the numbers $f(x)$ remain arbitrarily close to L_1 when we take x sufficiently close to a (but $x \ne a$) for values of $x > a$. We write $\lim_{x \to a^+} f(x) = L_1$.

(b) The limit as x approaches a from the left exists and is L_2 if the numbers $f(x)$ remain arbitrarily close to L_2 when we take x sufficiently close to a (but $x \ne a$) for values of $x < a$. We write $\lim_{x \to a^-} f(x) = L_2$.

(c) We say that $\lim_{x \to a} f(x)$ exists and is L if and only if $\lim_{x \to a^+} f(x)$ and $\lim_{x \to a^-} f(x)$ both exist and are *both* equal to L.

(2) *Continuity*

(a) A function f is continuous at a if

  (i) $a \in \text{dom} f$

  (ii) $\lim_{x \to a} f(x)$ exists and $\lim_{x \to a} f(x) = f(a)$.

(b) A continuous function (one that is continuous at every point in its domain) has a smooth graph, no jumps or sustained oscillations.

(3) *Examples of continuous functions*

(a) polynomials   (c) the exponential function

(b) rational functions   (d) the logarithm function

(4) *General comment*

If one knows what a continuous function is, the notion of limit can be handled in the following way. The $\lim_{x \to a} f(x) = L$ if the function is continuous at a when we set $f(a) = L$. (This definition allows the possibility that f is originally not defined at a or is defined at a but $f(a)$ was different from L originally.)

Example. We know that $f(x) = x^3 + 1$ is continuous at $x = 1$ and $f(1) = 2$. Hence, $\lim_{x \to 1} f(x) = 2$.

REVIEW EXERCISES

1. Decide by inspection whether the following functions in Figure 2.6.1–2.6.7
    (a)   have a right-hand limit at a
    (b)   have a left-hand limit at a
    (c) $\lim_{x \to a} f(x)$ exists
    (d) f is continuous at a.

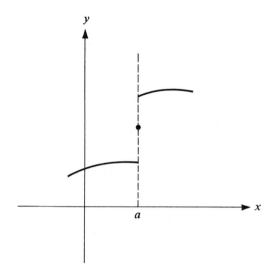

FIGURE 2.6.1

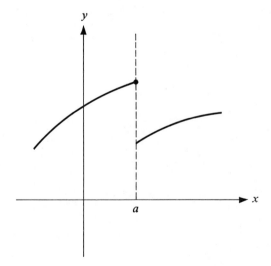

FIGURE 2.6.2

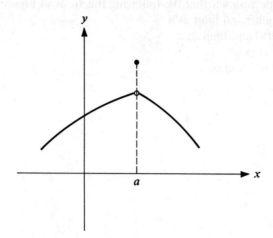

FIGURE 2.6.3

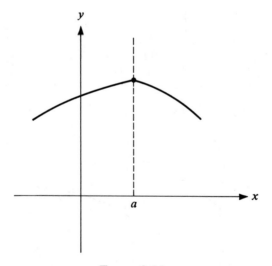

FIGURE 2.6.4

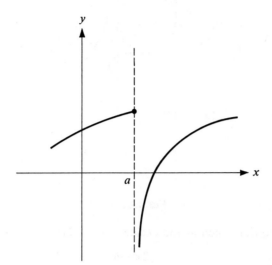

FIGURE 2.6.5

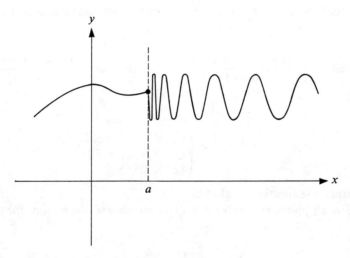

FIGURE 2.6.6

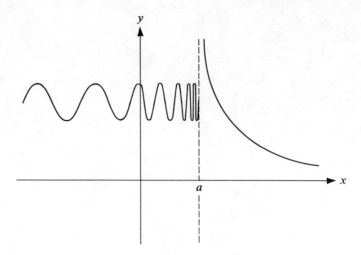

FIGURE 2.6.7

2. Graph the following function and decide whether $\lim_{x \to 3} f(x)$ exists.

$$f(x) = \begin{cases} 2x - 4, & x < 3 \\ 3, & x = 3 \\ 5 - x, & x > 3. \end{cases}$$

Is f continuous at 3?

3. Graph the function $f(x) = x^2 - |x|$. Is this function continuous at $x = 0$?

4. Find

(a) $\displaystyle \lim_{x \to 1^-} \frac{(1 + x + x^2)|x - 1|}{(x - 1)}$ (b) $\displaystyle \lim_{x \to 1^+} \frac{(1 + x + x^2)|x - 1|}{(x - 1)}$.

5. Let

$$f(x) = \begin{cases} 0 & \text{if } x < -1 \\ x^2 & \text{if } -1 \le x \le 1 \\ 1 & \text{if } 1 < x < 2 \\ \frac{1}{2}x & \text{if } 2 < x. \end{cases}$$

(a) Graph the function carefully.
(b) Give all points at which f is *not* continuous and the reasons for this.

6. Let

$$f(x) = \begin{cases} \dfrac{x^2 - 9}{x - 3} & \text{if } x \ne 3 \\ 3 & \text{if } x = 3. \end{cases}$$

(a) Graph $f(x)$.
(b) Is f continuous at $x = 3$? Why? If not, can you redefine f at $x = 3$ in order to make it continuous at that point?

7. Evaluate the following limits, if they exist.

(a) $\lim\limits_{x \to 2} \dfrac{x^2 - x - 2}{x(x - 2)}$

(d) $\lim\limits_{x \to 2} \dfrac{|x - 2|}{x - 2}$

(b) $\lim\limits_{x \to -1} \dfrac{(x - 1)^2 - 4}{x + 1}$

(e) $\lim\limits_{x \to 1} \dfrac{x^2 - x}{x^2 - 1}$

(c) $\lim\limits_{x \to 1} \dfrac{2x^2 - x - 1}{x(x - 1)}$

8. Suppose that $f(x) = 1/(x + 1)$. Evaluate the following limits.

(a) $\lim\limits_{x \to 2} \dfrac{f(x) - f(2)}{x - 2}$

(b) $\lim\limits_{x \to x_0} \dfrac{f(x) - f(x_0)}{x - x_0}$, where x_0 is any number not equal to -1.

9. (a) Draw a figure showing the graph of a function f such that for $x < 2$ the graph is part of a straight line, for $x > 2$ the same is true (though not the same straight line as when $x < 2$), and $f(2) = 1$, $\lim\limits_{x \to 2} f(x) = 0$. Is f continuous?
    (b) The same problem as (a) except that $f(2) = -1$, $\lim\limits_{x \to 2} f(x) = -1$.

10.* A continuous function $y = f(x)$ is known to be negative at $x = 0$ and positive at $x = 1$. Why is it true that the equation $f(x) = 0$ has at least one root [that is, there exists at least one point $x_0 \in (0, 1)$ such that $f(x_0) = 0$] between $x = 0$ and $x = 1$? Illustrate with a sketch. (Intuitive answer only.)

APPENDIX

2.7 The Definition of a Limit

As we pointed out in the first part of this chapter, it is essential in any careful study of calculus to make the terms "closer and closer" and "approaches" precise. If we return to the provisional definition of a limit in Section 2.3, we see that we could make this more precise by noting that the statement "$f(x)$ remains arbitrarily close to L when x is sufficiently close to a" means that the distance between $f(x)$ and L is made small or equivalently $|f(x) - L|$ is small whenever $|x - a|$ is sufficiently small and $x \neq a$. This leads us to adopt the following definition of a limit.

DEFINITION.   *The function f approaches the limit L at a, written $\lim_{x \to a} f(x) = L$ if for every number $e > 0$ there is some $d > 0$ such that $|f(x) - L| < e$ for $0 < |x - a| < d$.*

Observe that in general the choice of d depends upon the previous choice of e. In particular, we do *not* require a number d that works for *all* e, but rather that for *each* e there exists a number d that works for it.

Let us now look at a geometrical interpretation of the above definition. We do this in various steps. We begin with a number L and a number a on the y and x axes, respectively. (Figure 2.7.1.) Next we select a number $e > 0$ such that $L \in (L - e, L + e)$. (Figure 2.7.2.) We must then find a number $d > 0$ such that $x \in (a - d, a + d)$ and $x \neq a$, (Figure 2.7.3) implies that $|f(x) - L| < e$ or $L - e < f(x) < L + e$.

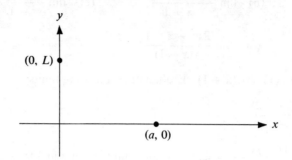

FIGURE 2.7.1

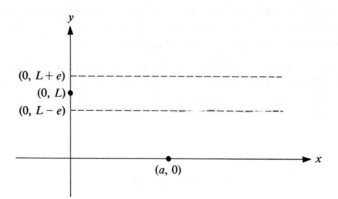

FIGURE 2.7.2

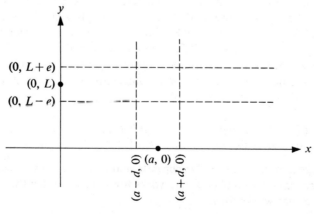

FIGURE 2.7.3

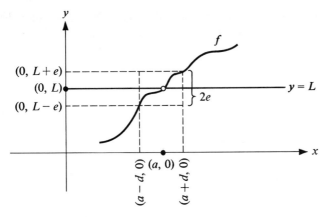

FIGURE 2.7.4

It should be evident to you that the definition of the limit requires that for each $e > 0$, we must find a $d > 0$ such that the graph of the function lying about the interval $(a - d, a + d)$ is contained in a rectangle of width $2d$ and height $2e$. The circle in the graph indicates that we do not care whether or not f is defined at a.

This definition offers us a precise way of verifying that the limit is a certain number. That is, we have some intuitive idea from our previous discussion of what the limit is and then we verify that this is indeed the case via the definition. However, if we want to show that a limit does not exist, we must negate the statement in the definition. That is, if it is *not* true that for every $e > 0$ there is some $d > 0$ such that for all x, if $0 < |x - a| < d$ then $|f(x) - L| < e$, then, there is *some* fixed $e > 0$ such that for *every* $d > 0$ there is *some* x that satisfies $0 < |x - a| < d$ but not $|f(x) - L| < e$. Let us now look at some examples.

Example 1.   Verify that $\lim_{x \to 1} (x + 1) = 2$.

Solution.   We must show that for every $e > 0$ there is some $d > 0$, such that if $0 < |x - 1| < d$, then $|x + 1 - 2| = |x - 1| < e$. It should be clear that corresponding to any e there is such a d, namely $d = e$, since $0 < |x - 1| < d = e$ implies that $|x - 1| < e$. Hence $\lim_{x \to 1} (x + 1) = 2$.

Example 2.   Verify that $\lim_{x \to 2} (3x + 1) = 7$.

Solution.   We must show that for every $e > 0$, there is some $d > 0$ such that if $0 < |x - 2| < d$, then $|3x + 1 - 7| = |3x - 6| = 3|x - 2| < e$. Again it should be clear that corresponding to any e, there is such a d—namely $d = e/3$, since $0 < |x - 2| < e/3$ implies that $3|x - 2| < e$, which yields $|3x - 6| < e$. Hence, $\lim_{x \to 2} (3x + 1) = 7$.

Now that we have rigorously defined the concept of limit, we can derive all of the Properties 1–8 given in Section 2.3. We illustrate how this is accomplished by deriving Property 4.

PROPERTY 4.   *Suppose that f and g are defined for x close to a number a, and assume that $\lim_{x \to a} f(x) = L$ and $\lim_{x \to a} g(x) = M$. Then $\lim_{x \to a} (f + g)(x) = L + M$.*

Proof. We must show that for every $e > 0$, there is some $d > 0$ such that if $0 < |x - a| < d$, then $|f(x) + g(x) - L - M| < e$. We first observe that $|f(x) + g(x) - L - M| = |(f(x) - L) + (g(x) - M)|$. Now, by the triangle inequality (Section 1.1, Property 3),

$$|(f(x) - L) + (g(x) - M)| \leq |f(x) - L| + |g(x) - M|.$$

Using the fact that $\lim_{x \to a} f(x) = L$ and $\lim_{x \to a} g(x) = M$, we know that for $e/2$, there is a d_1 and d_2 such that if $0 < |x - a| < d_1$, then $|f(x) - L| < e/2$, and if $0 < |x - a| < d_2$, then $|g(x) - M| < e/2$. Hence, if we choose d to be the smaller of the two numbers d_1, d_2, we see that if $0 < |x - a| < d$, then $|(f(x) - L) + (g(x) - M)| \leq |f(x) - L| + |g(x) - M| < e/2 + e/2 = e$. Thus, the d that "works" is the smaller of the two numbers d_1, d_2 obtained from the hypothesis that $\lim_{x \to a} f(x)$ and $\lim_{x \to a} g(x)$ exist.

The remaining properties could be derived in a similar way, although for Properties 5 and 6 a bit of algebraic manipulation is necessary. We end by observing that right- and left-hand limits could also be rigorously defined as follows.

DEFINITION. $\lim_{x \to a^+} f(x) = L$ *means that for every* $e > 0$ *there is some number* $d > 0$ *such that if* $0 < x - a < d$, *then* $|f(x) - L| < e$.

and

DEFINITION. $\lim_{x \to a^-} f(x) = M$ *means that for every* $e > 0$ *there is some number* $d > 0$ *such that if* $0 < a - x < d$, *then* $|f(x) - M| < e$.

It is clear from the above definitions that $\lim_{x \to a} f(x)$ exists if and only if $\lim_{x \to a^+} f(x)$ and $\lim_{x \to a^-} f(x)$ exist and they are equal.

The definition of a limit given in this section suggests the following.

Game.   A function $f(x)$ and a number L are furnished.
(1)   Player A chooses an $e > 0$.
(2)   Player B chooses a $d > 0$.
(3)   Player A chooses x, where $0 < |x - a| < d$.
Then $f(x)$ is evaluated and if $|f(x) - L| < e$, player B wins. If $|f(x) - L| \geq e$, then player A wins. If *every* time the game is played, B wins, then $\lim_{x \to a} f(x) = L$. If B is not able to win every time, then $\lim_{x \to a} f(x) \neq L$. (It is assumed that both players play to win.)

Exercises

1.   Using the rigorous definition of a limit, derive Property 1 of Section 2.4.

2.   Verify that:

(a) $\lim_{x \to 1} (x + 1) = 2$ (c) $\lim_{x \to \frac{1}{2}} (2x + 1) = 2$.

(b) $\lim_{x \to 3} (2x - 1) = 5$

3.   Prove that if c is any constant, then if $\lim_{x \to a} f(x)$ exists and is L, $\lim_{x \to a} [cf(x)] = cL$.

4. Show that, if m and b are constants, $\lim_{x \to k} (mx + b)$ exists for every k and that a single choice of d to match a given e will apply to all values of k.

5. Derive Property 5 of Section 2.4. [HINT: If $\lim_{x \to a} f(x) = L$ and $\lim_{x \to a} g(x) = M$, write

$$|f(x)g(x) - LM| = |f(x)g(x) - Lg(x) + Lg(x) - LM|.$$

Now

$$|f(x)g(x) - LM| \leq |g(x)|\,|f(x) - L| + L|g(x) - M|. \quad \text{(Why?)}$$

To complete the proof use the facts that $\lim_{x \to a} f(x) = L$ and $\lim_{x \to a} g(x) = M$.]

6. Prove that $\lim_{x \to a} f(x) = \lim_{h \to 0} f(a + h)$.

7. Suppose that $f(x) \leq g(x)$ for all x. Prove that $\lim_{x \to a} f(x) \leq \lim_{x \to a} g(x)$.

8. (a) If $\lim_{x \to a} f(x)$ exists and $\lim_{x \to a} [f(x) + g(x)]$ exists, must $\lim_{x \to a} g(x)$ exist?
   (b) If $\lim_{x \to a} f(x)$ exists and $\lim_{x \to a} f(x)g(x)$ exists, does it follow that $\lim_{x \to a} g(x)$ exists?

9. Prove using a rigorous definition of limit that the function
   (a) $f(x) = x^2$ is continuous at $x = 2$.
   (b) $f(x) = x^3$ is continuous at $x = 1$.

3

Differentiation

3.1 The Derivative

To motivate the concept of a limit in Section 2.1, we briefly discussed the idea of measuring the slope or steepness of a hill at any given point on the hill. We could obtain a rough value for the slope of the hill at P_1 by sending our friend up the hill to a point P and then calculating the slope of the line joining P_1 to P (a healthful outdoor activity indulged in by surveyors daily). (See Figure 3.1.1.) The value obtained in this way would

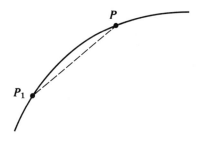

FIGURE 3.1.1

be only an approximation, but it is not unreasonable to think that the accuracy would improve as P was brought closer to P_1. Thus, it is tempting to define the slope at P_1 as the limit of the approximations as P approaches P_1 (assuming the limit exists). With the foregoing discussion as motivation let us consider a similar but more abstract problem.

How can we find the tangent to the curve $y = f(x)$ at a point P? (See Figure 3.1.2.) We have not yet defined the tangent to a curve, so this is the first problem to be overcome.

100

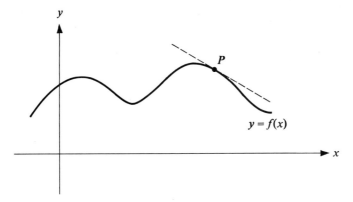

FIGURE 3.1.2

One might think that it suffices to say that the tangent is simply a line that touches the curve at the point P and nowhere else. (For example, this would appear to be a satisfactory definition of a tangent to a circle.) This is clearly unsuitable as can be seen in Figure 3.1.3. We would hope (relying on our intuition!) that we would call L the tangent at P, yet this line cuts the graph of f in three places.

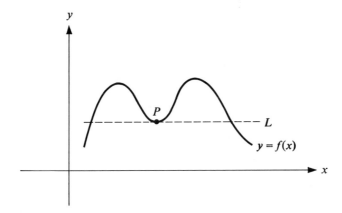

FIGURE 3.1.3

Now let us try a method for obtaining a tangent and see what happens. Suppose P_0 is a point on the curve $y = f(x)$ at which we wish to find the tangent. (We shall be referring to Figure 3.1.4 in the following discussion.) Let P_1 be another point on the curve and L_1 be the line through P_0 and P_1. This does not seem to be very close to anything we would be willing to call a tangent. Consider the lines $L_2, L_3, L_4, \ldots$ corresponding to the points $P_2, P_3, P_4, \ldots$, where the P's are getting closer to P_0. In this case at least the lines L_i appear to be approaching a line that is a geometrically appealing candidate for the tangent. Let us look more closely at this situation.

Think of $P_0 = (x_0, f(x_0))$ as a fixed point on the curve at which we want to find the tangent. Choose another point on the curve $(x, f(x))$ and let L_x be the line through these two points. (See Figure 3.1.5.) Now what we want to know is this: As x approaches x_0,

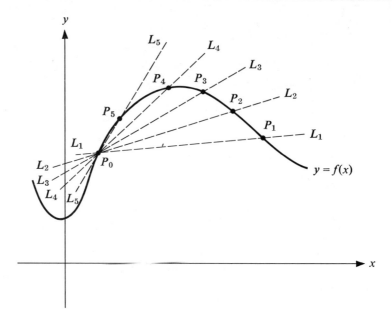

FIGURE 3.1.4

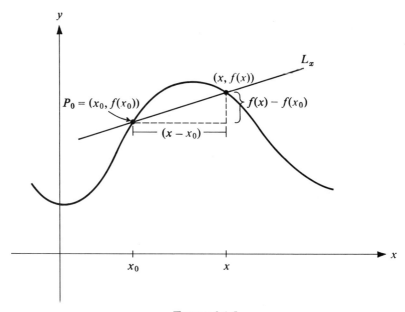

FIGURE 3.1.5

does the line L_x approach some line L? It is not very clear just what this last phrase is supposed to mean. But note that L_x passes through P_0 in all cases. Certainly our tangent line should pass through P_0. Since one point and the slope determine a line we will ask instead: Does the slope of L_x approach a number as x approaches x_0? The slope of the line L_x is given by

$$m_{L_x} = \frac{f(x) - f(x_0)}{x - x_0}.$$

Hence, we really are asking the following question: Does

$$\lim_{x \to x_0} \frac{f(x) - f(x_0)}{x - x_0}$$

exist? The last expression is so important, we shall incorporate it into a definition.

DEFINITION.   *If* $\lim_{x \to x_0} [f(x) - f(x_0)]/[x - x_0]$ *exists then we say that f is differentiable at* x_0 *and write*

$$f'(x_0) = \lim_{x \to x_0} \frac{f(x) - f(x_0)}{x - x_0}.$$

The number $f'(x_0)$ *is called the derivative of f at* x_0.

Since we have come this far with the tangent problem we will finish it by means of the following definition.

DEFINITION.   *Let* $y = f(x)$ *be a function differentiable at* x_0. *Let* $f'(x_0) = M$. *We define the tangent line to the curve* $y = f(x)$ *at* $(x_0, f(x_0))$ *to be the line whose equation is given by*

$$y = f(x_0) + M(x - x_0).$$

Some people might feel that another definition of the tangent would be preferable. We will not try to argue that the one given above is better in one way or another. We do remark that the given definition works very well in practice.

Let us look at some examples before proceeding further.

Example 1.   Consider the curve $y = x^2$. We wish to determine the equation of the tangent line T at $(1, 1)$. (See Figure 3.1.6.)

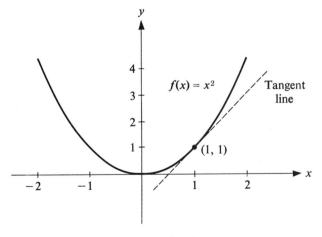

FIGURE 3.1.6

Solution. First we must determine whether or not the function $f(x) = x^2$ has a derivative at $x = 1$. That is, does $\lim_{x \to 1} [f(x) - f(1)]/[x - 1]$ exist? Observe that

$$\frac{f(x) - f(1)}{x - 1} = \frac{x^2 - 1}{x - 1} = \frac{(x - 1)(x + 1)}{(x - 1)} = (x + 1) \quad \text{if } x \neq 1.$$

Hence, $\lim_{x \to 1} [f(x) - f(1)]/(x - 1) = \lim_{x \to 1} (x + 1) = 2$ (from our work in Chapter 2). Thus f is differentiable at $x = 1$ and $f'(1) = 2$. The slope of the tangent is 2; thus its equation is

$$y = 1 + 2(x - 1) = 2x - 1, \quad \text{since } f(1) = 1.$$

Example 2. Find the equation of the line tangent to the curve $y = x^2$, at the point (x_0, x_0^2), where x_0 is any real number.

Solution. First we must determine the limit

$$\lim_{x \to x_0} \frac{x^2 - x_0^2}{x - x_0} = \lim_{x \to x_0} \frac{(x + x_0)(x - x_0)}{(x - x_0)} = 2x_0.$$

Hence, the slope of the line tangent to $y = x^2$ at (x_0, x_0^2) is $2x_0$. The equation of the line is

$$y = x_0^2 + 2x_0(x - x_0).$$

Example 3. Can we determine the tangent to the curve $y = |x|$ at $x = 0$? (See Figure 3.1.7.)

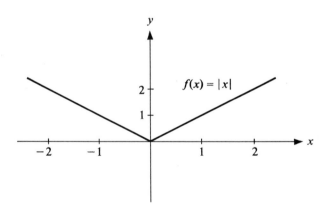

FIGURE 3.1.7

Solution. We will see that the answer is *No!*, since f does not have a derivative at $x = 0$. Now, $[f(x) - f(0)]/[x - 0] = |x|/x$. Hence we ask whether $\lim_{x \to 0} |x|/x$ exists? We know, since

$$|x| = \begin{cases} x, & x \geq 0 \\ -x, & x < 0 \end{cases}, \quad \text{that } \lim_{x \to 0^+} \frac{|x|}{x} = 1 \quad \text{and} \quad \lim_{x \to 0^-} \frac{|x|}{x} = -1.$$

Thus $\lim_{x \to 0} |x|/x$ does not exist and $f'(0)$ is not defined.

The limit, $\lim_{x \to 0^+} [f(x) - f(0)]/(x - 0)$, is called the right-hand derivative of f at 0 and the limit, $\lim_{x \to 0^-} [f(x) - f(0)]/(x - 0)$, is called the left-hand derivative of f at 0. In example 3, the right-hand derivative is 1 and the left-hand derivative is -1.

The derivative is the single most important concept in calculus (with the integral not far behind). In all areas of social and natural science, the study of changes is of the utmost importance; changes in population, the value of the dollar, changes in hourly wages, interest rate, and so on. Equally important are the *rates* at which these changes occur. For example, if we are told that the consumer price index has gone down by 0.4 points, this does not mean very much until we find out whether this change took place over a week, a month, or a year. The derivative measures the rate of change in the function f when there is a change in x. It can also be used to solve problems involving the maximizing and minimizing of quantities and curve tracing.

Example.   The total cost of producing and marketing x units of a commodity is assumed to be a function of x alone—that is, independent of time and overhead. Designate this function by $C(x)$. We have assumed that if no items are produced, there is no cost; that is, $C(0) = 0$. The average cost to produce x units, $A(x)$, is given by

$$A(x) = \frac{C(x) - C(0)}{x - 0} = \frac{C(x)}{x},$$

Now, the rate of change of the total cost with respect to the number of units produced is called the *marginal cost*, and is simply given by

$$\lim_{x \to x_0} \frac{C(x) - C(x_0)}{x - x_0},$$

which we recognize as $C'(x_0)$, the derivative of $C(x)$ at x_0.

Suppose the total cost is $C(x) = x^2$. The average cost of producing 100 units is

$$A(100) = \frac{(100)^2}{100} = \$100$$

and the marginal cost is

$$\lim_{x \to 100} \frac{x^2 - (100)^2}{x - 100} = \lim_{x \to 100} \frac{(x + 100)(x - 100)}{(x - 100)} = \$200.$$

Exercises

1. (a)   Let $f(x) = ax + b$. Convince yourself that $[f(x) - f(x_0)]/(x - x_0)$ is independent of the choice of x and x_0. Give a geometrical interpretation of this fact.
    (b)   Find the equation of the tangent line to the above curve.
2. Find the equation of the tangent line to the curve $y = 1/x$ at the point $(1, 1)$. Construct the graph of this function first.
3. Find the equation of the tangent line to the curve $y = x^2 - 1$ at the point $(-1, 0)$. First construct its graph.

4. Find the derivatives of the following functions at the indicated points, using the definition given in this section.

   (a) $f(x) = 3$ at $x = 2$.

   (b) $f(x) = 2x - 5$ at $x = 1$.

   (c) $f(t) = t + 3$ at $t = \frac{1}{2}$.

   (d) $f(x) = x^2 - 2$ at $x = 2$.

   (e) $f(x) = \dfrac{1}{x^2}$ at $x = 1$.

   (f) $f(x) = \sqrt{x}$ at $x = 1$.

5. (a) Graph the function defined by

$$f(x) = \begin{cases} x^2, & x \geq 0 \\ -x, & x < 0. \end{cases}$$

   (b) Is f continuous at $x = 0$? Why?

   (c) Does f have a derivative at $x = 0$? Why?

6. Suppose the demand for "Double Bubble Bath Soap" is expressed by means of

$$D(p) = 100 - 2p + p^2,$$

where p is the price per box (in cents) and D is the weekly demand (in thousands of boxes). What is the instantaneous (or *marginal*) rate of change of the demand for the detergent when the price per box is 25 cents?

7. Suppose that the total cost of producing widgets is given by

$$C(x) = x^2 + 1,$$

where x denotes the number of widgets produced. What is the average cost of producing 50 widgets? What is the marginal cost of producing 50 widgets?

8. A manufacturer, after studying the market behavior for his product, comes to the following conclusion: To sell his output of x tons per week he must charge a price of $P = F(x)$ dollars per ton. At a higher price he would sell less; at a lower price he could sell more. His revenue $R(x)$ is given by $R(x) = xF(x)$, and $R'(x_0)$ represents the *marginal revenue* for x_0 tons. (The marginal revenue gives the rate of increase of revenue per unit increase in output; more bluntly, it tells how fast revenues are going up.) Obviously the manufacturer's profit is given by the difference between revenue and cost $C(x)$. That is, $T(x) = xF(x) - C(x)$, where T is the profit, C the cost, and F the price he must charge per ton.

   (a) Suppose that a manufacturer finds he must charge a price of $P = F(x) = 1 + 1/x + x$ dollars per ton. What is his total revenue for 100 tons? What is marginal revenue for 100 tons?

   (b) Referring to (a), if his costs are given by $C(x) = x + 1$ dollars per ton, determine his total profit for selling 100 tons. What is the rate of increase of profit per unit increase of production for 100 tons? (This is called the *marginal profit*.)

3.2 The Relation Between Differentiability and Continuity

Let us now introduce some jargon that is commonly used by mathematicians. If a function f has a derivative at x_0, we say that it is differentiable at x_0. If f does not have a derivative at x_0, we say it is nondifferentiable at x_0. If f is differentiable at every point in the interval (a, b), we say f is differentiable on (a, b).

It would be convenient at this point to relate the concepts of continuity and differentiability. To say a function is continuous at a means roughly the function knows what its value will be at a and has that value. To say a function is differentiable at a means it also knows what its direction is at a. Note that these are different. Thus, it is one thing to predict the free market price of gold on October 10, 1968 and quite another to also predict whether the market will be rising or falling on that day (assume for the sake of the argument that it cannot do both by taking closing prices).

To illuminate the last remarks we quote the following result. (We call any results that are logically deduced from definitions and previously defined properties, theorems.)

THEOREM. *If f is differentiable at a, then f is continuous at a.*

Note that the converse is *not* true. Consider the function $f(x) = |x|$. We have seen that f is continuous at $x = 0$, but that f does not have a derivative at 0; that is, f is not differentiable at 0.

It is a simple matter to establish the validity of the above theorem. For those readers interested, we include a proof—the others can learn the result and omit the following discussion.

Proof of Theorem. If f is differentiable at $x = x_0$, then we know that

$$\lim_{x \to x_0}[f(x) - f(x_0)]/(x - x_0)$$

exists and is $f'(x_0)$. We must prove that $\lim_{x \to x_0} f(x) = f(x_0)$ in order to establish continuity. This is equivalent to showing that $\lim_{x \to x_0}[f(x) - f(x_0)] = 0$.

We may write

$$f(x) - f(x_0) = \frac{f(x) - f(x_0)}{(x - x_0)} \cdot (x - x_0) \qquad \text{for } x \neq x_0.$$

Hence,

$$\lim_{x \to x_0}[f(x) - f(x_0)] = \lim_{x \to x_0} \frac{f(x) - f(x_0)}{(x - x_0)} \cdot (x - x_0)$$

$$= \lim_{x \to x_0} \frac{f(x) - f(x_0)}{(x - x_0)} \cdot \lim_{x \to x_0}(x - x_0) = f'(x_0) \cdot 0 = 0,$$

by our properties of limits discussed in Chapter 2. Thus, $\lim_{x \to x_0} f(x) = f(x_0)$, and f is continuous at $x = x_0$.

Example. Let

$$f(x) = \begin{cases} x + 2 & \text{if } 0 \leq x < 2 \\ 5 & \text{if } x = 2 \\ 6 - x & \text{if } 2 < x \leq 6. \end{cases}$$

(a)   Sketch graph of f.
(b)   Is f continuous at $x = 2$?
(c)   Is f differentiable at $x = 2$?

   Solution.   (a) The graph of f is shown in Figure 3.2.1.

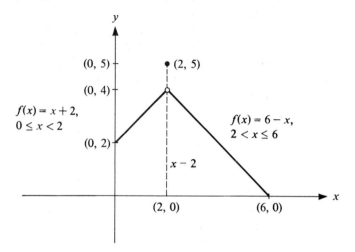

FIGURE 3.2.1

(b)   It is clear that $\lim_{x \to 2^+} f(x) = \lim_{x \to 2^-} f(x) = 4$.   However, $f(2) = 5 \neq 4$.   Thus, f is not continuous at $x = 2$.

(c)   The function is not differentiable at $x = 2$ since f is *not* continuous at $x = 2$. We could also see this by observing that $\lim_{x \to 2}[f(x) - f(2)]/(x - 2)$ does not exist. Note that

$$\lim_{x \to 2^+} \frac{f(x) - f(2)}{x - 2} = \lim_{x \to 2^+} \frac{6 - x - 5}{x - 2} = \lim_{x \to 2^+} \frac{1 - x}{x - 2}.$$

Now, $\lim_{x \to 2^+}(1 - x)/(x - 2)$ does not exist; the denominator tends to 0, while the numerator tends to -1. (If both numerator and denominator tend to zero, then it is possible for limit to exist—that is,

$$\lim_{x \to 2^+} \frac{2 - x}{x - 2} = \lim_{x \to 2^+} (-1) = -1,$$

since $x \neq 2$.) Also,

$$\lim_{x \to 2^-} \frac{f(x) - f(2)}{x - 2} = \lim_{x \to 2^-} \frac{x + 2 - 5}{x - 2} = \lim_{x \to 2^-} \frac{x - 3}{x - 2}.$$

This limit does not exist. Since both the left- and right-hand limits do not exist, it follows that $\lim_{x \to 2}[(f(x) - f(2)]/(x - 2)$ cannot exist.

   We end this section by briefly discussing the geometrical significance of a derivative. As has already been observed the derivative $f'(x_0)$ can be thought of as the slope of the tangent to the curve $f(x)$ at the point x_0. Thus the derivative tells you in what direction $f(x)$ is going at x_0. If $f'(x_0) > 0$, then $f(x)$ is increasing at x_0. If $f'(x_0) < 0$, then $f(x)$ is decreasing at x_0. The size of $f'(x_0)$ gives an indication of how steeply the function is increasing or decreasing. We have more to say about this in the next chapter.

Exercises

1. Let

$$f(x) = \begin{cases} 2, & x \neq 1 \\ 5, & x = 1. \end{cases}$$

    (a) Sketch a graph of f.
    (b) Is f differentiable at 1? Give two different arguments.

2. Let

$$f(x) = \begin{cases} -x, & x < 0 \\ x^{1/2}, & x \geq 0. \end{cases}$$

(See Figure 3.2.2.)

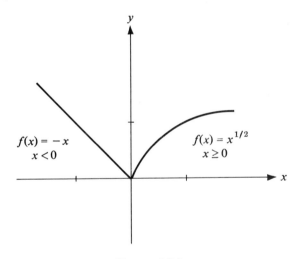

$f(x) = -x$
$x < 0$

$f(x) = x^{1/2}$
$x \geq 0$

FIGURE 3.2.2

    (a) Evaluate:

$$\text{(i)} \quad \lim_{x \to 0^-} \frac{f(x) - f(0)}{x - 0} \qquad \text{(ii)} \quad \lim_{x \to 0^+} \frac{f(x) - f(0)}{x - 0}.$$

    (b) Is f continuous at $x = 0$? Is f differentiable at $x = 0$? Why?

3. Consider the function defined by $f(x) = |x - 1|$. Is f differentiable at $x = 1$? Why?

4. Suppose f is defined by:

$$f(x) = \begin{cases} 0 & \text{if } x \leq 0 \\ x & \text{if } 0 < x \leq 1. \\ 1 & \text{if } x > 1 \end{cases}$$

   (a) Sketch a graph of f.

   (b) Is f continuous at all points in its domain?

   (c) At what points is f not differentiable? Why?

5. Consider the postage-stamp function discussed in Exercise 8 in Section 2.5. At what points is this function nondifferentiable? Why?

6. Sketch a graph of a function such that

   (a) $\displaystyle\lim_{x \to a^+} \frac{f(x) - f(a)}{x - a}$ exists but $\displaystyle\lim_{x \to a^-} \frac{f(x) - f(a)}{x - a}$ does *not* exist.

   (b) $\displaystyle\lim_{x \to a^-} \frac{f(x) - f(a)}{x - a}$ exists but $\displaystyle\lim_{x \to a^+} \frac{f(x) - f(a)}{x - a}$ does *not* exist.

   (c) For the functions you give in (a) and (b), can f be differentiable at a? Why?

7. Let

$$f(x) = \begin{cases} \dfrac{x^2 - 9}{x - 3}, & 0 \le x < 3 \\[2mm] 7, & x = 3 \\[2mm] 9 - x, & 3 < x \le 9. \end{cases}$$

   (a) Sketch a graph of f.

   (b) Is f continuous at $x = 3$? Why?

   (c) Is f differentiable at $x = 3$? Why?

8.* Let f be a function with the following property:

$$|f(x_2) - f(x_1)| \le |x_2 - x_1|^2$$

for *all* real numbers x_1, x_2. Find $f'(x_0)$ for x_0 an arbitrary real number.

3.3 The Derivative of Several Basic Functions

There are several different notations for the derivative of f at x_0 that are useful in applications. We may write

$$f'(x_0) = \frac{df}{dx}\bigg|_{x = x_0} = \frac{df}{dx}(x_0).$$

We have seen in Section 3.1 that if $f(x) = x^2$, then

$$f'(x_0) = \frac{df}{dx}\bigg|_{x = x_0} = \frac{df}{dx}(x_0) = 2x_0 \, ;$$

hence, $f'(3) = 6, f'(-1) = -2$, and $f'(175) = 350$.

   Instead of writing $f'(x_0) = 2x_0$, it is certainly just as clear to write $f'(x) = 2x$. Let us see why in more detail. Given a function f, let A be the set of points for which f is differentiable. Consider the rule that associates with x_0 the number $f'(x_0)$. Since to any number $x_0 \in A$ we are associating precisely one well-defined number $f'(x_0)$, this defines a function g that assigns to x_0 the value $f'(x_0)$. Instead of writing this function as $g(x)$, we usually write $f'(x)$. Is this consistent with the usual meaning of $f'(x)$ and the definition of g? Yes, since $g(x_0) = f'(x_0) = f'(x)|_{x=x_0}$, all we are really saying is that we use the symbol $f'(x)$ to define a function whose value at x_0 is the number $f'(x_0)$. Similarly, we may write

$$f'(x) = f' = \frac{df}{dx} = \frac{df(x)}{dx} = \frac{df}{dx}(x).$$

Note, the domain of f' consists of all numbers x_0 for which

$$\lim_{x \to x_0}[f(x) - f(x_0)]/(x - x_0)$$

exists. In general, the domain of f' is not the same as the domain of f. For example, if $f(x) = |x|$ then domain $f' = \{x : |x| \neq 0\}$ while domain $f = \{$all real numbers$\}$.

We will now give the derivatives of several well-known functions. For the sake of convenience, we first list them in tabular form and then briefly discuss how they are obtained.

Table 3.3.1

$f(x)$	$f'(x)$
1. k, where k is any constant	0
2. x^n, for *any n*	nx^{n-1}
3. e^x	e^x
4. $\log x$, $x > 0$	$\dfrac{1}{x}$

The following remarks refer to Table 3.3.1.

Remark 1. It is clear from a geometrical point of view that the slope of any line parallel to the x axis is 0. Hence, one would expect $f'(x) = 0$ if $f(x) = k$, a constant. (See Figure 3.3.1.)

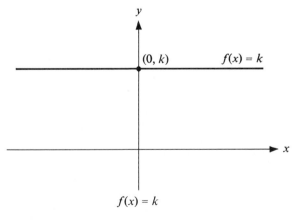

FIGURE 3.3.1

Moreover, if x_0 is any number, then

$$\lim_{x \to x_0} \frac{f(x) - f(x_0)}{x - x_0} = \lim_{x \to x_0} \frac{k - k}{x - x_0} = \lim_{x \to x_0} \frac{0}{x - x_0} = \lim_{x \to x_0} 0 = 0.$$

Thus, entry 1 can be verified from the definition.

Remark 2.   It is not difficult to verify entry 2 if n is a positive integer.

We include this as Exercise 11.

Remark 3.   Note that if $f(x) = e^x$, then $f'(x) = e^x$. This simplicity is not just an accident. Indeed, it is fair to say that e was defined with this in mind.

To see more clearly why this is so, let us consider the function $f(x) = a^x$. By definition,

$$f'(x_0) = \lim_{x \to x_0} \frac{f(x) - (fx_0)}{x - x_0}$$

$$= \lim_{x \to x_0} \frac{a^x - a^{x_0}}{x - x_0}$$

$$= \lim_{x \to x_0} a^{x_0} \frac{a^{(x - x_0)} - 1}{x - x_0}$$

$$= a^{x_0} \lim_{h \to 0} \frac{a^h - 1}{h},$$

where we let $h = x - x_0$. The expression $\lim_{h \to 0}(a^h - 1)/h$ is not easy to evaluate. However, the limit does exist and $\lim_{h \to 0}(a^h - 1)/h = \log_e a$. Thus, if $f(x) = a^x$, then $f'(x_0) = (\log_e a)a^{x_0}$.

When we set $a = e$, the expression becomes considerably simpler since $\log e = 1$. There is a geometric interpretation to the foregoing discussion. First, note that $\lim_{h \to 0}(a^h - 1)/h$ is just $f'(0)$, where $f(x) = a^x$. Thus, $\log a$ is the slope of the curve $y = a^x$ as it crosses the y axis. For a large, the number $\log a$ is large; for a small, it is small (in fact negative for $0 < a < 1$). In singling out e^x, we are selecting the curve that crosses the y axis with slope 1. Conversely, we can define e to be that real number such that the curve $y = e^x$ crosses the y axis with slope 1.

The formula for the derivative of the logarithm can be obtained by using implicit differentiation. This is done explicitly in Chapter 6, Section 6.2, Example 5.

Let us now consider some examples.

Example 1.   If $f(x) = x^3$, find $f'(5)$.

Solution.   Since $f(x) = x^3$, $f'(x) = 3x^2$. Thus $f'(5) = 3(5^2) = 75$.

Example 2.   If $f(x) = \log x$, find $f'(1/10)$.

Solution.   Since $f(x) = \log x$, $f'(x) = 1/x$. Thus

$$f(1/10) = \frac{1}{1/10} = 10.$$

Example 3.   If $f(x) = e^x$, find $f'(0)$.

Solution. Since $f(x) = e^x, f'(x) = e^x$. Thus $f'(0) = e^0 = 1$.

Note that you can *not* find $f'(x_0)$ by *first* substituting x_0. Thus, if $f(x) = x^5$ and we wish to find $f'(2)$, we *must first* write $f'(x) = 5x^4$, *then* substitute $x = 2$ to obtain $f'(2) = 5 \cdot 2^4 = 5 \cdot 16 = 80$. If you first substitute $x = 2$ in the function $f(x) = x^5$ to obtain $f(2) = 32$ (the constant function 32) and then differentiate, you get zero. In other words,

$$\frac{d}{dx} f(x_0) = \frac{d}{dx} (\text{constant}) = 0.$$

This is why we wrote

$$\frac{df}{dx} (x_0)$$

and deliberately avoided the more attractive but incorrect expression $df(x_0)/dx$.

Example 4. If $f(x) = x^{1/2}$, find $f'(9)$.

Solution. Since $f(x) = x^{1/2}, f'(x) = \frac{1}{2}x^{-1/2}$. Thus $f'(9) = \frac{1}{2}9^{-1/2} = \frac{1}{2} \cdot 1/\sqrt{9} = \frac{1}{6}$.

Example 5. If $f(x) = x^{3/2}$, find the tangent to the curve at the point $(4, 8)$.

Solution. Since $f(x) = x^{3/2}$, $f'(x) = \frac{3}{2}x^{1/2}$. Thus $f'(4) = \frac{3}{2}(4^{1/2}) = 3$. Hence, $y - 8 = 3(x - 4)$ is the equation of the tangent to the curve at the point $(4, 8)$.

Example 6. Does $f(x) = x^{1/3}$ have a derivative at $x = 0$?

Solution. Let us go ahead and simply calculate the derivative from the formula. Since $f(x) = x^{1/3}$,

$$f'(x) = \frac{1}{3}x^{-2/3} = \frac{1}{3} \cdot \frac{1}{x^{2/3}}.$$

Thus, if we substitute 0, we obtain $f'(0) = \frac{1}{3} \; 1/0$!!! This tells us something is amiss! Draw the graph of $f(x) = x^{1/3}$. (See Figure 3.3.2.) Notice the curve becomes steeper as we approach 0. It should be clear now that f does not have a derivative at $x = 0$.

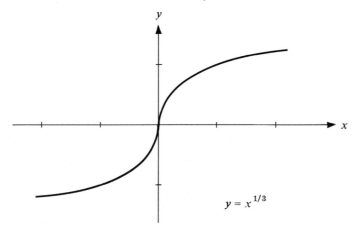

$$y = x^{1/3}$$

FIGURE 3.3.2

When a formula yields a bizarre result, stop and ask if there is something you are over-looking. We remark at this point that the function $f(x) = x^n$ does not have a derivative at $x = 0$ for $0 < n < 1$ and $-\infty < n < 0$.

Exercises

1. If $f(x)$ is given by
   (a) $f(x) = x^3$, find $f'(1)$ and $f'(-3)$.
   (b) $f(x) = x^{1/6}$, find $f'(2)$ and $f'(\frac{1}{4})$.
   (c) $f(x) = e^x$, find $f'(2)$ and $f'(0)$.
   (d) $f(x) = \log x$, find $f'(\frac{1}{4})$ and $f'(1)$.
   (e) $f(x) = x^4$, find $f'(1)$ and $f'(-1)$.
   (f) $f(x) = x^{3/2}$, find $f'(4)$ and $f'(9)$.
   (g) $f(x) = x^{4/3}$, find $f'(8)$ and $f'(27)$.

2. Find the equation of the tangent to the curve $y = x^{1/2}$ at the point $(9, 3)$. Draw the curve and its tangent at that point.

3. Find the equation of the tangent to the curve $y = \log x$ at the point $(e, 1)$. Draw the curve and its tangent at that point.

4. Consider the function defined by $f(x) = x|x|$.
   (a) Construct its graph.
   (b) Is f continuous at $(0, 0)$? Why?
   (c) Is f differentiable at $(0, 0)$? Why?
   (d) Find $f'(-2)$ and $f'(2)$.
   (e) Find the equation of the line tangent to the curve at the point $(1, 1)$.

5. Let
$$f(x) = \begin{cases} x, & x < 0 \\ x^2, & x \geq 0. \end{cases}$$
   (a) Is f differentiable at $(0, 0)$? Why?
   (b) Is f differentiable at $(1, 1)$? $(-1, -1)$? $(2, 4)$? Why?
   (c) Calculate f' and carefully indicate the domain of f.

6. If $f(x)$ is given by the expression below, find df/dx at the point indicated.
   (a) x^{21} at $x = 5$.   (d) $x^{-1/3}$ at $x = 8$.
   (b) $\log x$ at $x = 4$.   (e) x^{-2} at $x = -4$.
   (c) e^x at $x = -\frac{1}{2}$.

7. Show that the rate of change of the area of a square field with respect to the length of a side is one-half the perimeter of the field. [HINT: Let x denote the length of a side; the area is then given by the function $A(x)$, where $A(x) = x^2$.]

8. Let A be the area of a circle of radius r. Show that the rate of change of A with respect to the radius is given by the circumference of the circle.

9. Suppose that the total costs of producing x boxes of greeting cards is given by $S(x) = 2x^3$. Find the average and marginal costs of producing 100 boxes of greeting cards.

10. The volume of a sphere of radius r inches is given by $V = \frac{4}{3}\pi r^3$. If the radius is changing, find the rate of change of V with respect to r when $r = 2, 4, 6,$ and 8 inches, respectively.

11. If $f(x) = x^n$, where n is a positive integer, show that $f'(x) = nx^{n-1}$. [HINT: Observe that if x_0 is any real number, then

$$\frac{f(x) - f(x_0)}{x - x_0} = \frac{x^n - x_0^n}{x - x_0}.$$

We recall from high school algebra that $x^n - x_0^n$ may be factored as follows:

$$x^n - x_0^n = (x - x_0)(x^{n-1} + x^{n-2}x_0 + x^{n-3}x_0^2 + \cdots + xx_0^{n-2} + x_0^{n-1}).$$

Hence, for $x \neq x_0$,

$$\frac{x^n - x_0^n}{x - x_0} = x^{n-1} + x^{n-2}x_0 + x^{n-3}x_0^2 + \cdots + xx_0^{n-2} + x_0^{n-1}.$$

Thus,

$$f'(x_0) = \lim_{x \to x_0} \frac{x^n - x_0^n}{x - x_0} = \underbrace{x_0^{n-1} + x_0^{n-1} + \cdots + x_0^{n-1}}_{n \text{ times}} = nx_0^{n-1}. \Bigg]$$

3.4 Some Basic Differentiation Formulas

So far, while we can differentiate several elementary functions, we do not know how to differentiate such functions as $5x^2$ or $x^2 + x^3$ or $x^2/(x-1)$. This difficulty will now be overcome. First, we consider the simplest case—that of differentiating a constant times another function. We state this result in the form of a theorem. (A theorem is a statement that can be derived from definitions and previous results.)

THEOREM 1. *Let f be a differentiable function. If $h(x) = Cf(x)$, where C is a constant, then $h'(x) = Cf'(x)$. (In d/dx notation, this becomes $(d/dx)[Cf(x)] = C\,df(x)/dx$.)*

It is an easy matter to verify Theorem 1. See Exercise 9 in this section.
We turn immediately to an application of Theorem 1.

Example 1. If $h(x) = 5x^2$, find $h'(x)$.

Solution. Let $h(x) = Cf(x)$, where $C = 5$ and $f(x) = x^2$. Then $f'(x) = 2x$ and so $h'(x) = Cf'(x) = 5(2x) = 10x$; or

$$\frac{dh}{dx} = \frac{d}{dx}(5x^2) = 5\frac{d}{dx}(x^2) = 5(2x) = 10x.$$

As will be seen shortly, the d/dx notation affords a very neat, clear and concise way of simplifying and keeping track of the various steps when differentiating complicated expressions. For this reason it is often preferable.

Next, we consider the derivative of the sum and difference of two functions.

THEOREM 2.   *Let f and g be two differentiable functions. If* $h(x) = f(x) \pm g(x)$, *then h is differentiable and* $h'(x) = f'(x) \pm g'(x)$. *(Using the d/dx notation, we have*

$$\frac{d}{dx}[f(x) \pm g(x)] = \frac{df(x)}{dx} \pm \frac{dg(x)}{dx}.\bigg)$$

We could verify the above result directly from the definition of a derivative. Those interested should refer to Exercise 10 of this section. We now consider some examples.

Example 2.   If $h(x) = x^3 + \log x$, find $h'(x)$.

Solution.   Write $h(x) = f(x) + g(x)$, where $f(x) = x^3$ and $g(x) = \log x$. Then $f'(x) = 3x^2$ and $g'(x) = 1/x$. Hence, $h'(x) = f'(x) + g'(x) = 3x^2 + 1/x$. Alternatively,

$$\frac{dh(x)}{dx} = \frac{d}{dx}(x^3) + \frac{d}{dx}(\log x) = 3x^2 + \frac{1}{x}.$$

Example 3.   If $h(x) = 4x^{1/2} + 2e^x$, find $h'(x)$.

Solution.

$$\frac{dh(x)}{dx} = \frac{d}{dx}(4x^{1/2}) + \frac{d}{dx}(2e^x) = 4\frac{d}{dx}(x^{1/2}) + 2\frac{d}{dx}(e^x)$$

$$= 4(\tfrac{1}{2}x^{-1/2}) + 2(e^x) = 2x^{-1/2} + 2e^x.$$

Note that the *d/dx* notation eliminates the need to introduce auxiliary functions while permitting us to do the problem *one step* at a time.

Exercises

1.   Find the derivative of the following functions

(a) $2x^3 + \log x$ (f) $e^x - \log x + x^3 - x^4$

(b) $4e^x + x^{-10}$ (g) $2x^{1/2} + x^4 + e^x$

(c) $x + x^2 + x^3$ (h) $5x^{-3/2} - \log x$

(d) $4 \log x - 30x^2$ (i) $x^4 + 2x^3 - x^2 + 1$

(e) $-e^x + x^{-1} + \log x$ (j) $1/x + x^{4/3} + 3x^5 + e^x$

2.   If $f(x) = -x^2 + x^3$, find the tangent to the curve at $(2, 4)$.

3.   If $f(x) = x^{1/2} + 2x^2$, find the tangent to the curve at the point $(1, 3)$.

4.   The relation between sales and advertising cost x for a product is given by the formula $S(x) = 200x^2 - 3x$. Find the rate of change in sales for $x = \$500$.

5.   Suppose that $f(x) = 3x^4 - 6x^2$. For what values of x is $f'(x) = 0$?

6.   A company's total sales revenue is given by the equation $R(x) = \tfrac{1}{2}x + 2x^2$, where x is the number of years the company has been in business and $R(x)$ is in millions of dollars. At what rate is the company's total sales revenue growing at the end of 5 years?

7. Find $f'(x)$ at the indicated point, if

(a) $f(x) = x^3 + 3x^{1/3} + 1$ at $x = 9$.

(b) $f(x) = x^2 + e^x + x + 1$ at $x = 0$.

(c) $f(x) = 3 \log x + 2e^x + x^3 + x^2$ at $x = 1$.

(d) $f(x) = 10e^x - (x^2 + 1)^2 + 5$ at $x = 0$.

(e) $f(x) = -3x^3 + 9x^2 + \log x - 5e^x$ at $x = 1$.

8. If $f(x) = a_0 + a_1 x + a_2 x^2 + \cdots + a_n x^n$, find $f'(x)$.

9. If $h(x) = Cf(x)$, where f is differentiable, show that $h'(x) = Cf'(x)$.
[HINT:

$$h'(x_0) = \lim_{x \to x_0} \frac{h(x) - h(x_0)}{x - x_0} = \lim_{x \to x_0} \frac{Cf(x) - Cf(x_0)}{x - x_0}.\Bigg]$$

10. Prove Theorem 2 of this section.
[HINT: Since $h(x) = f(x) + g(x)$, it follows that

$$\frac{h(x) - h(x_0)}{x - x_0} = \frac{[f(x) + g(x)] - [f(x_0) + g(x_0)]}{x - x_0}$$

$$= \frac{f(x) - f(x_0)}{x - x_0} + \frac{g(x) - g(x_0)}{x - x_0}.\Bigg]$$

3.5 Differentiation of Products and Quotients

We begin with the formula for the derivative of a product of two functions. The result is stated as Theorem 3.

THEOREM 3. *Let f and g be two differentiable functions and suppose $h(x) = f(x)g(x)$. Then h is also differentiable and $h'(x) = f'(x)g(x) + f(x)g'(x)$. (In d/dx notation, we have*

$$\frac{d}{dx}[f(x)g(x)] = \frac{df(x)}{dx} \cdot g(x) + \frac{dg(x)}{dx} \cdot f(x).\Big)$$

The proof of this theorem is a bit tricky and is outlined in Exercise 12 of this section. We immediately proceed to some examples.

Example 1.   If $h(x) = x^3 \log x$, find $h'(x)$.

Solution.   Let $h(x) = f(x)g(x)$, where $f(x) = x^3$ and $g(x) = \log x$. Then $f'(x) = 3x^2$ and $g'(x) = 1/x$. Hence

$$h'(x) = f'(x)g(x) + f(x)g'(x) = 3x^2 \cdot \log x + x^3 \left(\frac{1}{x}\right).$$

Alternatively,

$$\frac{dh}{dx} = \frac{d}{dx}(x^3) \cdot \log x + x^3 \cdot \frac{d}{dx}(\log x) = 3x^2 \log x + x^3 \left(\frac{1}{x}\right) = 3x^2 \log x + x^2.$$

Example 2.   If $h(x) = 4e^x(x^6 - x^{-6})$, find $h'(x)$.

Solution.

$$\frac{dh(x)}{dx} = 4\frac{d}{dx}\left[e^x(x^6 - x^{-6})\right]$$

$$= 4\left\{\frac{d}{dx}(e^x) \cdot (x^6 - x^{-6}) + e^x\frac{d}{dx}(x^6 - x^{-6})\right\}$$

$$= 4\left\{e^x \cdot (x^6 - x^{-6}) + e^x\left[\frac{d}{dx}(x^6) - \frac{d}{dx}(x^{-6})\right]\right\}$$

$$= 4\{e^x(x^6 - x^{-6}) + e^x[6x^5 - (-6x^{-7})]\}$$

$$= 4e^x\{x^6 - x^{-6} + 6x^5 + 6x^{-7}\}.$$

(The last step was not really necessary, but there are advantages to algebraic simplification.)

Note that Theorem 1 can be obtained from Theorem 3 by applying the product rule to $Cf(x)$. (See Exercise 8 of this section.)

We conclude this section with the formula for the derivative of the quotient of two functions.

THEOREM 4.   *Let f and g be two differentiable functions such that $g(x) \neq 0$. Then the function defined by $h(x) = f(x)/g(x)$ is differentiable and we have*

$$h'(x) = \frac{f'(x)g(x) - f(x)g'(x)}{[g(x)]^2}.$$

(*In d/dx notation,*

$$\frac{dh(x)}{dx} = \frac{(df(x)/dx)g(x) - f(x)(dg(x)/dx)}{[g(x)]^2}.)$$

We observe that if we know the derivative of $1/g(x)$, then the result follows from the product rule. In Exercise 11 of this section, it is shown that

$$\frac{d}{dx}\left[\frac{1}{g(x)}\right] = -\frac{dg(x)/dx}{[g(x)]^2}.$$

Hence,

$$\frac{d}{dx}\left[\frac{f(x)}{g(x)}\right] = \frac{df(x)}{dx}\cdot\frac{1}{g(x)} + f(x)\cdot\frac{d}{dx}\left[\frac{1}{g(x)}\right]$$

$$= \frac{df(x)/dx}{g(x)} - \frac{f(x)(dg(x)/dx)}{[g(x)]^2} = \frac{(df(x)/dx)g(x) - f(x)(dg(x)/dx)}{[g(x)]^2}.$$

Thus, we need only know $(d/dx)[1/g(x)]$ and the product rule for differentiation in order to obtain the quotient rule.

We consider some examples.

Example 3.   If $h(x) = x^3/(1 + x^2)$, find $h'(x)$.

Solution.   Note that $1 + x^2 \neq 0$ for any real number x; thus, we can apply the quotient rule. Let $h(x) = f(x)/g(x)$, where $f(x) = x^3$ and $g(x) = 1 + x^2$. Then $f'(x) = 3x^2$ and $g'(x) = 2x$. Hence,

$$h'(x) = \frac{f'(x)g(x) - f(x)g'(x)}{[g(x)]^2} = \frac{3x^2(1 + x^2) - x^3(2x)}{(1 + x^2)^2}.$$

It is possible to simplify this expression, but we leave that and the d/dx approach to the reader. Alternatively, we could use the remark after the theorem and obtain

$$\frac{d}{dx}\left[\frac{x^3}{1 + x^2}\right] = \frac{d}{dx}[x^3] \cdot \frac{1}{(1 + x^2)} + x^3 \cdot \frac{d}{dx}\left[\frac{1}{1 + x^2}\right] = \frac{3x^2}{1 + x^2} - \frac{x^3 \cdot 2x}{(1 + x^2)^2},$$

which is equivalent to the above.

Exercises

1.   Find the derivative of the following expressions.

(a) $x^2 e^x$

(b) $\log x - x^{10} \log x$

(c) $\dfrac{x^3}{e^x}$

(d) $\dfrac{e^x}{x^3}$

(e) $\dfrac{\log x}{x^5}$

(f) $5x e^x \log x$

(g) $10(\log x) \cdot \log x$

(h) $e^x \cdot e^x$

(i) $\dfrac{1}{1 + \dfrac{1}{1 + (1/x)}}$

(j) $e^x \log x + 5x^{1/2}(x^2 + 2x + 1)$

(k) $\dfrac{e^x}{1 + e^x}$

(l) $\dfrac{x^2}{1 + x}$

(m) $\dfrac{\log x}{1 + e^x}$

2.   Find the equation of the tangent line to the following curves at the indicated points.
   (a) $f(x) = x^2 e^x$ at $(1, e)$
   (b) $f(x) = 2x + x \log x$ at $(1, 2)$
   (c) $f(x) = \sqrt{1 + x^2} \cdot e^x$ at $(0, 1)$
   (d) $f(x) = x/(x^2 + 1)$ at $(0, 0)$ and $(1, \frac{1}{2})$.

3.   If the total cost of producing x cases of beer is given by

$$C(x) = x^3 - \tfrac{1}{2}x^2 + 7x,$$

find the marginal cost when 10 cases of beer have been produced.

4. Find the points on the graph of the function f, where

$$f(x) = 2x^3 - 3x^2 - 12x + 20,$$

where the tangent line is parallel to the x axis.

[HINT: It should be clear that at those points x_i, where the tangent line is parallel to the x axis, the slope of the tangent line must be zero. (Why?) Hence, to find those points x_i, we must find those x_i such that $f'(x_i) = 0$.]

5. (Refer to Exercise 8, Section 3.1.) Suppose that a New York publishing firm must charge a price of $F(x) = 1/x + \frac{1}{2}x + 0.01x^2$ dollars per ton of paperbacks.
    (a) What is his total revenue for 10 tons?
    (b) What is his marginal revenue for 10 tons?
    (c) Suppose the firm's costs are given by $C(x) = 0.30x + 0.001x^2$ dollars per ton. Determine his total profit for selling 10 tons. What is his marginal profit?
    (d) For what values of x does the marginal cost equal the marginal revenue? (We shall see in Chapter 4 that total profit is largest when marginal revenue equals marginal cost.)

6. Find the values of the constants a, b, and c if the curve $f(x) = ax^2 + bx + c$ is to pass through the point $(1, 2)$ and is to be tangent to the line $y = x$ at the origin.

7. Find $f'(x)$ at the indicated point if

    (a) $f(x) = \dfrac{x^3 + 5x^2 - 7x}{x^2 + 1}$ at $x = 1$.

    (b) $f(x) = (x - 3)(2x + 1)(5x - 7)$ at $x = 2$.

    (c) $f(x) = (x^3 + 7x - 1)(x^2 + 2x + 5) - \dfrac{3x + 1}{x + 2}$ at $x = 3$.

    (d) $f(x) = 7e^x \log x = \dfrac{x - 1}{e^x} + x^{5/2}$ at $x = 1$.

    (e) $f(x) = \dfrac{x^3 - 5x + 1}{1 + e^x} + x^2 e^x \log x$ at $x = 1$.

8. Use the product rule (Theorem 3) to show that if $h(x) = Cf(x)$, then $h'(x) = Cf'(x)$. (C is a constant.)

9. If $u(x) = f(x) \cdot g(x) \cdot h(x)$, find $u'(x)$ in terms of $f, f', g, g', h,$ and h'.

10. If $h(x) = f(x) \cdot f(x)$, find $h'(x)$.

11. Let f be differentiable at x_0 when $f(x_0) \neq 0$, and suppose that $h(x) = 1/f(x)$. Show from first principles that $h'(x_0) = -f'(x_0)/[f(x_0)]^2$. [HINT: Observe that

$$\frac{h(x) - h(x_0)}{x - x_0} = \frac{1/f(x) - 1/f(x_0)}{x - x_0} = \frac{f(x_0) - f(x)}{x - x_0} \cdot \frac{1}{f(x)f(x_0)}.$$

Now take the

$$\lim_{x \to x_0} \frac{h(x) - h(x_0)}{x - x_0}$$

and use the fact that f is continuous at $x = x_0$.]

12. Prove Theorem 3 of this section.
    [HINT: Let x_0 be in the domain of $f \cdot g$. Then,

$$\frac{h(x) - h(x_0)}{x - x_0} = \frac{f(x)g(x) - f(x_0)g(x_0)}{x - x_0}$$

$$= \frac{f(x)g(x) - f(x_0)g(x) + f(x_0)g(x) - f(x_0)g(x_0)}{x - x_0}.$$

(We have simply added and subtracted the quantity $f(x_0)g(x)$, leaving the numerator unchanged.) Thus,

$$\frac{h(x) - h(x_0)}{x - x_0} = \frac{f(x) - f(x_0)}{x - x_0} g(x) + f(x_0) \frac{g(x) - g(x_0)}{x - x_0}.$$

Now take

$$\lim_{x \to x_0} \frac{h(x) - h(x_0)}{x - x_0}$$

and use the fact that g is continuous at x_0.]

3.6 Differentiation of Composite Functions

The preceding sections in this chapter have greatly increased our repertory of functions that we are able to differentiate. However, we still cannot differentiate such functions as $(x^2 + 1)^5$, e^{x^2}, $\log x^2(x^2 + 1)$, etc. Recall from Section 1.6 that these are really functions of functions or composite functions. In that section we said that h is the composition of two functions f and g, written $h = f \circ g$, if $h(x) = f[g(x)]$. Thus, if $g(x) = x^2 + 1$ and $f(x) = x^2$, $(x^2 + 1)^2 = f[g(x)] = [f \circ g](x)$. Although we did not state it explicitly in Chapter 2, it is not difficult to verify the fact that if f and g are continuous then the composition function $f \circ g$ is also continuous. Now we ask a similar question concerning differentiability. Namely, if f and g are differentiable, is $f \circ g$ differentiable? The answer is yes—although the verification of this is a bit subtle. Let us now see how we form the derivative of $f \circ g$.

The composite functions we shall wish to differentiate will all be of the following form

$$h(x) = [f(x)]^n, \tag{3.6.1}$$

$$h(x) = e^{f(x)}, \tag{3.6.2}$$

$$h(x) = \log f(x). \tag{3.6.3}$$

The derivatives for these are given in Table 3.6.1.

Table 3.6.1

$h(x)$	$h'(x)$
1. $[f(x)]^n$	$n[f(x)]^{n-1}f'(x)$
2. $e^{f(x)}$	$e^{f(x)}f'(x)$
3. $\log f(x)$	$\dfrac{1}{f(x)} f'(x)$

Using the d/dx notation we may also write

$$\frac{d}{dx}[f(x)]^n = n[f(x)]^{n-1}\frac{df}{dx},$$ (3.6.1′)

$$\frac{d}{dx}e^{f(x)} = e^{f(x)}\frac{df}{dx},$$ (3.6.2′)

$$\frac{d}{dx}\log f(x) = \frac{1}{f(x)}\frac{df}{dx}$$ (3.6.3′)

from Table 3.6.1.

We will now illustrate the use of this table by several examples.

Example 1.   If $h(x) = (1 - x^2)^{1/2}$, find $h'(x)$.

Solution.   The function h is of the form $[f(x)]^n$, where $f(x) = 1 - x^2$ and $n = \frac{1}{2}$. Therefore,

$$h'(x) = \tfrac{1}{2}(1 - x^2)^{1/2-1} \cdot f'(x)$$
$$= \tfrac{1}{2}(1 - x^2)^{-1/2} \cdot (-2x).$$

Example 2.   If $h(x) = e^{(4x+9x^2)}$, find $h'(x)$.

Solution.   The function h is of the form $e^{f(x)}$, where $f(x) = 4x + 9x^2$. Thus,

$$\frac{dh}{dx} = \frac{d}{dx}e^{(4x+9x^2)} = e^{(4x+9x^2)} \cdot \frac{d}{dx}(4x + 9x^2)$$

$$= e^{(4x+9x^2)} \cdot (4 + 18x).$$

Example 3.   If $h(x) = \log(1 + 2x)$, find $h'(x)$.

Solution.   The function h is of the form $\log f(x)$, where $f(x) = (1 + 2x)$. Thus,

$$h'(x) = \frac{1}{(1 + 2x)} \cdot f'(x) = \frac{1}{(1 + 2x)} \cdot 2.$$

The d/dx notation is very convenient for handling more complicated expressions. It allows you to keep track of what you are doing and to proceed one step at a time with a minimum of confusion. Consider the next example.

Example 4. If $h(x) = e^{[x \log (x^2+1)]^3}$, find dh/dx.

Solution. Let us simply proceed one step at a time. Thus,

$$\frac{dh}{dx} = \frac{d}{dx} e^{[x \log(x^2+1)]^3}$$

$$= e^{[x \log(x^2+1)]^3} \cdot \frac{d}{dx} [x \log(x^2+1)]^3$$

(although the expression $e^{[x \log (x^2+1)]^3}$ is complicated, it is just $e^{f(x)}$)

$$= e^{[x \log(x^2+1)]^3} \cdot 3[x \log(x^2+1)]^2 \cdot \frac{d}{dx} [x \log(x^2+1)]$$

(since $[x \log(x^2+1)]^3$ is of the form $[f(x)]^n$)

$$= e^{[x \log(x^2+1)]^3} \cdot 3[x \log(x^2+1)]^2 \cdot$$

$$\left\{ \frac{d}{dx}(x) \cdot \log(x^2+1) + x \frac{d}{dx} \log(x^2+1) \right\}$$

(we are just differentiating a product)

$$= e^{[x \log(x^2+1)]^3} \cdot 3[x \log(x^2+1)]^2 \cdot$$

$$\left\{ 1 \cdot \log(x^2+1) + x \cdot \frac{1}{x^2+1} \cdot \frac{d}{dx}(x^2+1) \right\}$$

(Why?)

$$= e^{[x \log(x^2+1)]^3} \cdot 3[x \log(x^2+1)]^2 \cdot$$

$$\left\{ \log(x^2+1) + \frac{x}{x^2+1}(2x) \right\}.$$

The last example should convince you that you can differentiate even a very complicated expression! Just hammer away at it, one step at a time.

The formulas in Table 3.6.1 are special cases of a very general theorem.

THEOREM 5 (THE CHAIN RULE). *If g and f are differentiable, then so is the function $h = g \circ f$. In addition,*

$$h'(x) = g'[f(x)]f'(x).$$

(This holds, of course, for x in the domain of $g \circ f$.)

The proof of this theorem has been placed in an appendix. If we use the d/dx notation, then the conclusion of Theorem 5 may be written as

$$\frac{dh}{dx} = \frac{dg}{du} \cdot \frac{du}{dx},$$

where $u = f(x)$. Thus, if $h(x) = [f(x)]^n$, then $u = f(x)$ and $g(u) = u^n$;

$$\frac{dh}{dx} = \frac{dg}{du} \cdot \frac{du}{dx} = nu^{n-1} \cdot \frac{du}{dx} = n[f(x)]^{n-1} \frac{df}{dx}.$$

This checks with entry 1 in Table 3.6.1.

Theorem 5 can be used to derive the formulas in Table 3.6.1. Let us do this for $h(x) = e^{f(x)}$. In this case, $h(x) = g[f(x)]$, where $g(x) = e^x$ and $f(x)$ is just $f(x)$. By Theorem 5, $h'(x) = g'[f(x)] \cdot f'(x)$. Since $g(x) = e^x$, $g'(x) = e^x$. Hence, $g'(f(x)) = e^{f(x)}$. Substituting, we find $h'(x) = e^{f(x)} \cdot f'(x)$ as expected.

We end this section with the following example.

Example 5.   Suppose that

$$h(x) = (x^4 - 3)^{1/3}.$$

Find $h'(1)$.

Solution.   Let us use Theorem 5 for this example.

Let

$$g(x) = x^{1/3} \quad \text{and} \quad f(x) = x^4 - 3.$$

Hence,

$$g'(x) = \tfrac{1}{3} x^{-2/3} \quad \text{and} \quad f'(x) = 4x^3,$$

$$h'(x) = g'[f(x)] \cdot f'(x) = \tfrac{1}{3}[x^4 - 3]^{-2/3} \cdot 4x^3 = \tfrac{4}{3}[x^4 - 3]^{-2/3} \cdot x^3,$$

and

$$h'(1) = \frac{4}{3} \cdot [-2]^{-2/3} = \frac{4}{3} \cdot \frac{1}{\sqrt[3]{4}}.$$

Exercises

1.   Find the derivatives of the following expressions.

(a) $\log(2x^2 + x + 1)$

(b) $\dfrac{x}{\sqrt{(x^2 - 1)}}$

(c) $\dfrac{x^3 + 2x}{(x^2 + 2x + 1)^{3/2}}$

(d) $x^2 e^{4x}$

(e) $\log(e^x + x)$

(f) $(x^2 - 5)^{3/2}$

(g) $(x + \log x)^{-4}$

(h) $\left(e^x - \dfrac{1}{x}\right)^7$

(i) $\log(xe^x)$

(j) $[e^{(x^3 - 2x)}]^{100}$

(k) $[\log x]^2 + e^{x^3}$

(l) $e^{x^2} + [\log x]^3$

(m) $\log(x^2 + 2x + 1)^2$

(n) $\log\left[\dfrac{x^3 - 9}{e^{5x}}\right]^{1/2}$

(o) $e^{\left[\frac{\log x}{x}\right]^{10}}$

(p) $\log[\log x]$

(q) $\log[\log(\log x)]$

(r) $\left[\dfrac{x + \sqrt{1 - x^2}}{x^2 - \sqrt{1 + x^2}}\right]^{1/2}$

2.   Let $f(x) = x^2 - 4$. Does $f'(x_0) = 0$ for any x_0? If so, for what x_0?

3.   Let $f(x) = 3x^4 + 2x^3 + 3x^2 + 1$. Find all values of x such that $f'(x) = 0$.

4. Let $h(x) = 1/f(x)$. This can also be written as $h(x) = [f(x)]^{-1}$. Find dh/dx by the rule for differentiation of composite functions. Find dh/dx by the method for quotients. Compare your answers.

5. Let $f(x) = \log(e^x)$. Find $f'(x)$ by the composite function method. Note that $\log(e^x) = x$. Check your answer.

6. Let $h(x) = f(x)/g(x)$. Rewrite this as $h(x) = f(x) \cdot [g(x)]^{-1}$. Find dh/dx by using the chain rule. Check your answer against the standard method of differentiating f/g.

7. Find the equation of the tangent to the curve

$$f(x) = \frac{2}{\sqrt{x^2 - 1}}$$

at the point on the curve, where $x = 10$.

8. If $f'(x) = \sqrt{3x^2 - 1}$ and $y = f(x^2)$, find dy/dx.

9.* Previously we have verified that if n is *any* positive integer, then the derivative of x^n is nx^{n-1}. Now we can, with the use of the chain rule, extend this result to any rational number n, where $n = p/q$ with p, q integers.

(a) First, use the quotient rule to show that if $n < 0$,

$$\frac{d}{dx}(x^n) = nx^{n-1}.$$

[HINT: Let $n = -m$, where m is positive. Then,

$$\frac{d}{dx}[x^{-m}] = \frac{d}{dx}\frac{1}{x^m} = -\frac{mx^{m-1}}{x^{2m}} = -mx^{-m-1}.]$$

(b) Now, we show that $d/dx[x^n] = nx^{n-1}$ if $n = 1/p$, p an integer and nonzero. [HINT: Let $g(x) = x^{1/p}$, thus (see Chapter 1) $[g(x)]^p = x$. Since $[g(x)]^p = x$, the derivative of $[g(x)]^p$ must be 1. (Why?) Thus,

$$\frac{d}{dx}[g(x)]^p = 1$$

and thus $p[g(x)]^{p-1} \cdot g'(x) = 1$. Hence,

$$g'(x) = \frac{1}{p}[g(x)]^{1-p} = \frac{1}{p}(x^{1/p})^{1-p} = \frac{1}{p} \cdot x^{(1/p)-1}.]$$

(c) Next show that $d/dx[x^n] = nx^{n-1}$, where $n = p/q, p, q$ integers and $q \neq 0$. (Use the chain rule.)
[HINT:

$$\frac{d}{dx}[x^{p/q}] = \frac{d}{dx}[x^{1/q}]^p = p[x^{1/q}]^{p-1} \cdot \frac{d}{dx}[x^{1/q}]$$

$$= px^{p/q - 1/q} \cdot \frac{1}{q}x^{(1/q)-1}.$$

Complete the verification.]

10*. (a)　Consider the function $f(x) = x^x$ for $x > 0$. One might think that $f'(x) = x \cdot x^{x-1}$. Explain why the chain rule (Theorem 5) does *not* yield this result.

(b)　If $f(x) = x^x$ for $x > 0$, find $f'(x)$.
　　[HINT: Recall that $x^x = e^{x \log x}$.]

(c)　If $f(x) = x^{(x^2)}$, find f'.

(d)　If $f(x) = x^{\log x}$, find f'.

3.7　Higher Derivatives

Let f be a differentiable function. Since f' is a function in its own right, one can ask whether $g(x) = f'(x)$ has a derivative at x_0, say. If g is differentiable at x_0, then we write $g'(x_0) = f''(x_0)$. In this case one says that f has a second derivative at x_0 that can be denoted by

$$f''(x_0) = f''(x)\bigg|_{x=x_0} = \frac{d^2 f}{dx^2}\bigg|_{x=x_0}.$$

It is obviously possible (formally) to define derivatives of higher orders in this manner. The third derivative is usually written as $f'''(x)$ or $d^3 f(x)/dx^3$. It is unlikely that we will have a need for more than first and second derivatives. We could also define the second and higher derivatives in terms of limits, but there is really no advantage to doing this. Let us now consider some examples.

Example 1.　If $f(x) = -4x^3 + 3x^2 + x - 1$, find $f''(x)$.

　Solution.　There is no fancy method for finding second derivatives; you simply differentiate and then differentiate again. Thus,

$$f'(x) = -12x^2 + 6x + 1 \quad \text{and} \quad f''(x) = -24x + 6.$$

Example 2.　If $f(x) = x^2 e^x$, find $f''(x)$.

　Solution.

$$\frac{df}{dx} = 2xe^x + x^2 e^x.$$

$$\frac{d^2 f}{dx^2} = \frac{d}{dx}[2xe^x + x^2 e^x] = \frac{d}{dx}(2xe^x) + \frac{d}{dx}(x^2 e^x)$$

$$= (2e^x + 2xe^x) + (2xe^x + x^2 e^x) = e^x(2 + 4x + x^2).$$

We can also give a geometrical significance to the second derivative. The first derivative, roughly speaking, tells you whether the curve is going up or down and how steeply. The second derivative tells you how fast the curve is bending (turning) and whether it is turning up or down. We illustrate this via Figures 3.7.1–3.7.4.

This is discussed in much greater detail in the next chapter. We also see some applications of second derivatives in the next chapter.

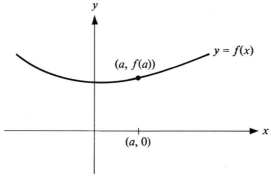

$f''(a) > 0$ and small

FIGURE 3.7.1

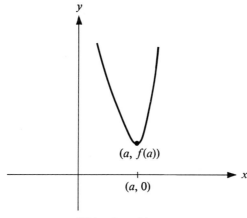

$f''(a) > 0$ and large

FIGURE 3.7.2

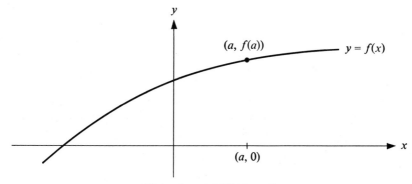

$f''(a) < 0$ and $|f''(a)|$ small

FIGURE 3.7.3

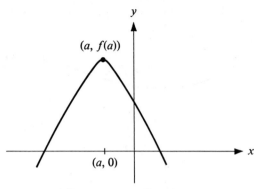

$f''(a) < 0$ and $|f''(a)|$ large

FIGURE 3.7.4

Exercises

1. Find the second derivatives of the the following expressions:

(a) $x^3 - 9x + 1$ (g) $\log(\log x)$

(b) $(x^2 - 1)^{1/2}$ (h) $e^{(1-x)/(1+x)}$

(c) $x \log x$ (i) $\dfrac{x}{(x^2 - 1)^{1/3}}$

(d) $(\log x)e^x$ (j) $\log(x^3 + x^2 e^x + 1)$

(e) $(x + 1)e^{4x^2}$ (k) $e^{\log(x^2 + 1)}$

(f) $\dfrac{x + 1}{x - 1}$ (l) $[x^3 - 9x^2 + 1]^3$

2. If $f(x) = x^4 - x^3 + x^2 - x + 1$, find all points where $f''(x)$ equals 0.

3. If $y = (1 + x)/(1 - x)$, find the third derivative of y, $y'''(x)$ at the point $x = \frac{1}{2}$.

4. (a) If $y = a_0 + a_1 x + a_2 x^2 + a_3 x^3$, find $y'''(x)$ and $y^{(iv)}(x)$ (that is, the fourth derivative of y).

   (b) Suppose that $y(x) = a_0 + a_1 x + a_2 x^2 + \cdots + a_n x^n$, where n is a positive integer. Find $y'(x)$, $y''(x)$, $y'''(x)$, $y^{(iv)}(x)$. Can you guess what $y^{(n)}(x)$, where $y^{(n)}(x)$ denotes the nth derivative of x, might be? What do you think is the value of $y^{(n+1)}(x)$?

   (c) Referring to (b), what is $y(0)$? $y'(0)$? $y''(0)$? ... $y^{(n)}(0)$? [in terms of the a_i's $(i = 0, 1, 2, \ldots, n)$, of course]. Rewrite $y(x)$ by substituting the various derivatives of y, evaluated at $x = 0$.

5. If $h(x) = 1/f(x)$, find $h''(x)$, in terms of f, f', f''.

6. If $h(x) = f(x) \cdot g(x)$, find $h''(x)$.

7. Let $f(x) = \sqrt{1 - x^2}$ for $-1 < x < 1$ and $g(x) = x^2/2$. Show that for $-1 < x < 1$, $(g \circ f)''(x) = -1$. What is $(f \circ g)''(x)$?

3.8 Summary of Chapter 3

(1) (a)   The derivative $f'(x_0)$ of f at a point x_0 in the domain of f is defined by

$$\lim_{x \to x_0} \frac{f(x) - f(x_0)}{x - x_0} = f'(x_0)$$

provided that this limit exists. Other notations for the derivative:

$$\frac{df}{dx}(x_0) \quad \text{and} \quad \frac{df(x)}{dx}\bigg|_{x = x_0}.$$

(b)   If f has a derivative at a point x_0, we say that f is differentiable at x_0.

(c)   If f is differentiable at x_0, then f is continuous at x_0. The converse does *not* hold. However, if f is *not* continuous at x_0, then f is not differentiable at x_0.

(2) The derivative of a function f measures the rate of change in $f(x)$ when there is a change in x. Thus, if f represents: (a) cost; f' denotes marginal cost (b) position; f' denotes velocity.

(3) We list here all the important formulas discussed in this chapter. We assume all functions are differentiable.

(a-1) $\dfrac{d}{dx}[C] = 0$ when C is a constant.

(a-2) $\dfrac{d}{dx}[x^n] = nx^{n-1}$ for any n.

(a-3) $\dfrac{d}{dx}[e^x] = e^x$.

(a-4) $\dfrac{d}{dx}[\log x] = \dfrac{1}{x}$ for $x > 0$.

(b-1) $\dfrac{d}{dx}[Cf] = C\dfrac{df}{dx}$ when C is a constant.

(b-2) $\dfrac{d}{dx}[f \pm g] = \dfrac{df}{dx} \pm \dfrac{dg}{dx}$.

(b-3) $\dfrac{d}{dx}[f \cdot g] = g\dfrac{df}{dx} + f\dfrac{dg}{dx}$.

(b-4) $\dfrac{d}{dx}\left[\dfrac{f}{g}\right] = \dfrac{g\, df/dx - f\, dg/dx}{[g(x)]^2}$.

(b-5) $\dfrac{d}{dx}[(f \circ u)(x)] = \dfrac{d}{dx}\{f[u(x)]\} = \dfrac{df}{du} \cdot \dfrac{du}{dx}$.

(c-1) $\dfrac{d}{dx}[f(x)^n] = n[f(x)]^{n-1} \cdot \dfrac{df}{dx}$.

(c-2) $\dfrac{d}{dx}[e^{g(x)}] = e^{g(x)} \cdot \dfrac{dg}{dx}.$

(c-3) $\dfrac{d}{dx}[\log g(x)] = \dfrac{1}{g(x)} \cdot \dfrac{dg}{dx}.$

REVIEW EXERCISES

1. (a) If $f(x) = x^5$, then evaluate

$$\lim_{x \to 2} \frac{f(x) - f(2)}{x - 2}.$$

[HINT: Do it the easy way, using something you know from Section 3.1.]

  (b) Suppose that $f(x) = 1/(x + 2)$. Find, using the *definition* of a derivative, $f'(2)$. (You may not use the quotient rule, but must use the definition given in Section 3.1.)

2. If $f(x)$ is given as follows, determine $f'(x)$.

  (a) $f(x) = \dfrac{x^2}{\sqrt{x^2 - 4}}$

  (f) $f(x) = (\log x)^8$

  (b) $f(x) = \dfrac{x^3 - 1}{x - 1}$

  (g) $f(x) = \log(x^8)$

  (c) $f(x) = 2x^2 + x^3 e^x$

  (h) $f(x) = \left(\dfrac{x + 1}{x - 1}\right)^3$

  (d) $f(x) = e^{(x^3 + x^2 + 1)^2}$

  (i) $f(x) = x^2 e^{-x^2} + e^{-x} \log x.$

  (e) $f(x) = \log(x^2 + 2x + 1)^5$

  (j) $f(x) = \sqrt{x + \sqrt{x}}$

3. Find the equation of the line tangent to the curve $y = \log x/(x + 1)$ at the point $(1, 0)$.

4. Determine the constant c such that the straight line joining the points $(0, 3)$ and $(5, -2)$ is tangent to the curve $y = c/(x + 1)$.

5. Find the points on the curve

$$y = \tfrac{1}{3}x^3 + \tfrac{1}{2}x^2 + x + 1,$$

where the tangent is parallel to the x axis.

6. Consider the funtion defined by

$$f(x) = \begin{cases} \dfrac{x^2 - 9}{x - 3} & \text{if } 0 \le x < 3 \\ 2 & \text{if } x = 3 \\ 9 - x & \text{if } 3 < x \le 9. \end{cases}$$

(a)   Sketch the graph of f and indicate the domain and range of the function.

(b)   Does $\lim_{x \to 3} f(x)$ exist? Why?

(c)   Is f continuous at $x = 3$? Why? If not, how can f be defined at $x = 3$ in order to make it continuous at that point?

(d)   Is f differentiable at $x = 2$? Why?

7.   A Budd car will hold 100 people. If the number x of persons per trip who use the Budd car is related to the fare charged (p dollars) by $p = (5 - x/20)^2$, write the function expressing the total revenue per trip received by the Budd car company. What is the number x, of people per trip that will make the marginal revenue equal to zero? What is the corresponding fare? (See Exercise 8, Section 3.1.)

8.   If s represents the distance a body moves in time t seconds, determine

(a)   the velocity, $v = ds/dt$, and (b) the acceleration, $a = d^2s/dt^2$

if $s(t) = 250 + 40t - 16t^2$.

9.   If $f'(x) = \sqrt{3x^2 - 1}$ and $y = f(x^2)$, find dy/dx.

10.* Suppose we are given a function f satisfying the following conditions for all x and y.

(i) $f(x + y) = f(x) \cdot f(y)$

(ii) $f(x) = 1 + xg(x)$, where $\lim_{x \to 0} g(x) = 1$.

Prove that (a) the derivative $f'(x)$ exists and (b) $f'(x) = f(x)$.

[HINT: Let $x - x_0 = h$ in the definition of a derivative, so that as $x \to x_0$, $h \to 0$. Thus,

$$\lim_{x \to x_0} \frac{f(x) - f(x_0)}{x - x_0} = \lim_{h \to 0} \frac{f(x_0 + h) - f(x_0)}{h}.$$

Now, use Properties (i) and (ii).]

11. The demand for Superlux, the wonder detergent, is expressed by means of the equation

$$d(p) = 520 - 45p + p^2,$$

where p is the price per box (in cents) and d is the weekly demand (in thousands of boxes). What is the marginal rate of change of the demand for Superlux when the price per box is 25 cents?

12. Let

$$f(x) = \begin{cases} x^2, & x \geq 0 \\ x, & x < 0. \end{cases}$$

Does $f'(0)$ exist? Why? Sketch a graph of f.

13. Find $f'(x)$ if

$$f(x) = \left[\frac{(x - 5)^{2/5} \cdot (1/x + 2)^{1/3}}{(x - 5)^2} \right]^{1/2}.$$

14. Consider the function

$$f(x) = \begin{cases} \dfrac{x^2}{2}, & x \le 0 \\[2mm] \dfrac{x^2}{4}, & x > 0. \end{cases}$$

Determine the functions $f'(x)$ and $f''(x)$ and state the domain for each. Sketch the graphs of $f(x), f'(x),$ and $f''(x)$.

15. Determine where the first and second derivatives of $y = 2x/(1 + x^2)$ vanish.

16. If $f(x) = e^{x^2 - 2x}$, what is $f'(0)$? $f''(0)$? $f'''(0)$? $f^{(iv)}(0)$?

17. Given that $F'(x) = G(x)$, show that

$$\frac{d}{dx} F(ax + b) = aG(ax + b),$$

provided that F is differentiable at $ax + b$. (a, b are arbitrary constants.) [HINT: Use the chain rule.]

APPENDIX

3.9 Proof of the Chain Rule

We shall now verify Theorem 5—the chain rule. It is necessary for us to show that if $x_0 \in$ domain of $f \circ g$, then

$$\frac{d}{dx} [f \circ g(x_0)] = f'[g(x_0)]g'(x_0).$$

If we let $h(x) = f[g(x)]$, then

$$\frac{h(x) - h(x_0)}{x - x_0} = \frac{f[g(x)] - f[g(x_0)]}{x - x_0}.$$

Now we multiply the right-hand side of the preceding by $[g(x) - g(x_0)]/[g(x) - g(x_0)]$ obtaining

$$\frac{h(x) - h(x_0)}{x - x_0} = \frac{f[g(x)] - f[g(x_0)]}{g(x) - g(x_0)} \cdot \frac{g(x) - g(x_0)}{x - x_0}.$$

Taking limits,

$$\lim_{x \to x_0} \frac{h(x) - h(x_0)}{x - x_0} = \lim_{x \to x_0} \frac{f[g(x)] - f[g(x_0)]}{g(x) - g(x_0)} \cdot \lim_{x \to x_0} \frac{g(x) - g(x_0)}{x - x_0}.$$

It looks as though we might be finished since

$$\lim_{x \to x_0} \frac{f[g(x)] - f[g(x_0)]}{g(x) - g(x_0)} = f'[g(x_0)]$$

and thus

$$h'(x) = f'[g(x_0)] \cdot g'(x_0).$$

However, there is a flaw in this argument. The above method is valid *only if* $g(x) - g(x_0)$ $\neq 0$. If $g(x) = C$, a constant, then clearly the above method is in error since

$$g(x) - g(x_0) = 0.$$

Thus, we must somehow overcome this difficulty.

We must define an auxiliary function $F(t)$ as follows:

$$F(t) = \begin{cases} \dfrac{f(t) - f[g(x_0)]}{t - g(x_0)}, & \text{if } t \neq g(x_0) \\ f'[g(x_0)], & \text{if } t = g(x_0). \end{cases}$$

First observe that F is continuous at $g(x_0)$ since

$$\lim_{t \to g(x_0)} F(t) = \lim_{t \to g(x_0)} \frac{f(t) - f[g(x_0)]}{t - g(x_0)} = f'[g(x_0)],$$

the value of F at $g(x_0)$!

For $y \neq x_0$, we have

$$\frac{f[g(y)] - f[g(x_0)]}{y - x_0} = F[g(y)] \cdot \frac{g(y) - g(x_0)}{y - x_0}$$

since if $g(y) = g(x_0)$, then both sides are 0, and if $g(y) \neq g(x_0)$,

$$F[g(y)] = \frac{f[g(y)] - f[g(x_0)]}{g(y) - g(x_0)}.$$

Now,

$$\lim_{y \to x_0} \frac{f[g(y)] - f[g(x_0)]}{y - x_0} = F[g(x_0)] \cdot g'(x_0)$$

since F is continuous at $g(x_0)$ *and g* is differentiable at that point.

Substituting for $F[g(x_0)]$ and for the left-hand side of the above, we have

$$h'(x_0) = f'[g(x_0)] \cdot g'(x_0).$$

3.10 Why We Study Proofs

You are a systems analyst for Yoyodyne, a West Coast electronics firm. To impress a client you are preparing a report with a four-color pencil, 200-day moving averages, and third derivatives. This is being done on a flight from Los Angeles to New York City. Just past Albuquerque, New Mexico, you need to know the derivative of e^{-x^2} and suddenly realize you do not remember it. What to do? The man on your left is a professor of classics at Columbia. He tells you that calculus is the Greek word for pebble. The man on your right is a doorknob salesman from Sheboygan, Wisconsin. No help there either. Gritting your teeth and keeping in mind the words of your old high school football coach, you decide to derive the chain rule. By the time you pass Wheeling, West Virginia, the derivative of e^{-x^2} is in your possession.

THE MORAL: Next time, fly American. They have a stewardess who speaks calculus like a native.

4

Applications of Derivatives

4.1   Maxima and Minima.
Definitions and Basic Results

In this chapter we consider the problem of determining maximum and minimum values
of a function. Such problems frequently arise in applications. For example, consider
the following classic problem. A farmer has 6000 ft of fence. He wishes to fence in a
rectangular portion of land along a river in such a way as to enclose the largest possible
area. (See Figure 4.1.1.) How long should the sides of the rectangle be to accomplish

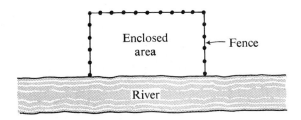

FIGURE 4.1.1

this? The first thing to do is rephrase the problem in a more mathematical but equivalent
form. Let x and y be the length of the sides of the rectangle. (See Figure 4.1.2.) Since
the perimeter of the fence is 6000 ft, we see that $2x + y = 6000$ or $y = 6000 - 2x$. Let
$A(x)$ represent the area of the fenced-in plot when the length of the land is x. Then,
$A(x) = xy = x(6000 - 2x) = 6000x - 2x^2$. By the nature of the problem, $x \geq 0$ and
$x \leq 3000$. (There are only 6000 ft of fence.)

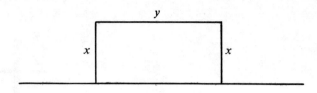

FIGURE 4.1.2

A problem equivalent to the original one is the following: Find the maximum value of the function $A(x) = 6000x - 2x^2$ for $0 \le x \le 3000$. This problem probably does not look any easier than the first. Let us suggest one way of obtaining an approximate solution. Simply graph the function $A(x) = 6000x - 2x^2$ at the integers between 0 and 3000. (This would be time-consuming but not impossible.) What you would get would look like Figure 4.1.3. However, it is not clear that the graph might not look like that in Figure 4.1.4 if more points were plotted.

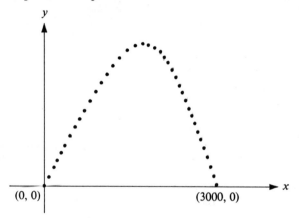

Function graphed for integer values

FIGURE 4.1.3

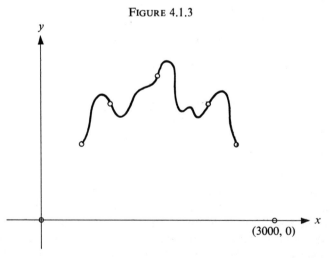

FIGURE 4.1.4

We now present a method for attacking a wide range of problems of a similar nature. First, it is necessary to carefully define the terms "maximum" and "minimum."

DEFINITION 1. *A function f, defined on an open interval (a, b) or a closed interval $[a, b]$, has a local maximum at $x_0 \in (a, b)$ if there exists a number $d > 0$ such that $f(x_0) \geq f(x)$ for $x \in (x_0 - d, x_0 + d)$.*

In general, f has a local maximum at x_0 if $f(x_0)$ is greater than or equal to f evaluated at *any* nearby point. (See Figure 4.1.5.)

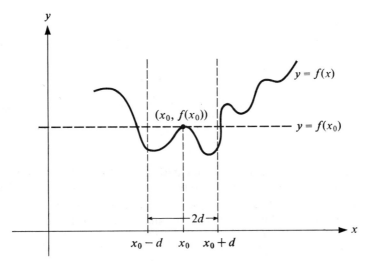

A local maximum at $(x_0, f(x_0))$

FIGURE 4.1.5

DEFINITION 2. *A function f, defined on an open interval (a, b) or a closed interval $[a, b]$, has a local minimum at $x \in (a, b)$ if there exists a $d > 0$ such that $f(x_0) \leq f(x)$ for $x \in (x_0 - d, x_0 + d)$.* (See Figure 4.1.6.)

A word of caution is in order at this point. Note that we have defined local maximum and local minimum *only* at interior points of an interval. We have *not* defined them for the *end points* of an interval. We could do this, but it is not useful for our purposes.

Observe that the function $f(x) = 2$ for all real x has a local maximum and a local minimum at every point.

The next theorem tells us how to go about finding local maxima and minima. Before stating the theorem, we clarify what is meant by saying a function is differentiable at an end point of a closed interval. If f is defined on $[a, b]$, then to say that f is differentiable at a means that $\lim_{x \to a^+} [f(x) - f(a)]/(x - a)$ exists; f is differentiable at b means that $\lim_{x \to b^-} [f(x) - f(b)]/(x - b)$ exists. In this case, we write $f'(a)$ and $f'(b)$ to denote those limits respectively.

THEOREM 1. *Let f be a function differentiable on the closed interval $[a, b]$. Let f have either a local maximum or a local minimum at $x_0 \in (a, b)$. Then $f'(x_0) = 0$.*

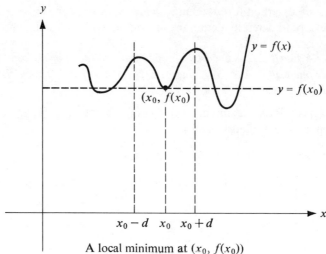

A local minimum at $(x_0, f(x_0))$

FIGURE 4.1.6

The verification is not difficult and is included as Exercise 5 in this section. Note that the theorem does *not* say that if $f'(x_0) = 0$ then x_0 is a local maximum or local minimum as can be seen by Example 1. It does say that if f has a local maximum (or minimum) the point where it has the local maximum (or minimum) will be included in the set of points where $f'(x) = 0$.

Example 1.   Find the local maximum and minimum of $f(x) = x^3$ on $[-1, 2]$.

Solution. $f'(x) = 3x^2$ so that 0 is the only place where $f'(x) = 0$. You might think that 0 must be a local maximum or local minimum, but it is not. It is impossible to find any $d > 0$ such that $f(0) \geq f(x)$ for $x \in (-d, d)$. In like manner we cannot find a $d_1 > 0$ such that $f(0) \leq f(x)$ for any $x \in (-d_1, d_1)$. For example, if $d = \frac{1}{2}$, then we see that for $x \in (-\frac{1}{2}, 0)$ we have $f(0) = 0 > x^3$, while for $x \in (0, \frac{1}{2}), 0 < x^3$. Thus, even though $f'(0) = 0$, 0 is still not a local maximum nor a local minimum. Consider the graph of $f(x)$ as shown in Figure 4.1.7. The fact that $f'(0) = 0$ means only that the function is "flat" at zero.

We now define the maximum and minimum of a function over a closed interval $[a, b]$. Sometimes this is called the absolute maximum and absolute minimum for emphasis.

DEFINITION 3.   *The function f has a maximum [minimum] on $[a, b]$ at x_0 if $f(x_0) \geq f(x)$ $[f(x_0) \leq f(x)]$ for **all** $x \in [a, b]$.*

Thus, in Example 1, $x = 2$ and $x = -1$ are the maximum and minimum points respectively, since $f(-1) = -1 \leq f(x)$ for $x \in [-1, 2]$ and $f(2) = 8 \geq f(x)$ for $x \in [-1, 2]$.

We quote the following results, which will be useful for further applications.

THEOREM 2.   *Let f be a function differentiable on* [a, b]. *Then*
  (i)   *f has a maximum and a minimum on* [a, b], *and*
  (ii) *this maximum* [minimum] *occurs at a, b or one of the numbers* x_i, $i = 1, 2, \ldots, n$,
       where $f'(x_i) = 0$.

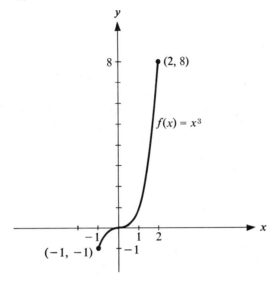

FIGURE 4.1.7

Since the verification of this theorem is fairly difficult, we omit it. Most people
would be all too willing to believe part (i) of the theorem. Actually, it is the subtle half.
Stop and ask yourself if the function in Figure 4.1.11 has a maximum on the interval
[0, 4]. A certain amount of reflection should convince you that it does not. (Note first

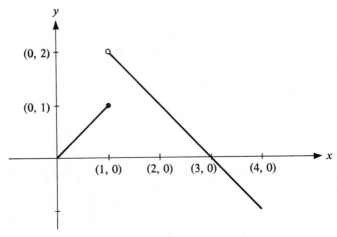

FIGURE 4.1.8

that $f(x) \leq 2$ for $x \in [0, 4]$, and second that although the functional values get arbitrarily close to 2, there is no x_0 for which $f(x_0) = 2$.) Clearly, f is **not** differentiable at 1. Indeed, f is not even continuous at 1. This example shows that just because a function is defined on $[a, b]$ does not mean it has a maximum (or a minimum) on $[a, b]$.

Let us now complete the problem we began in this chapter. We were faced with the task of maximizing the function $f(x) = 6000x - 2x^2$ for $0 \leq x \leq 3000$. Since the function is differentiable, it *must* have a maximum on the closed interval in question. Now we will locate the zeros of the derivative $f'(x) = 6000 - 4x$. Thus, setting the derivative equal to zero, $6000 - 4x = 0$ or $x = 1500$. Hence the maximum must occur at either 0, 1500, or 3000. Since $f(0) = 0$, $f(3000) = 0$, and $f(1500) = 4,500,000$, it is clear that the maximum occurs at $x = 1500$. Thus the area to be fenced in should have the dimensions shown in Figure 4.1.9.

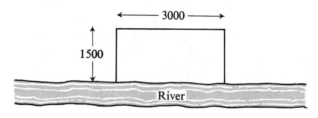

FIGURE 4.1.9

Example 2.   Find the maximum and minimum for the function

$$f(x) = x^2 + 2x + 2, \qquad -2 \leq x \leq 2.$$

Solution.   Clearly, f is differentiable on $[-2, 2]$. Hence, the maximum and minimum are attained at either $x = -2$, $x = 2$, or those points x for which $f'(x) = 0$. Now, $f'(x) = 2x + 2$ and $f'(x) = 0$ imply that $x = -1$. We see that $f(-2) = 2$; $f(-1) = 1$ and $f(2) = 10$. Hence the maximum occurs for $x = 2$ and the minimum for $x = -1$. The graph of f is shown in Figure 4.1.10.

Example 3.   Find the maximum and minimum of the function $f(x) = 2x^3 - 15x^2 + 36x + 1$ on the interval $1 \leq x < 5$.

Solution.   Note that since f is differentiable on $[1, 5]$, it must have both a maximum and a minimum. Since $f'(x) = 6x^2 - 30x + 36 = 6(x - 2)(x - 3)$, the zero's of f' are $x = 2$ and $x = 3$. By simply evaluating f we find that $f(1) = 24$, $f(2) = 29$, $f(3) = 28$, and $f(5) = 56$. Hence the minimum occurs at $x = 1$ and the maximum at $x = 5$.

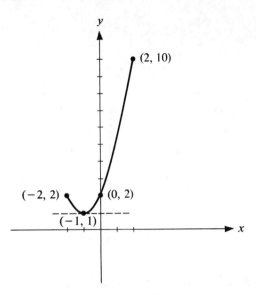

FIGURE 4.1.10

Exercises

1. Find all the local and absolute maxima and minima for the following functions by graphing and inspection.

(a)
$$f(x) = \begin{cases} |x|, & x \neq 0 \\ 2, & x = 0 \end{cases} \quad -1 \leq x \leq 1.$$

(b)
$$f(x) = \begin{cases} x + 2, & -5 \leq x \leq -1 \\ x^2, & -1 \leq x \leq 1 \\ -2x + 3, & 1 \leq x \leq 4. \end{cases}$$

(c)
$$f(x) = \begin{cases} 0, & x \neq \pm 1, \pm 2, \pm 3, \pm 4, \ldots \\ x, & x = \pm 1, \pm 2, \pm 3, \pm 4, \ldots. \end{cases}$$

(d)

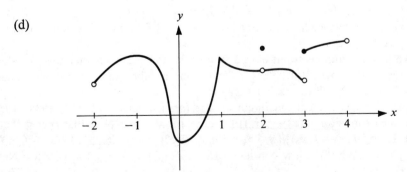

FIGURE 4.1.11

(e)

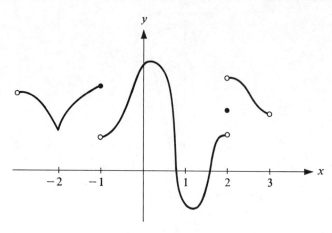

FIGURE 4.1.12

2. Find the maxima and minima for the following functions.

(a) $f(x) = x^2 + 1$, $-1 \le x \le 10$.

(b) $f(x) = x^3/3 + x^2 + x + 11$, $|x| \le 4$.

(c) $f(x) = x^3/3 - x^2 - 15x + 9$, $|x| \le 6$.

(d) $f(x) = \log x$, $1 \le x \le 10$.

(e) $f(x) = \log x + (1 - x)^2$, $1 \le x \le 4$.

(f) $f(x) = (x + 1)/(1 + x^2)$, $|x| \le 4$.

(g) $f(x) = (x^2 + 2x + 1)/(1 + x^2)$, $|x| \le 4$.

(h) $f(x) = x + 1$, $|x| \le 8$.

(i) $f(x) = (\sqrt{x+1} - \sqrt{x})/x$, $1 \le x \le 4$.

(j) $f(x) = e^{-x^2}$, $|x| \le 1$.

(k) $f(x) = x^2 e^x$, $-10 \le x \le 1$.

(l) $f(x) = x - \log x$, $\frac{1}{2} \le x \le 4$.

3. A photographer has a thin piece of wood 32 in. long. How should he cut the wood to make a rectangular picture frame that encloses the maximum area? [HINT: Call the length of the required frame x and the width y. We wish to maximize the area enclosed, $A = xy$. Now, $2x + 2y = 32$. Thus, $x + y = 16$, $y = 16 - x$, and $A = x(16 - x)$. Note that we have obtained the area as a function of x alone. Clearly, $0 \le x \le 16$.]

4. Find two positive numbers that have a sum of 20 and are such that their product is as large as possible. [HINT: Let x denote one number and y the other. Then, $x + y = 20$ and we wish to maximize xy.]

5. Show that if f has a local maximum at $x_0 \in (a, b)$, then $f'(x_0) = 0$. [HINT: First consider the expression $[f(x) - f(x_0)]/[x - x_0]$ for $x < x_0$. Note that $f(x) \le f(x_0)$. (Why?) Thus, $[f(x) - f(x_0)]/[x - x_0] \ge 0$. (Why?) Thus, $\lim_{x \to x_0^-} [f(x) - f(x_0)]/[x - x_0] \ge 0$. (Why?) Now consider $[f(x) - f(x_0)]/[x - x_0]$ for $x_0 < x$. Note that $f(x) \le fx(_0)$. (Why?) Thus, $[f(x) - f(x_0)]/[x - x_0] \le 0$. (Why?) Thus, $\lim_{x \to x_0^+} [f(x) - f(x_0)]/[x - x_0] \le 0$. (Why?) We may then conclude that $f'(x_0) = 0$. (Why?)] In this exercise, we assume f is differentiable on (a, b).

4.2 Some Applications of Maxima and Minima

We now illustrate how the material developed in Section 4.1 is used for solving problems involving the maximization or minimization of functions.

Example 1. A man wishes to travel from point A to point B. (See Figure 4.2.1.) To do this he must rent a car and drive to the railroad and then take the train the rest of the way. It costs \$2 a mile to travel by car and \$1 a mile to travel by rail. Find the least expensive route from A to B and assume he drives in a straight line and meets the railroad some place between B and C. We also assume that the distance between A and C is 3 miles and between B and C is 5 miles.

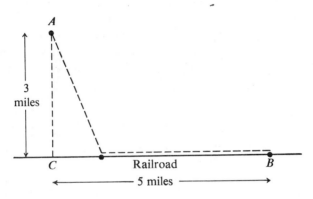

FIGURE 4.2.1

Solution. Since the problem asks us to minimize the cost of the trip, we should set up a function for the cost in terms of some variable. In this problem a convenient variable is the distance from C to the point when he meets the railroad. Thus, if $f(x)$ is the cost of traveling from A to B, then $f(x) = 2\sqrt{9 + x^2} + 1(5 - x)$ for $0 \le x \le 5$. (See Figure 4.2.2.) Now we determine $f'(x)$.

$$f'(x) = \frac{2x}{\sqrt{9 + x^2}} - 1.$$

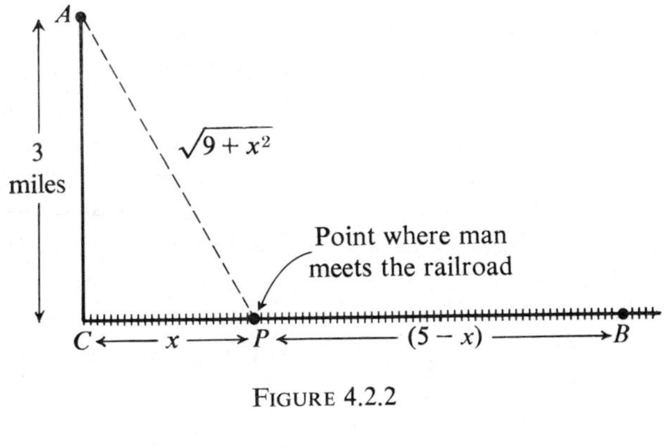

FIGURE 4.2.2

Next, we find the zeros of $f'(x)$; that is, find all solutions to the equation,

$$0 = \frac{2x}{\sqrt{9 + x^2}} - 1.$$

Multiplying both sides by $\sqrt{9 + x^2}$ we obtain, $\sqrt{9 + x^2} = 2x$, and thus, $9 + x^2 = 4x^2$ yielding $x^2 = 3$ and $x = \sqrt{3} \cong 1.732$. Note that we reject the solution $x = -\sqrt{3}$, since the distance from C to P must be positive. Thus the minimum *must* occur at either $x = 0$, $x = \sqrt{3}$, or $x = 5$. A simple calculation reveals that $x = \sqrt{3}$ yields the minimum cost.

If we look at the first examples in Sections 4.1 and 4.2, we should be able to extract a general schema for attacking problems of this kind.

Given a Max–Min problem, we proceed as follows:

(1) Draw a clear picture describing the situation and label it carefully.

(2) Make up a function $f(x)$ for the quantity you are trying to maximize (minimize) in terms of a convenient variable x. (It may be necessary to express other quantities in terms of x to do this.)

(3) Find $f'(x)$.

(4) Find all points x_i such that $f'(x_i) = 0$, $i = 1, \ldots, n$. That is, find all the zeros of f'.

(5) Since we are assuming the variable x ranges over some closed interval $[a, b]$, the maximum (minimum) must occur at a, b or one of the x_i's.

Evaluate f at a, b and x_i, $i = 1, \ldots, n$. Pick out the x values that yield the largest and smallest values for $f(x)$.

Let us now look at some additional examples.

Example 2. The Indiana University real estate office handles an apartment house with 100 units. When the rent of each of the units is $80 per month, all of the units are filled. Experience shows that for each $5 per month increase in rent, five units become vacant. The cost of servicing a rented apartment is $20 per month. What rent should be charged to maximize profit?

Solution. Let us choose for our unknown, x, the number of $5 per month increases in rent. Thus, the amount of rent charged per month is $80 + 5x$ dollars. We see that the number of units rented for each value of x is $100 - 5x$ units. Hence, in order to determine the profit $f(x)$, we must subtract the cost of servicing from the amount taken in from the rented apartments—$(80 + 5x)(100 - 5x)$. Thus, we wish to maximize the function, $f(x) = (80 + 5x)(100 - 5x) - 20(100 - 5x)$, where $0 \le x \le 20$. (We can leave the rent unchanged, or at the other extreme after 20 such increases all apartments are empty.) Now $f(x) = (60 + 5x)(100 - 5x)$ and hence, $f'(x) = 5(100 - 5x) - 5(60 + 5x) = 5$ $(40 - 10x)$. Letting $f'(x) = 0$, we obtain $x = 4$. We claim that this value of x maximizes the profit. By our theorem, since f is differentiable, it must have a maximum at either $x = 0$, $x = 20$, or $x = 4$. Now, $f(0) = \$6000$, $f(20) = 0$, and $f(4) = \$6400$. Thus the rent that should be charged is $80 + $20 = $100 per month. At that rent, 80 units will be rented producing a profit of $6400 per month.

Example 3.   Consider a firm that produces a single product. The firm would like to decide how much of this product should be produced during the coming year in order to maximize its profits over this period. How can this be done?

Solution.   Let x be the quantity of the product produced and suppose that the revenues generated on selling x units amount to $R(x)$. Let $C(x)$ be the total cost of producing the x units. Then, it is clear that the profit obtained as a result of producing x units is

$$P(x) = R(x) - C(x).$$

Suppose that M is the maximum quantity that can be produced. Then we wish to find an x satisfying $0 \leq x \leq M$ that maximizes $P(x)$. If we assume that R and C are differentiable on $[0, M]$, then the maximum occurs at either 0, M or those points x such that $P'(x) = 0$. Now,

$$P'(x) = R'(x) - C'(x)$$

and $P'(x) = 0$ implies that

$$R'(x) = C'(x).$$

Thus, if x_1, the quantity of the product produced that yields maximum profit, is not 0 or M, it satisfies the equation $R'(x_1) = C'(x_1)$; that is, the quantity x_1, such that marginal revenue is equal to marginal cost. (See Exercise 8, Section 3.1.) In textbooks in economics, the statement is often found that production should be adjusted to the point where marginal revenue is equal to marginal cost.

Exercises

1. A triangular area is enclosed on two sides by a fence and on the third side by the straight edge of a river. The two sides of the fence have equal length, 50 ft. Find the maximum area enclosed. [HINT: The area enclosed is given by $\frac{1}{2}hx$. Thus, you must find h in terms of x. (See Figure 4.2.3.)]

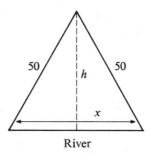

River

FIGURE 4.2.3

2. A retailer knows that if he charges x dollars for an alarm clock, he will be able to sell $480 - 40x$ clocks. The alarm clock costs him \$4 per clock. How much should he charge per clock in order to maximize his total profit? [HINT: His profit per clock is obviously $x - 4$ dollars. Hence, his total profit is $(480 - 40x)(x - 4)$. Clearly, $4 \leq x \leq 12$ and it is the above quantity that must be maximized.]

3. A rectangular trough is formed from a piece of metal 12 in. wide and of indeterminate length. (See Figure 4.2.4.) What should the dimensions be so as to maximize the amount of water carried? Thus, what dimensions maximize the cross-sectional area of the trough?

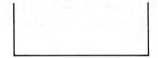

FIGURE 4.2.4

4. A trough is to be formed as in Exercise 3, but this time in the shape of an isosceles triangle. What should the dimensions be so as to maximize the amount of water carried? Again the metal is 12 in. wide. A convenient variable is the distance labeled x in Figure 4.2.5. Again you must find the dimensions that make the cross-sectional area of the trough largest.

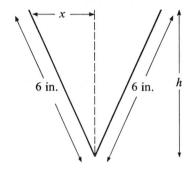

FIGURE 4.2.5

5. A box is to be formed from a piece of paper 16 in. square by cutting out squares in the corners and folding as shown in Figure 4.2.6. What should the dimensions of the box be so as to maximize its volume?

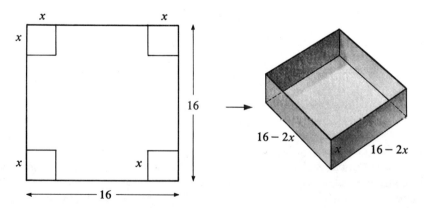

FIGURE 4.2.6

6. The sum of two positive integers is 75. How large can the square of the first times the second be?

7. According to U.S. Post Office regulations the girth plus length of a package cannot exceed 72 in. If a package has square ends, what is the maximum volume permissible under this regulation? (See Figure 4.2.7.)

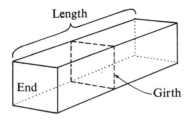

FIGURE 4.2.7

8. You wish to construct a tomato can out of $100\,\pi$ sq in. of metal. What is the volume of the largest can that can be so constructed? [HINT: Let x = radius of can, h = height of can. Then

$$\text{Area of top} = \pi x^2,$$

$$\text{Area of side} = 2\pi xh,$$

$$\text{Volume of can} = \pi x^2 h.$$

(You may assume that $x \le \sqrt{50}$ or else all the metal would be used in the construction of the top and bottom.)] See Figure 4.2.8.

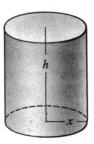

FIGURE 4.2.8

9. In the manner of Exercise 8, you now wish to construct a beer can. This means the top and bottom must be double thickness. What is the largest can that can be constructed out of the $100\,\pi$ sq in. of metal?

10. An Arab wishes to walk from his present point to the river and then to his horse. Assuming he walks in a straight line and meets the river somewhere between A and B, find the shortest distance from Arab to river to horse. See Figure 4.2.9.

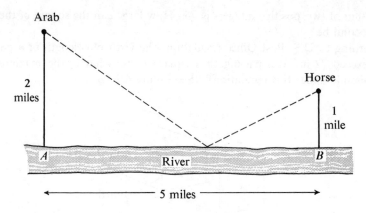

FIGURE 4.2.9

11. Find the area of the largest rectangle that can be inscribed in a semicircle as shown in Figure 4.2.10.

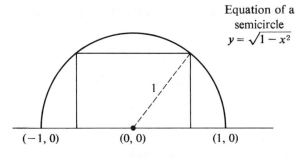

FIGURE 4.2.10

12. A firm's revenue and cost functions are given by

$$R(x) = -\tfrac{1}{4}x^2 + 2x - 1, \qquad 2 \le x \le 5$$

and

$$C(x) = \tfrac{3}{4}x + 1, \qquad 2 \le x \le 5,$$

respectively; x is the output in thousands of units. Find the quantity of output which minimizes the cost, the quantity that maximizes the revenue, and the quantity that maximizes the profit. Is there any relationship between these values?

4.3   Increasing and Decreasing Functions. The Geometrical Significance of the Derivative

So far we have only used information on the zeros of f' to locate maxima and minima. However, it is possible to learn a great deal more about a function by studying the behavior of its derivative. We begin by defining what is meant by an increasing or decreasing function.

DEFINITION.   *Let f be defined on* (a, b). *Then f is strictly increasing on* (a, b) *if* $x_1 < x_2$, *where* $x_1, x_2 \in (a, b)$, *implies that* $f(x_1) < f(x_2)$. *If* $x_1 < x_2$, *where* $x_1, x_2 \in (a, b)$, *implies that* $f(x_1) > f(x_2)$ *we say that f is strictly decreasing on* (a, b). (*See Figure* 4.3.1.)

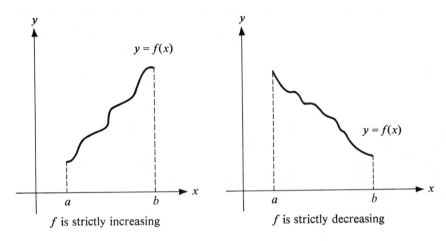

FIGURE 4.3.1

It is easy to see that if f is a strictly increasing function on (a, b) and if f' exists at $x_0 \in (a, b)$, then $f'(x_0) \geq 0$. This seems reasonable from a geometric standpoint and is not hard to see analytically. (Consider $[f(x) - f(x_0)]/[x - x_0]$; if $x > x_0$, then $f(x) - f(x_0) \geq 0$; thus $[f(x) - f(x_0)]/[x - x_0] \geq 0$. Hence, $\lim_{x \to x_0^+} [f(x) - f(x_0)]/[x - x_0] \geq 0$ and a similar argument shows that $\lim_{x \to x_0^-} [f(x) - f(x_0)]/[x - x_0] \geq 0$. Hence, $f'(x_0) \geq 0$.)

It is important for us that a converse to the preceding statement is true. Namely,

THEOREM 3.   *Let f be differentiable on* (a, b).
  (i)   *If* $f'(x) \geq 0$ *for* $x \in (a, b)$, *then f is strictly increasing on* (a, b).
  (ii)   *If* $f'(x) < 0$ *for* $x \in (a, b)$, *then f is strictly decreasing on* (a, b).

Before discussing the verification of the above result, let us look at some examples.

Example 1.   If $f(x) = 2x^3 - 21x^2 + 60x - 9$, find all intervals on which f is strictly increasing or strictly decreasing.

Solution.   First we differentiate f to obtain

$$f'(x) = 6x^2 - 42x + 60 = 6(x^2 - 7x + 10) = 6(x - 2)(x - 5).$$

Now a simple analysis reveals that

$$f'(x) \geq 0 \text{ for } x < 2$$
$$f'(x) < 0 \text{ for } 2 < x < 5$$
$$f'(x) \geq 0 \text{ for } 5 < x.$$

Clearly $f'(x) = 0$ for $x = 2, 5$. The results are displayed in Figure 4.3.2. Thus f is strictly increasing on $(-\infty, 2)$ and $(5, \infty)$ and strictly decreasing on $(2, 5)$. (By $(-\infty, a)$ we mean $\{x \mid x < a\}$, while (b, ∞) is $\{y \mid y > b\}$, where a and b are real numbers.)

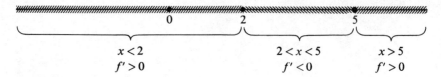

$$\begin{array}{ccc} x < 2 & 2 < x < 5 & x > 5 \\ f' > 0 & f' < 0 & f' > 0 \end{array}$$

FIGURE 4.3.2

We can use the information in Example 1 to make a very rough sketch of the graph of f. (See Figure 4.3.3.) First, we find that $f(2) = 43$ and $f(5) = 16$. We sketch the graph in a wiggly manner to indicate that we do not know what it is doing more precisely. However, just from the information we have, it is clear that f has a local maximum at 2 and a local minimum at 5.

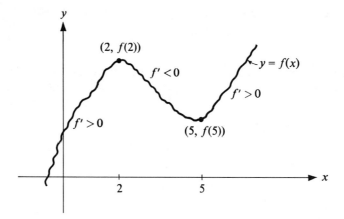

FIGURE 4.3.3

Theorem 3 is a corollary of the mean-value theorem. This is one of the most important and useful theorems of all calculus and we now state it.

THEOREM 4 (MEAN-VALUE THEOREM).   *Let f be a function continuous on $[a, b]$ and differentiable on (a, b). Then, there exists a number $c \in (a, b)$ such that*

$$f'(c) = \frac{f(b) - f(a)}{b - a}.$$

At first glance the theorem may appear technical and uninteresting but it really says a very simple and intuitively appealing thing. Suppose you join the points $P_1 = (a, f(a))$ and $P_2 = (b, f(b))$ in Figure 4.3.4 with a straight line L. The slope of L is

$$m = [f(b) - f(a)]/[b - a].$$

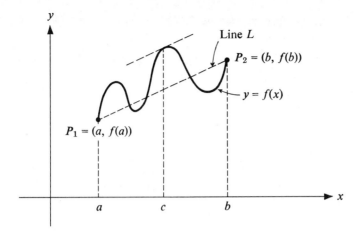

FIGURE 4.3.4

Then, some place between a and b the tangent to f has the same slope m. (There may, in fact, be several points where the tangent to f has slope m, but the important thing is that there is *always* at least one such point.)

If we take a few liberties with the notion of function, the theorem could be paraphrased as follows: You cannot get from P_1 to P_2 without going in the right direction at least once.

Someone tells you he left Chicago at 1:00 PM and arrived in Indianapolis 170 miles away at 3:00 PM. When you tell him that he must have been going 85 miles per hour at some point, you are using a disguised version of the mean-value theorem. (This becomes clearer in a later chapter.)

We omit the proof of this theorem since it requires tools out of the mainstream of our course. The interested reader can find a proof in any of a number of calculus books. [For example, T. Apostol, *Calculus* Vol. I, Blaisdell Publishing Company, Waltham, Mass. (1967), p. 83.]

Let us briefly outline how Theorem 3 follows from the mean-value theorem. Suppose x_1 and x_2 are *any* two numbers in (a, b) with $x_2 > x_1$. If f is differentiable on (a, b), f is certainly differentiable on $[x_1, x_2]$. Hence, there exists a number $c \in (x_1, x_2)$ such that

$$\frac{f(x_2) - f(x_1)}{x_2 - x_1} = f'(c).$$

But in the hypothesis (i) of Theorem 3, $f'(c) > 0$. Hence,

$$\frac{f(x_2) - f(x_1)}{x_2 - x_1} > 0$$

and $x_2 > x_1$ implies that $f(x_2) > f(x_1)$. Thus f is strictly increasing on (a, b). The second part follows in a similar manner.

As another corollary of the mean-value theorem, we have the following result, which is extremely useful in applications—particularly in Chapter 5.

THEOREM 5.   *Let f be differentiable on (a, b) and let $f'(x) = 0$ for all $x \in (a, b)$. Then $f(x) = k$ (k constant) for $x \in (a, b)$.*

This result is easily verified. Let x_1, x_2 be any elements of (a, b). Then the mean-value theorem states that there exists a $c \in (x_1 x_2)$ such that

$$\frac{f(x_2) - f(x_1)}{x_2 - x_1} = f'(c) = 0.$$

Hence, for any $x_1, x_2 \in (a, b), f(x_1) = f(x_2)$ implying $f(x) = k$ on (a, b).

Let us summarize what we have so far.

(1)   If $f'(x) > 0$ for $x \in (a, b)$, then $f(x)$ is strictly increasing on (a, b).

(2)   If $f'(x) < 0$ for $x \in (a, b)$, then $f(x)$ is strictly decreasing on (a, b).

(3)   If $f'(x) = 0$ on (a, b) then f is constant.

Exercises

1.   Determine by inspection when f' is positive and negative.   (See Figures 4.3.5–4.3.8.)

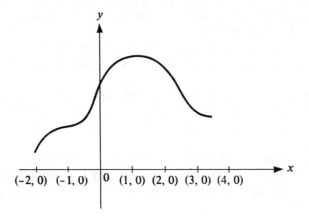

FIGURE 4.3.5

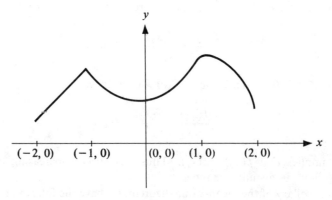

FIGURE 4.3.6

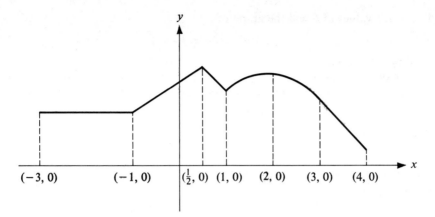

FIGURE 4.3.7

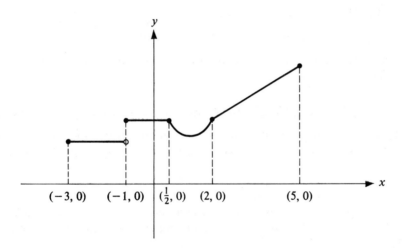

FIGURE 4.3.8

2. Determine where the following functions are increasing and where they are decreasing. Also decide where all relative maxima and minima occur. Sketch *as much* of the graph as you can. (Similar to Figure 4.3.3)

   (a) $f(x) = x^2 - 5x + 6$ (f) $f(x) = \sqrt{1 - x^2}$

   (b) $f(x) = x^3 + x^2 - 8x + 1$ (g) $f(x) = x - \log x$

   (c) $f(x) = x - x^2$ (h) $f(x) = xe^{-x^2}$

   (d) $f(x) = x + \dfrac{1}{x}$ (i) $f(x) = \dfrac{x}{1 + x^2}$

   (e) $f(x) = x^4 + 32x + 32$ (j) $f(x) = \sqrt{x^2 - 1}$.

3. For which values of b will the graph of

$$y = 2x^3 - 3x^2 + bx + 1$$

be always increasing?

4. In each of the following problems, a, b, c refer to the equation $[f(b) - f(a)]/[b - a]$ $= f'(c)$ that expresses the mean-value theorem. Given $f(x)$, a, and b, find c.

(a) $f(x) = x^2 + 2x - 1$, $a = 0, b = 1$.
(b) $f(x) = x^3$, $a = 0, b = 3$.
(c) $f(x) = x^{2/3}$, $a = 0, b = 1$.
(d) $f(x) = \sqrt{x - 1}$, $a = 1, b = 3$.

[HINT TO (a):

$$\frac{f(b) - f(a)}{b - a} = \frac{2 + 1}{1} = 3. \qquad f'(x) = 2x + 2. \qquad \therefore \quad f'(c) = 2c + 2.$$

Hence, by the mean-value theorem,

$$2c + 2 = 3, \qquad 2c = 1, \qquad \text{and} \qquad c = \tfrac{1}{2}].$$

5. Can the mean-value theorem be applied to the function $f(x) = |x|$ on the interval $[-1, 1]$? Why?

4.4   Concavity and Graphing. The Significance of the Second Derivative

By looking at the second derivative, we can obtain more information on the behavior of the function. First we introduce the concept of concavity, which is of great use in graphing functions.

DEFINITION.   *A curve is concave upward on (a, b) if*
  (i)   *the chord joining any two points on the curve lies above the curve, or equivalently,*
  (ii)   *at every point on the curve there is a tangent line that lies below the curve.* (See Figure 4.4.1.)

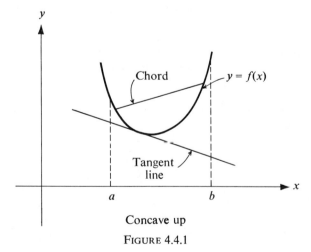

Concave up

FIGURE 4.4.1

DEFINITION.   *A curve is concave downward on* (a, b) *if*

(i) *the chord joining any two points on the curve lies below the curve, or equivalently,*

(ii) *at every point on the curve there is a tangent line that lies above the curve.* (See Figure 4.4.2.)

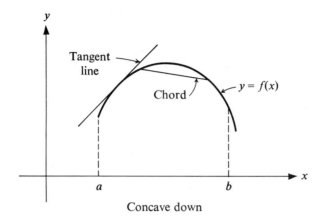

Concave down

FIGURE 4.4.2

Example 1.   Consider the graph of the function $f(x) = x^2$ on $[-1, 1]$. (See Figure 4.4.3.) It is intuitively clear that this curve is concave upward on $(-1, 1)$, since at every point there is a tangent line that lies below the curve.

Example 2.   Consider the graph of the following function $f(x) = x^3$ on $[-1, 1]$. (See Figure 4.4.4.) Observe that in this case the curve is concave upward in $(0, 1]$ and concave downard in $[-1, 0)$. Thus, a change in concavity occurs at $x = 0$.

Now, the fortunate thing is that the second derivative will tell us when a curve is concave up or down. We have the following result.

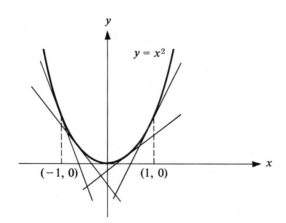

FIGURE 4.4.3

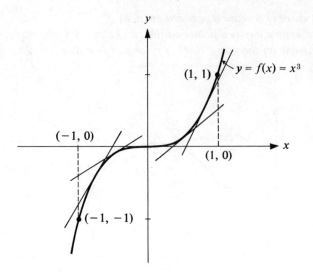

FIGURE 4.4.4

THEOREM 6.   *If $f''(x) > 0$ on (a, b), then $f(x)$ is concave up on (a, b).*

Although we do not give a rigorous proof of this theorem, we remark that geometrically it is easy to observe its verification. If $f''(x) > 0$ on (a, b), then by Theorem 3, $f'(x)$ is strictly increasing on (a, b) and the curve is bending upward (since the slope of the tangent is always increasing). In like manner if $f''(x) < 0$, $f'(x)$ is strictly decreasing on (a, b) and the curve is bending downward. For example, if $f(x) = x^2$, then $f'(x) = 2x$ and $f''(x) = 2$. In this case, the curve must be concave up since $2 > 0$. We show the graph of f, f', and f'' in Figure 4.4.5. You can see that the derivative is strictly increasing for all x. However, the function itself is increasing for $x \geq 0$ and is decreasing for $x < 0$.

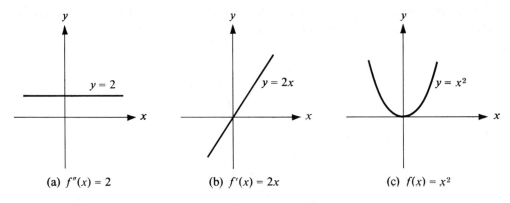

(a) $f''(x) = 2$ (b) $f'(x) = 2x$ (c) $f(x) = x^2$

FIGURE 4.4.5

Let us now consider the function in Example 1 of Section 4.3 in more detail. The function under discussion is

$$f(x) = 2x^3 - 21x^2 + 60x - 9.$$

So far we know it looks like that in Figure 4.4.6. Now, $f'(x) = 6x^2 - 42x + 60$ and

$f''(x) = 12x - 42 = 12(x - \frac{7}{2})$. Thus $f''(x) < 0$ for $x < \frac{7}{2} = 3\frac{1}{2}$ and $f''(x) > 0$ for $x > \frac{7}{2} = 3\frac{1}{2}$. Therefore, the curve is concave down for $x < \frac{7}{2}$ and concave up for $x > \frac{7}{2}$. At $x = 3\frac{1}{2}, f''(x) = 0$, which means that at $x = 3\frac{1}{2}$ the curve "changes" from being concave down to concave up. This point is called a point of inflection. Using the information we have just obtained we are able to draw the more detailed graph (see Figure 4.4.7).

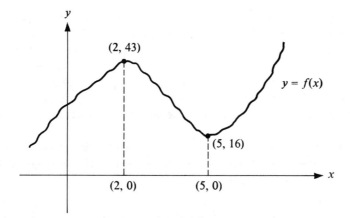

FIGURE 4.4.6

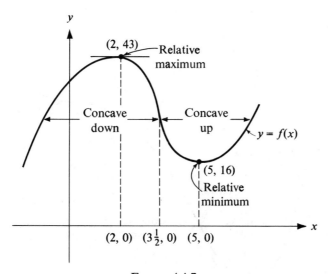

FIGURE 4.4.7

Example 3. Draw the graph of $f(x) = x^4 - 8x^3 + 18x^2 + 1$.

Solution. First we find $f'(x)$ and $f''(x)$ and write them in a convenient form.

$$f'(x) = 4x^3 - 24x^2 + 36x = 4x(x^2 - 6x + 9)$$
$$= 4x(x - 3)^2$$
$$f''(x) = 12x^2 - 48x + 36 = 12(x^2 - 4x + 3)$$
$$= 12(x - 3)(x - 1).$$

It is easy to see that $f'(0) = f'(3) = 0$ and 0, 3 are the only places where $f'(x)$ equals zero. Also

$$f'(x) < 0 \quad \text{for} \quad x < 0$$
$$f'(x) > 0 \quad \text{for} \quad x > 0.$$

Thus, the curve is decreasing for $x < 0$ and increasing for $x > 0$. Now how about $f''(x)$. It is easy to see that $f''(x) > 0$ for $x < 1$, $f''(x) < 0$ for $1 < x < 3$ and $f''(x) > 0$ for $x > 3$. Hence, for $x < 1$ and $x > 3$ the curve is concave up, and for $x \in (1, 3)$, the curve is concave down. We summarize the results on the real line in Figure 4.4.8. In Figure 4.4.9, we draw the graph of the function. We remark that even though $f'(3) = 0$, $x = 3$ does not yield a local maximum or minimum. Actually, the curve is said to have a "point of inflection" at 3; that is, the curve changes its concavity at this point. It is

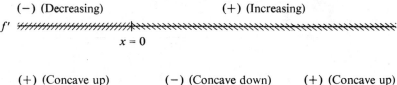

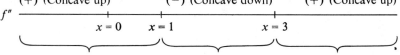

FIGURE 4.4.8

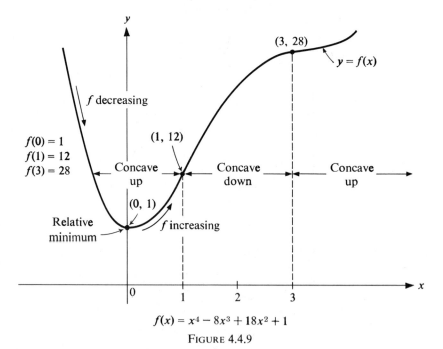

$$f(x) = x^4 - 8x^3 + 18x^2 + 1$$

FIGURE 4.4.9

important to remember that a point x_1 for which $f'(x_1) = 0$ *and* where the curve changes its concavity is neither a minimum or a maximum.

We summarize our results by giving the following rules for graphing $y = f(x)$.

RULE 1. *Find $f'(x)$ and $f''(x)$.*

RULE 2. *Find the zero's of f' and f''.*

RULE 3. *Find the intervals in which* (i) $f' < 0$ *and* (ii) $f' > 0$. (Two algebra methods for doing this are described in an appendix to the chapter.)

RULE 4. *Find the intervals on which* (i) $f'' < 0$ *and* (ii) $f'' > 0$.

RULE 5. *Use Rules 1–4 with the table below to graph the function.*

Table 4.4.1

1. $f'(x) > 0$ on (a, b)	f is increasing
2. $f'(x) < 0$ on (a, b)	f is decreasing
3. $f''(x) > 0$ on (a, b)	f is concave up
4. $f''(x) < 0$ on (a, b)	f is concave down

We end this section by considering one more example.

Example 4. One of the most common functions used in the application of statistics to many fields is the so-called *normal density function* defined by

$$f(x) = e^{-x^2}.$$

Find all relative maximum and minimum points; all intervals where the function is increasing and decreasing; all intervals where it is concave up and concave down. Also, sketch a graph of the function for $-2 \le x \le 2$.

Solution. Now, $f'(x) = -2xe^{-x^2}$ and

$$f''(x) = -2e^{-x^2} + 4x^2e^{-x^2} = 2(2x^2 - 1)e^{-x^2}.$$

Since e^{-x^2} is always positive, $f'(x) < 0$ for $x > 0$ and $f'(x) > 0$ for $x < 0$. Now, $f'(x) = 0$ for $x = 0$. Observe that at 0 the function changes from an increasing function to a decreasing function. Thus, since $f(0) = 1, f(+2) = e^{-4} = 0.02$ and $f(-2) = e^{-4} = 0.02$, we see that on $[-2, 2]$, $(0, 1)$ is a *relative maximum*. It is not hard to see that $(0, 1)$ is also an absolute maximum point of the curve. Now $f''(x) > 0$ for $2x^2 - 1 > 0$, which means that for $x^2 > \frac{1}{2}$ or $|x| > 1/\sqrt{2}$, the curve is concave upward while for $2x^2 - 1 < 0$ or $|x| < 1/\sqrt{2}$, the curve is concave down. On $[-2, 2]$, the absolute minima occur at $x = -2$ and $x = +2$. We summarize as follows:

(1)　　$x < 0$:　　f is increasing

(2)　　$x > 0$:　　f is decreasing

(3) $x = 0$:   a maximum point

(4) $|x| < \frac{1}{2}$:   curve is *concave down*

(5) $|x| > \frac{1}{2}$:   curve is *concave up*

(6) $x = \pm \dfrac{1}{\sqrt{2}}$:   inflection points

(7) $x = -2, x = +2$ absolute minima on $[-2, 2]$.

This information is used to sketch the graph of the function in Figure 4.4.10.

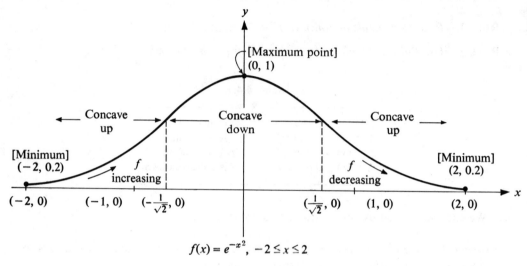

$$f(x) = e^{-x^2}, \quad -2 \le x \le 2$$

FIGURE 4.4.10

Exercises

1. Find the maxima and minima, the intervals where the graph is increasing, decreasing, concave up, concave down, of the following functions. Use the preceding information to graph the functions.

(a) $x^2 - 5x + 6$

(b) $x^3 + x^2 - 8x + 1$

(c) $2x^3 - 15x^2 - 36x - 1$

(d) $4x^3 + 7x^2 - 10x + 2$

(e) $(x - 2)^2(x + 3)$

(f) $x^3 - 3x^2 - 9x + 1$

(g) $x^2 - x^3$

(h) $\dfrac{x}{x^2 + 1}$

(i) $(x - 1)^2(x^2 + 1)$

(j) $x^2 + \dfrac{1}{x}$

*(k) $x - \log x$ for $0 < x < \infty$

*(l) $(x - 1)^3(x + 4)^2$

*(m) $\dfrac{x - 1}{x - 2}$

*(n) xe^{-x^2}

2. It is conjectured from experience that the ability A to memorize in the early years obeys a law of the form

$$A(x) = x \log x + 1, \qquad \text{when } x \in (0, 4)$$
$$A(0) = 1,$$

where x is measured in years. Where does A attain its local minimum? Sketch a graph of A.

3. Sketch a graph of a function f, for $x > 0$, if $f(1) = 0$ and $f'(x) = 1/x$ for all $x > 0$. Is such a curve necessarily concave upward or concave downward?

4. Sketch a smooth curve $y = f(x)$, illustrating $f(1) = 0$, $f''(x) < 0$ for $x < 1$, $f''(x) > 0$ for $x > 1$.

5. Make a diagram showing how the graph of $y = f(x)$ might appear if f has a second derivative for each x, given that $f(-3) = 4$, $f(-1) = 1$, $f(0) = 2$, $f''(x) > 0$ when $x < 0$ and $f''(x) < 0$ when $x > 0$, supposing in addition (a) that $f(2) = 0$ and (b) that $f'(x) > 0$ and $f(x) < 4$ when $x > 0$. Why is $f'(0) = 0$ impossible in both cases?

4.5   Tests for Maxima and Minima

The second derivative can be used in conjunction with the first derivative to determine local maxima and minima. The following theorem yields a very useful test.

THEOREM 7.   *Let $f'(x_0) = 0$, where f is defined on (a, b) and $x_0 \in (a, b)$.*
   (i)   *If $f''(x_0) > 0$, then f has a local minimum at x_0.*
   (ii) *If $f''(x_0) < 0$, then f has a local maximum at x_0.*
   (iii) *If $f''(x_0) = 0$, then it is unclear whether x_0 is a local maximum or minimum (it may be neither) and other tests must be made.*

We omit the proof of this theorem but note from some of our previous examples that when a maximum occurred the curve was always concave down in an interval about the maximum and always concave up near a minimum. Thus, one would expect a theorem like the present one.

Example 1.   If $f(x) = x^3 - x^2 - 8x + 1$, find all local maxima and minima of f on $(-\infty, \infty)$ and graph the function

$$[(-\infty, \infty) = \{x \,|\, x \text{ is a real number}\}].$$

Solution.   First, we find out where $f' = 0$. Now, $f'(x) = 3x^2 - 2x - 8 = (3x + 4)(x - 2)$ and thus $f'(x) = 0$ at $x = 2$ and $x = -\frac{4}{3}$. It is clear that $f''(x) = 6x - 2 = 2(3x - 1)$, $f''(2) = 10$ and $f''(-\frac{4}{3}) = -10$. Hence, by Theorem 7, 2 is a local minimum and $-\frac{4}{3}$ is a local maximum. To sketch the graph we observe the following facts:

(1) $f'(x) > 0$ for $x > 2$ and $x < -\frac{4}{3}$. Therefore, f is increasing for these values of x.
(2) $f'(x) < 0$ for $-\frac{4}{3} < x < 2$. Hence for $x \in (-\frac{4}{3}, 2)$, f is decreasing.
(3) $f''(x) > 0$ for $x > \frac{1}{3}$. Thus, f is concave up for $x > \frac{1}{3}$.

(4) $f''(x) < 0$ for $x < \frac{1}{3}$ implying that f is concave down for $x < \frac{1}{3}$.

(5) f has three zeros x_1, x_2, x_3, which we need not find exactly, and the curve must cross the y axis at the point $(0, 1)$.

Combining (1)–(5) and the facts that 2 is a local minimum and $-\frac{4}{3}$ is a local maximum, we obtain the graph shown in Figure 4.5.1.

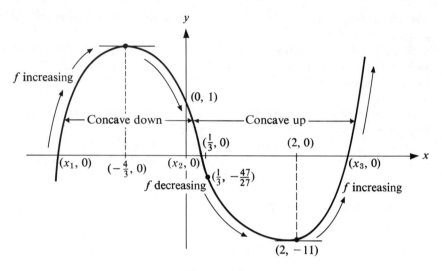

FIGURE 4.5.1

So far, we have been considering maximum–minimum (as opposed to local max and min) problems on closed intervals only. There, the situation is relatively simple. Recall that if f is continuous on a closed interval, f has both a maximum and minimum there! On open intervals there are more possibilities. (In particular, f may have a max and not a min or conversely.) Let us look at an example.

Example 2.   Suppose we consider the function in Example 4 of Section 4.4: $f(x) = e^{-x^2}$. This time, we consider the entire domain of the function—the open interval $(-\infty, \infty)$. Does this function have a maximum and minimum on $(-\infty, \infty)$?

Solution.   From our work in Example 4, Section 4.4, we see that a maximum occurs at $(0, 1)$. The second derivative test also gives this immediately. Since $f''(x) = 2e^{-x^2}$ $(2x^2 - 1)$, $f''(0) = -2$ and thus by Theorem 7, $x = 0$ yields a maximum. We claim that f does *not* have a minimum any place. Consider any point x_1. If $x_1 > 0$, then for $x_1 < x_2$ we see that $f(x_1) > f(x_2)$. (Why?) Thus x_1 is not a minimum. On the other hand, if $x_1 < 0$, then for $x_3 < x_1$ we recognize that $f(x_3) < f(x_1)$. (Why?) So again x_1 is not a minimum. By collecting the information we have and adding a little more, it is not hard to graph $f(x) = e^{-x^2}$ on $(-\infty, \infty)$. Figure 4.4.10 shows $f(x) = e^{-x^2}$ on $[-2, 2]$.   Now,

(1) $f > 0$ for all x. (Why?)
(2) f is symmetric about the y axis.
(3) f is increasing on $(-\infty, 0)$ and f is decreasing on $(0, \infty)$.
(4) $f''(x) = -2e^{-x^2} + 4x^2e^{-x^2} = 2e^{-x^2}(2x^2 - 1)$.

Note that $2e^{-x^2} > 0$ for all x and $2x^2 - 1 > 0$ for $|x| > 1/\sqrt{2}$ and $2x^2 - 1 < 0$ for $|x| < 1/\sqrt{2}$. Hence $f''(x) > 0$ for $|x| > 1/\sqrt{2}$ and $f''(x) < 0$ for $|x| < 1/\sqrt{2}$. Therefore, f is concave up on $|x| > 1/\sqrt{2}$ and f is concave down on $|x| < 1/\sqrt{2}$. The graph is shown in Figure 4.5.2.

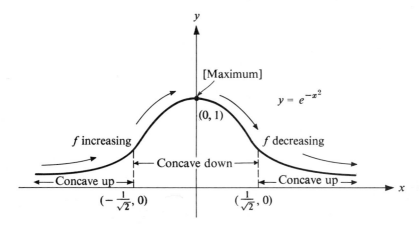

FIGURE 4.5.2

The clearest way to decide whether a function on an open interval has a maximum or minimum (and where) is to carefully graph the function as we have done here. Then these questions can be answered by inspection.

We will, however, state one rule that is simple and very useful for open interval problems. This is sometimes referred to as the first derivative test for determining maxima or minima.

THEOREM 8. *Let f be a differentiable function on (a, b) and $f'(p) - 0$ where $a < p < b$.*

(i) *If $f'(x) < 0$ for $x < p$ and $f'(x) > 0$ for $x > p$, then f has a minimum at p (and no maximum).*

(ii) *If $f'(x) > 0$ for $x < p$ and $f'(x) < 0$ for $x > p$, then f has a maximum at p (and no minimum).*

This theorem is easily verified. For example, if $f'(x) < 0$ for $x < p$, then f is *decreasing* there and $f(p) < f(x)$ for every $x < p$. Since $f'(x) > 0$ for $x > p$, f is increasing there and $f(x) > f(p)$ for $x > p$. Hence, in either case, $f(p) < f(x)$ for every x. Thus, $f(p)$ represents a true minimum. Part (ii) is verified in exactly the same way. We now give an example that illustrates the use of this rule.

Example 3. Find the shortest distance from the point $(5, 1)$ to the parabola $y = 2x^2$.

Solution. In keeping with our recipe for finding maxima, we first draw a picture. If $p = (x, 2x^2)$ is a general point on the parabola, then the distance from $(5, 1)$ to the point p is given by

$$d[(5, 1), p] = \sqrt{(5 - x)^2 + (1 - 2x^2)^2}.$$

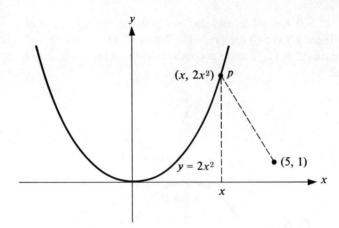

FIGURE 4.5.3

Let $f(x) = \sqrt{(5-x)^2 + (1-2x^2)^2}$. We wish to minimize f. Now,

$$f'(x) = \tfrac{1}{2}[-2(5-x) - 8x(1-2x^2)] \cdot [(5-x)^2 + (1-2x^2)^2]^{-1/2}.$$

Observe that $[(5-x)^2 + (1-2x^2)^2]^{-1/2} \neq 0$, and moreover this quantity is always positive by definition of the square root. Thus $f' = 0$ precisely when $-2(5-x) - 8x(1-2x^2) = 0$ or equivalently $16x^3 - 6x - 10 = 0$. It is easy to see that 1 is a root of this equation and factoring yields $(x-1)(16x^2 + 16x + 10) = 0$. The expression $16x^2 + 16x + 10$ has no real roots. (The roots are in fact $-16 \pm \sqrt{16^2 - 4 \cdot 10 \cdot 16}/32$.) Thus, this expression must always be positive. (Since there is no real x such that $16x^2 + 16x + 10 = 0$ and if $x > 0$, $16x^2 + 16x + 10 \geqslant 0$, it follows that $16x^2 + 16x + 10$ can never be negative; otherwise its graph must cross the axis and we would have a contradiction.) Since

$$f'(x) = \tfrac{1}{2}[(x-1)(16x^2 + 16x + 10)][(5-x)^2 + (1-2x^2)^2]^{-1/2},$$

$$\text{positive}$$

it is clear that $f'(x) < 0$ for $x < 1$ and $f'(x) > 0$ for $x > 1$. Hence, f has a minimum at $x = 1$ and no maximum. This should be clear since there are points on $y = 2x^2$ that are arbitrarily far from $(5, 1)$. Thus, the minimum distance is equal to

$$f(1) = \sqrt{(5-1)^2 + (1-2)^2} = \sqrt{17}.$$

In this example, it would be far more complicated to use the second derivative test.

In our discussion of maxima and minima, we have considered only differentiable functions. It is possible to modify the present techniques to handle functions that are differentiable except at a finite set of points.

Thus, let f be a continuous function on $[a, b]$, which is differentiable except at $r_1, \ldots, r_n$. Let $f'(x) = 0$ precisely for $x = s_1, \ldots, s_n$. Then f has an absolute maximum and minimum on $[a, b]$. These each occur at one of the following points.

(1) $a, b,$

(2) $r_1, \ldots, r_n,$

(3) $s_1, \ldots, s_n.$

We illustrate this by an example.

Example 4.   Find the maximum and minimum of $f(x) = |x|$ on $[-1, 2]$.

Solution.   The function f is differentiable except at $x = 0$. For $-1 \le x < 0, f'(x) = -1$. For $0 < x \le 2, f'(x) = 1$. Thus, the maximum and minimum must occur at $-1, 0$, or 2. Since $f(-1) = 1, f(0) = 0$, and $f(2) = 2$, it is clear that 0 is where the minimum occurs and 2 is where the maximum occurs. The graph of f is shown in Figure 4.5.4.

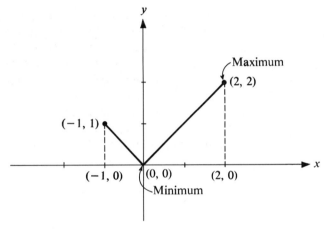

FIGURE 4.5.4

We do not discuss this situation for open intervals but the same method is valid. The interested student may wish to write it out in more detail on his own.

Exercises

1.   Use the second derivative test to find the local maxima and minima of the following functions and then sketch a graph of the functions.

(a) $f(x) = x^2 - 6x + 9$

(b) $f(x) = \dfrac{x^3}{3} + \dfrac{x^2}{2} - 12x + 9$

(c) $f(x) = x^3 - 3x^2 + 3x - 1$

(d) $f(x) = x^4 + 4x^3 + 6x^2 + 4x + 1$

(e) $f(x) = e^{-(x/2)^2}$

(f) $f(x) = 2x^3 + 3x^2 - 60x + 9$

(g) $f(x) = x^4 - 8x$

(h) $f(x) = x^2(x - 1)$

2.   The slope of a curve at any point (x, y) is given by the equation

$$\frac{dy}{dx} = 6(x - 1)(x - 2)^2(x - 3)^3(x - 4)^4.$$

(a) For what value (or values) of x is y a maximum? Why?

(b) For what value (or values) of x is y a minimum? Why?

3. Find the shortest distance from the point $(0, 1)$ to the parabola $y = 4x^2$.

4. A rectangular field to contain a given area A is to be be fenced off along a straight river. If no fencing is needed along the river, show that the least amount of fencing will be required when the length of the field is twice its width.

5. Determine a, b, c, d so that the curve whose equation is $y = ax^3 + bx^2 + cx + d$ has a minimum at $(-1, -3)$ and a maximum at $(0, -5)$.

6. Find two *positive* numbers that have a sum of 50 and are such that their product is as large as possible. Can the problem be solved if the product is to be as small as possible? Explain.

7.* Given $f(x) = ax^2 + bx + c$ with $a > 0$. By considering the minimum, show that $f(x) \geq 0$ for all real x, if and only if $b^2 - 4ac \leq 0$. [HINT: Suppose that $b^2 - ac \leq 0$. Now, $f'(x) = 2ax + b$ and $f'(x) = 0$ yields $x = -b/2a$. Now, $f''(x) = 2a > 0$ for all x. Thus, by the second derivative test, the point $(-b/2a, f(-b/2a))$ is a minimum point. But, $f(-b/2a) = -(b^2 - 4ac)/4a \geq 0$ and $f(x) \geq f(-b/2a)$ for all x. Hence, $f(x) \geq 0$ for all x. The converse is not difficult to show.]

8. Let $f(x) = (\log x)/x$, if $x > 0$. Describe intervals in which f is increasing, decreasing, concave up, and concave down. Sketch a graph of f.

9. Find the maximum and minimum points for the following functions. Also sketch a graph of the functions.

(a) $f(x) = 5|x|$ on $[-1, 3]$.

(b) $f(x) = |x - x^2|$ on $[-2, 2]$.

(c) $f(x) = \dfrac{x}{1 + |x|}$ on $(-\infty, \infty)$.

(d) $f(x) = \dfrac{|x|}{1 + |x|}$ on $(-\infty, \infty)$.

(e) $f(x) = |x^3|$ on $[-2, 3]$.

(f) $f(x) = |x^3|$ on $(-2, 3)$.

(g) $f(x) = \dfrac{1}{x - x^2}$ on $(0, 1)$.

10. Find the point on the curve $y = \sqrt{x}$ nearest the point $(a, 0)$

(a) if $a \geq \frac{1}{2}$.

(b) if $a < \frac{1}{2}$.

4.6 More Applications of Maxima and Minima

In this section, we simply further illustrate the theorems dealing with maxima and minima by considering several examples that appear in various applications.

Example 1. Suppose that firm A has a contract to supply 1000 items a month at a uniform daily rate, and that each time a production run is started, it costs $50. In order to avoid high production costs, the firm decides to produce a large quantity at a time and to store it until the contract calls for delivery. Unfortunately, even storage can be expensive; hence firm A does not want to store so much that the storage costs exceed the savings due to large production runs. If the cost of storage is 1 cent per item per month, how many should be made per run so as to minimize total average costs?

Solution. Let x denote the number of items made in each run and assume that a graph of inventory against time looks like that in Figure 4.6.1. The time period required

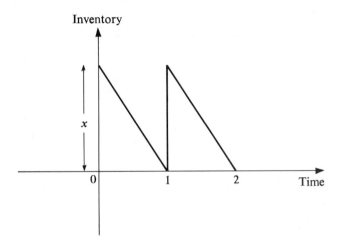

FIGURE 4.6.1

to sell x items is called an *inventory* cycle. If we assume that demand—and sales—are constant throughout the cycle, then the average inventory during any cycle is $x/2$ and the average cost of holding inventory is $(x/2)(0.01)$ dollars per month. Now, the batches of x items will last $x/1000$ months (since 1000 items must be supplied per month). The average set-up cost will be $50 \div x/1000$. (This is because each time a production run is started it costs $50 and every $x/1000$ months we have a production run.) Thus, the total average cost $C(x)$ is given by

$$C(x) = \frac{x}{2}(0.01) + 50 \div \frac{x}{1000}$$

$$= \frac{x}{200} + \frac{50{,}000}{x}.$$

Now,

$$C'(x) = \frac{1}{200} - \frac{50,000}{x^2} \quad \text{and} \quad C'(x) = 0$$

means that

$$\frac{1}{200} = \frac{50,000}{x^2}, \quad x^2 = 10^7, \quad \text{and} \quad x = 3160.$$

Also observe that $C''(x) = 100,000x^{-3}$ and $C''(3160) > 0$. Thus, by Theorem 7 of Section 4.5, the cost is minimized when 3160 items are made. This means that firm A should make a little over a 3-months' supply at a time.

Example 2.   A sheet of paper for a poster contains 18 sq ft. The margins at the top and bottom are 9 in. and at the sides 6 in. What are the dimensions if the printed area is a maximum?

Solution.   First, we draw a picture and let x = length of the poster and y = the width of the poster. (See Figure 4.6.2.)

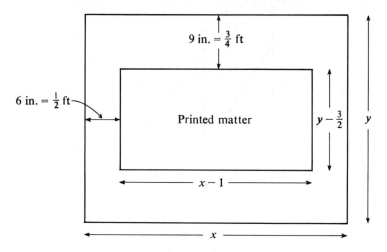

FIGURE 4.6.2

Now the area of the total poster is xy; however, we are told that this is 18 sq ft and thus, $xy = 18$ or $y = 18/x$. We wish to maximize the area A of the printed material. From Figure 4.6.2, we see that

$$A = (x - 1)\left(y - \frac{3}{2}\right) \quad \text{or} \quad A(x) = (x - 1)\left(\frac{18}{x} - \frac{3}{2}\right).$$

Thus, our problem has been reduced to maximizing the function $A(x)$, where $1 \le x \le 18$. Now,

$$A'(x) = \frac{18}{x} - \frac{3}{2} + (x - 1)\left(-\frac{18}{x^2}\right)$$

$$= \frac{18}{x} - \frac{3}{2} - \frac{18}{x} + \frac{18}{x^2} = -\frac{3}{2} + \frac{18}{x^2}$$

and $A'(x) = 0$ implies that

$$3x^2 = 36, \qquad x^2 = 12, \qquad \text{or} \qquad x = \pm 2\sqrt{3}.$$

It is also clear that $A''(x) = -36x^{-3}$ and $A''(2\sqrt{3}) < 0$ and $A''(-2\sqrt{3}) > 0$. Hence, the point $x = 2\sqrt{3}$ maximizes $A(x)$ by Theorem 7 of Section 4.5. The dimensions should be $x = 2\sqrt{3}$ ft and $y = 9/\sqrt{3}$ ft $= 3\sqrt{3}$ ft.

Example 3. The cost of fuel in running a locomotive is proportional to the square of the speed and is \$25 per hour for a speed of 25 miles per hour. Other costs amount to \$100 per hour, regardless of the speed. Find the speed that will make the cost per mile a minimum.

   Solution. Let $v =$ the required speed and C denote the total cost per mile. The fuel cost per hour is kv^2, where k is a constant to be determined. Since the cost is \$25 per hour for a speed of 25 miles per hour, we see that $25 = k(25)^2$ or $k = \frac{1}{25}$. Thus, the fuel cost per hour $C_1 = \frac{1}{25}v^2$. The total cost C in dollars per mile is given by

$$C(v) = \frac{\text{cost in dollars per hour}}{\text{speed in miles per hour}}$$

$$= \frac{\frac{1}{25}v^2 + 100}{v}.$$

Thus, we want to minimize the quantity $C(v) = v/25 + 100/v$, where $v > 0$. Now,

$$C'(v) = \frac{1}{25} - \frac{100}{v^2} \qquad \text{and} \qquad C'(v) = 0$$

implies that $v^2 = 2500$ or $v = \pm 50$. (We reject -50 since speed is positive.) Since $C''(v) = +200v^{-3}$, $C''(50) > 0$ and $v = 50$ produces a minimum value for C by Theorem 7 of Section 4.5. Thus, the most economical speed is 50 miles per hour.

Example 4. In radioactive experiments it is necessary to consider the *exponential distribution function f* defined by

$$f(t) = 1 - e^{-kt},$$

where the time $t > 0$ and where $k > 0$. Find the local maxima and minima of f, if any, and draw a graph of the function when $k = 1$. Also construct a graph of the derivative of f.

   Solution. We first find that $f'(t) = ke^{-kt}$. Now $f'(t) \neq 0$ for any value of t. We do note that as t becomes very large $f'(t)$ tends toward 0. Since f' exists for all t, the local maxima and minima must occur at a point where $f'(t) = 0$. Thus, *no* local maxima and minima exist. We observe that $f(0) = 0$ and if $k = 1$, $f'(t) = e^{-t}$. Since $e^{-t} > 0$ for all t, the function $f(t) = 1 - e^{-t}$ is increasing for all values of t. Also as $t \to \infty$, $f(t) \to 1$. The graph of $f(t)$ for $k = 1$ is shown in Figure 4.6.3. For $k = 1$, $f'(t) = e^{-t}$, and $f''(t) = -e^{-t}$. Notice that $f''(t) < 0$ for *all* values of t; hence the curve f' is strictly decreasing and does not have any local maximum or minimum points. The graph of f' is shown in Figure 4.6.4.

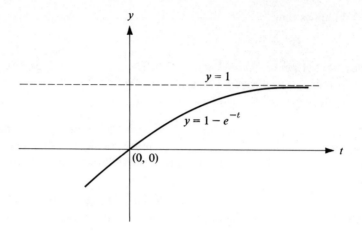

FIGURE 4.6.3

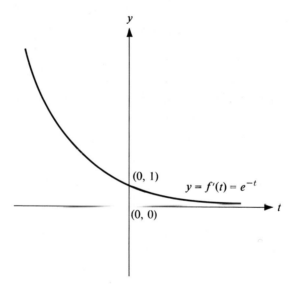

FIGURE 4.6.4

Exercises

1. The total cost of producing x radio sets per day is $\$(\frac{1}{4}x^2 + 35x + 25)$ and the price per set at which they may be sold is $\$(50 - \frac{1}{2}x)$.
   (a) What should be the daily output for a maximum total profit?
   (b) Show that the cost of producing a set has a local minimum.
2. A piece of wire 100 in. long is to be cut in two and the two pieces bent to form a square and a circle. (See Figure 4.6.5.)
   What is the smallest area that can be formed in this manner? What is the largest area (be careful)?

FIGURE 4.6.5

3. What is the largest area that can be achieved by a rectangle fitted under the curve $y = e^{-x^2}$ as in the picture? (See Figure 4.6.6.)

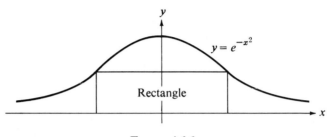

FIGURE 4.6.6

4. A flower bed is to be formed in the shape of a semicircle on top of a rectangle. The flower bed is to then be fenced in with the 200 ft of fence, which is available. What should the dimensions of the flower bed be to yield the largest area? (See Figure 4.6.7.)

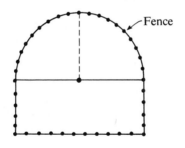

FIGURE 4.6.7

5. A printing company plans to have 40 sq in. of printed matter per page in a particular book. Each page is to have margins of $1\frac{1}{4}$ in. on the sides and $1\frac{1}{2}$ in. at the top and bottom. What are the most economical dimensions for the pages in terms of the cost of the paper?

6. A school group makes plans for an excursion from their home town in New Jersey to New York City. The bus company agrees to take the contract if they are guaranteed that at least 30 students will go. There is a possibility that as many as 200 could go. The fare is to be $50 a person if 30 go, and will decrease by 40 cents per person for every person above the minimum of 30 who goes. What number of people will give the bus company maximum revenue? Will they be able to get maximum revenue? What is the revenue if everyone eligible goes?

7. An American lawyer in order to evade the Internal Revenue Service places his money in a number of Swiss bank accounts via a Jamaican bank. This way he saves $20x$ dollars every year, where x is the number of accounts. Because he is absent minded, he forgets about these accounts at the rate of $x^2/40$ dollars per year. Thus, his profit per year is $f(x) = 20x - x^2/40$. What is the most profitable number of Swiss bank accounts for him to maintain?

8. We wish to consider a simple model of inventory. Let D equal the total number of items to be ordered in a year (known). Let x be the number of items in each order. Let A, B, C be positive constants. Then

$$\text{purchase cost} = AD,$$

$$\text{order cost} = B\frac{D}{x},$$

$$\text{holding cost} = Cx.$$

    The total cost of inventorying for a year, $T(x)$, is equal to purchase cost plus order cost plus holding cost, when $1 \leq x \leq D$. Find the order size (in terms of A, B, C, D) that makes the total cost smallest.

9. A manufacturer of widgets does a cost control analysis on his product and discovers the following. Since the government is furnishing the widget press and depreciation allowances are generous, there is no overhead. If x is the number of widgets produced, then the cost per widget $f(x)$ due to labor costs, raw materials, maintenance, and miscellaneous production costs is $f(x) = 400 \log x + (30 - x)^2$. If production capacity is 25 widgets, at what point is the cost per widget lowest? (Assume 1 is the minimum production level.)

$$\log 10 \cong 2.30$$

$$\log 20 \cong 3.00$$

$$\log 25 \cong 3.22$$

10. An architect wishes to incorporate 100 ft of existing stone wall, laid in a straight line, into a boundary around a rectangular garden. He has an additional 200 ft of fence. How should he lay out the remaining fence in order to maximize the area of the garden? Be careful.

11. The "information content" or "entropy" of a binary source (such as a telegraph that submits dots and dashes) whose two values occur with probability p and $1 - p$ is defined as

$$f(p) = -p \log p - (1 - p) \log (1 - p),$$

    where $0 < p < 1$. Show that f has a maximum at $p = \frac{1}{2}$. The practical significance of this result is that for maximum flow of information per unit time, dots and dashes should, in the long run, appear in equal proportions.

12. The demand for bell-bottom trousers varies inversely as the cube of the selling price. If the article costs 20 cents apiece to manufacture, find the selling price that yields a maximum profit.

13. Show that the sum of the square of any number and the square of its reciprocal is always greater than or equal to 2.

14. Find the area of the largest rectangle with lower base on the x axis and the upper vertices on the curve $y = 1 - x^2$.

15.* Show that for fixed a, the function

$$y = \left(a - \frac{1}{a} - x\right)(4 - 3x^2)$$

has just one maximum and one minimum, and that the difference between them is

$$\left(\frac{4}{9}\right)\left(a + \frac{1}{a}\right)^3 .$$

If a is allowed to vary, find the least value of this difference.

4.7 Summary of Chapter 4

In this section we summarize the most important results of this chapter. Let f be defined on the interval specified below.

(1) If f is continuous on $[a, b]$, it has both a maximum and a minimum on $[a, b]$. Moreover, the maximum and minimum occur either at
   (a) points where f' does not exist, or
   (b) points where $f'(x) = 0$, or
   (c) the end points a, b.

(2) (a) If $f'(x) > 0$, then f is increasing.
   (b) If $f'(x) < 0$, then f is decreasing.

(3) (a) If $f''(x) > 0$, then f is concave up.
   (b) If $f''(x) < 0$, then f is concave down.

(4) *Second derivative test.* Let $f'(x_0) = 0$.

   (a) If $f''(x_0) < 0$, then $(x_0, f(x_0))$ is a local maximum.
   (b) If $f''(x_0) > 0$, then $(x_0, f(x_0))$ is a local minimum.
   (c) If $f''(x_0) = 0$, no information. (See 5.)

(5) *First derivative test.* Let f be differentiable on (a, b), where $f'(x_0) = 0$ and $a < x_0 < b$.

   (a) If $f'(x) < 0$ for $x < x_0$ and $f'(x) > 0$ for $x > x_0$, then f has a minimum at x_0.
   (b) If $f'(x) > 0$ for $x < x_0$ and $f'(x) < 0$ for $x > x_0$, then f has a maximum at x_0.

REVIEW EXERCISES

In all exercises, you must use some test to justify whether or not you have a maximum or minimum.

1. For each of the following functions (i) find the intervals where the function is

increasing and decreasing, (ii) find the intervals where the function is concave up and concave down, (iii) find all maxima and minima, and (iv) sketch a graph of the function.

(a) $f(x) = x^3 + 6x^2 - 15x + 5$

(b) $f(x) = 3 + x^{2/3}$

(c) $f(x) = 1 - x^{2/3}, \quad -1 \leq x \leq 1$

(d) $f(x) = x^4/4 - x^2/2 + 1, \quad -2 \leq x \leq 2$

(e) $f(x) = e^{-(x/3)^2}$

(f) $f(x) = x^{-1}\log x, \quad x > 0$.

2. Suppose $P(x)$, the price for the quantity x, is given by $P(x) = 6 - x^2$. Find the maximum total revenue if demand varies so that $0 \leq x \leq \frac{5}{2}$. [HINT: Recall from Exercise 8, Section 3.1 that the total revenue is given by $xP(x)$.]

3. Determine the maximum total revenue if $P(x) = 8/(4 + x^2)$, $x \geq 0$. Sketch the graph of $P(x)$.

4. A man on a bank of a straight river 4 miles wide wishes to reach a point on the opposite shore that is 4 miles downstream. He can walk 5 mph and row 4 mph. Determine how the man should proceed in order to reach his destination in minimum time. [Recall that distance = (rate) (time)].

5. A manufacturer finds that his costs to produce x articles a day $0 \leq x \leq 100,000$, break down into (i) a fixed cost to employees of $1200 per day, (ii) a fixed unit production cost of $1.20 per day per article, and (iii) a maintenance cost of x^2 dollars per day.

(a)   Find the total cost $C(x)$ per article produced.

(b)   How many articles should be produced each day to minimize cost?

6. If all inputs (that is, labor, material, machinery, plant) to the production process are held fixed in amount except one, called the *variable input* x, and if x is increased over the interval $a \leq x \leq b$, than according to the *law of variable proportions* the total output $Q(x)$ will increase first with positive acceleration (that is, $d^2Q/dx^2 > 0$) and then with negative acceleration (that is, $d^2Q/dx^2 < 0$). That is, the graph of Q will be concave upward as x increases from 0 to an *inflection* point, from which point it will be concave downward. This inflection point is called, in economics, *the point of diminishing returns.*

Suppose that

$$Q(x) = 0.1x + 0.35x^2 - 0.01x^3, \qquad 0 \leq x \leq 23.$$

How many units of x can be employed before diminishing returns set in?

7. At a certain point x_0, where y' and y'' are continuous, we have $y' = 0$ and $y'' = 2$. Must $y = f(x)$ have a local maximum or local minimum at x_0? What if $y'' = 0$? What if $y'' = -2$? Explain.

8. Determine the coefficients a, b, c, d so that the curve whose equation is

$$y = ax^3 + bx^2 + cx + d$$

has a maximum at $(-1, 10)$ and an inflection point at $(1, -6)$.

9. A company wishes to manufacture at minimum cost a closed wooden box with square base and a volume of 12 cu ft. Find the dimensions of the box if the lumber is 5 cents per sq ft. [HINT: Volume of box = length × width × height.]

10. A farmer wishes to fence off a piece of land in the shape of a triangle using the river as one side. Assuming that the two sides of the fence have equal length and that the farmer has only 100 ft of fence, find the maximum area that he can enclose.

11. The S.D.S. has found that for each demonstration staged, they receive $3000 in contributions. They also notice that the cost for a demonstration rises cubically (that is, x demonstrations cost $250 x^3). What is the most profitable number of demonstrations to hold, given that they can hold at most 5?

12. Suppose that it costs a manufacturer $y = 2 + 3x$ dollars to produce x units per week. Assume that the price, P dollars per item, at which he can sell x items per week is $P = 15 - x$.
    (a) What level of production maximizes his profits?
    (b) What is the corresponding price?
    (c) What is his profit (per week) at this level of production?
    (d) If a tax of t dollars per item sold is imposed on this product, and the manufacturer still wishes to maximize his profit, at what price should he sell each item? Comment on the difference between this price and the price before tax.

13. (a) Let $f(x) = x^{1/2}$. Determine the equation of the secant line passing through the points $(0, 0)$ and $(4, 2)$. Find the point c, $0 < c < 4$, such that the slope of the curve at c is the slope of this secant line. Illustrate geometrically.
    (b) Can the mean-value theorem be applied to the function $f(x) = |2x - 1|$ over the interval $[-3, 3]$? Why?

APPENDIX

4.8   Note on Determining the Sign of a Function

In the graphing problems, one is faced with the task of deciding where f' and f'' are positive and negative. We present two methods for doing this.

4.8.1   A method for polynomials

Let $g(x) = (x - r_1)^{n_1}(x - r_2)^{n_2} \ldots (x - r_k)^{n_k}$. We are assuming that the r's are all distinct (different) and that they have been placed in increasing order, that is, $r_1 < r_2 < \cdots < r_k$. Then the sign of g changes at r_1 if n_1 is odd and does not change if n_1 is even. The same is true at a general r_i. The sign of g changes as we cross r_i if n_i is odd and does not change if n_i is even. Let us consider an example. Let $g(x) = (x - (-1))^5$ $(x - 2)^6(x - 7)(x - 12)^3(x - 14)^4$. It is easy to see that $g(-10) < 0$ and that $g(x) < 0$ for $x < -1$. (If $x < -1$, the terms $(x - 2)^6$ and $(x - 12)^4$ are positive and the three remaining terms negative, hence the result.) Since $n_1 = 5$ is odd, the sign of g changes at (-1), and hence $g(x) > 0$ for $-1 < x < 2$. The sign of $n_2 = 6$ is even and so the sign of g does not change at 2. Thus, $g(x) > 0$ for $2 < x < 7$. The reader can check the rest of the points and see if he comes out as in Figure 4.8.1.

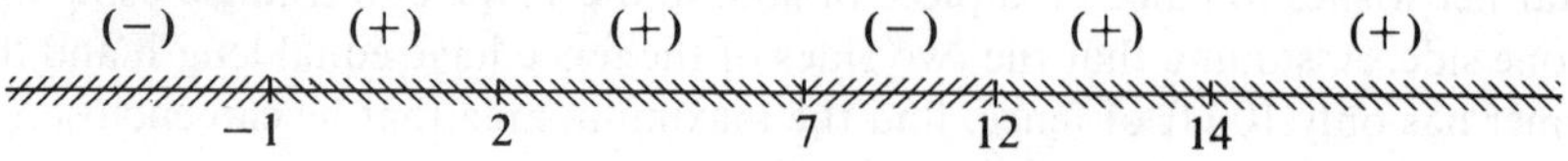

$$(-) \qquad (+) \qquad (+) \qquad (-) \qquad (+) \qquad (+)$$

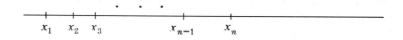

$$-1 \qquad\quad 2 \qquad\qquad\quad 7 \quad\ 12 \qquad\ 14$$

Sign of $g(x)$

FIGURE 4.8.1

If this means of determining the sign seems complicated, we may use a second method that involves a little more work but is simpler to apply.

4.8.2   A method for general functions

Let $g(x) = 0$ for $x = x_1, \ldots, x_n$ and at no other points. We assume g is continuous. Moreover, we assume the zeros of g are in increasing order, that is, $x_1 < x_2 < \cdots < x_n$. Now simply evaluate g at points a_i, where $x_i < a_i < x_{i+1}$; that is, evaluate g somewhere between each adjacent pair of x_i's. (See Figure 4.8.2.) The sign of g does *not necessarily* change at each x_i, but it cannot change any place else. So to decide whether g is positive or negative between x_i and x_{i+1}, we need to check it at a single point. [You also have to check the interval $(-\infty, x_i)$ and the interval (x_n, ∞).]

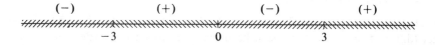

FIGURE 4.8.2

Example.   If $g(x) = (e^x - 1) \log [(1 + x^2)/10]$, determine the sign of $g(x)$ on $(-\infty, \infty)$.

Solution.   First we note that g is defined and continuous on $(-\infty, \infty)$. It should be clear by inspection and a simple calculation that $g(x) = 0$ precisely for $x = -3, 0, 3$. Since $g(-4) = (1/e^4 - 1) \log (\frac{17}{10}) < 0$; $g(-1) = (1/e - 1) \log (\frac{1}{5}) > 0$, $g(1) = (e - 1) \log (\frac{1}{5}) < 0$; and $g(5) = (e^5 - 1) \log (\frac{13}{5}) > 0$, we have the following behavior for $g(x)$ (see Figure 4.8.3).

$$(-) \qquad\qquad (+) \qquad\qquad (-) \qquad\qquad (+)$$

$$-3 \qquad\qquad\quad 0 \qquad\qquad\quad 3$$

FIGURE 4.8.3

5

Integration

Until now, we have been dealing with the problem of finding the derivative f' of a given function f. It is also natural to ask the following question: Given any function f, can we find a function g whose derivative is f? For example, if $f(x) = x^2$, does there exist a function g such that $g'(x) = x^2$? Clearly, if $g(x) = x^3/3$, then $g'(x) = x^2 = f(x)$. Now, is this the only function g satisfying $g'(x) = x^2$? A little thought yields a *negative* answer. Note that if $g(x) = x^3/3 + 8$, then $g'(x) = x^2$ (since the derivative of the constant function is zero). In fact, we observe that there are infinitely many functions satisfying $g'(x) = x^2$—namely functions of the form $x^3/3 + c$, where c is an arbitrary constant. This problem brings us into the realm of *integral calculus*. As we shall see in later sections, the techniques we are now developing have applications to a wide range of important problems in physics, biology, economics, psychology, and business. For example, in a sense that we shall make clear, the antiderivative of acceleration is velocity; of velocity is distance; of length is area; of marginal revenue is total sales revenue; of marginal cost is total cost; of area is volume; of learning rate is a knowledge function.

Let us make the following definition.

DEFINITION. *The function F is called an antiderivative of the function f on an interval I if and only if F is differentiable on I and $F'(x) = f(x)$ for $x \in I$. (I may be a closed, open, or half-open interval—see Section 1.1.)*

Example 1. Let $f(x) = 4x^2$. We assert that $F(x) = 4x^3/3$ is an antiderivative of $f(x)$, since $F'(x) = 4x^2$. Note that if c is any arbitrary constant, then $F(x) = 4x^3/3 + c$ is also an antiderivative of $4x^2$ since $F'(x) = 4x^2$

$$\left[\frac{d}{dx}(c) = 0! \right]$$

177

Example 2. Let $f(x) = ax^n$, $n \neq -1$. It is clear that $F(x) = ax^{n+1}/(n + 1)$ is an antiderivative of $f(x)$ since $F'(x) = (n + 1)ax^n/(n + 1) = ax^n$. In like manner if c is any arbitrary constant, then $F(x) = ax^{n+1}/(n + 1) + c$ is also an antiderivative. Thus there are infinitely many antiderivatives of $f(x)$.

Example 3. Let $f(x) = 1/x$. Recall from Section 3.3 that if $y = \log|x|$ then $(dy/dx) = 1/x$. Hence $F(x) = \log x$ is an antiderivative of $1/x$. In fact, if c is any arbitrary constant, then $F(x) = \log x + c$ is an antiderivative of $1/x$.

Example 4. Let $f(x) = e^x$. Then $F(x) = e^x$ is an antiderivative of $f(x)$ since $F'(x) = e^x = f(x)$. Clearly $e^x + c$, where c is an arbitrary constant, is also an antiderivative of e^x.

Example 5. Let $f(x) = x^2 + x^3 + 1/x + e^x$. Find an antiderivative of $f(x)$. Note that $F(x) = x^3/3 + x^4/4 + \log x + e^x$ is an antiderivative of $f(x)$ since

$$F'(x) = x^2 + x^3 + 1/x + e^x.$$

If we add any arbitrary constant to $F(x)$, we still have an antiderivative of f. Observe that we could find F by first finding an antiderivative of x^2, x^3, $1/x$, and e^x, and then adding them together. In the next section, we show that this is a general property for antiderivatives—namely if F, G, H are antiderivatives of f, g, h, then $F + G + H$ is an antiderivative of $f + g + h$.

It is now natural to ask if there is a relation between the antiderivatives of a function. For example, $x^3/3 + \pi$ and $x^3/3 + 2$ are both antiderivatives of x^2. Notice that these two antiderivatives differ by a constant—namely $\pi - 2$. We shall see now that this is true in general; any two antiderivatives of a function can differ only by a constant. Thus, apart from constant terms, a function can have only one antiderivative on a given interval. We state this result in the form of a theorem.

THEOREM.   *If F and G are two antiderivatives of a function f on an interval I, then there is a constant C such that*

$$F(x) = G(x) + C, \qquad x \in I.$$

This theorem simply states that any two antiderivatives of a function differ by a constant. Observe that it also tells us that if F is any antiderivative of f, then *all* antiderivatives of f must be of the form $F + C$.

The verification of the above theorem is quite straightforward. Let

$$H(x) = F(x) - G(x),$$

where $x \in I$. Since $F'(x) = G'(x) = f(x)$, then $H'(x) = 0$. By the corollary to the mean-value theorem (see Section 4.3) $H(x) = C$, where C is a constant. Hence, $H(x) = F(x) - G(x) = C$, the desired result.

At this stage, we point out why there is good reason for us to be concerned with intervals when discussing antiderivatives. Consider the following function:

$$f(x) = \begin{cases} \dfrac{1}{x} + 5, & x > 0 \\[2mm] \dfrac{1}{x} - 8, & x < 0. \end{cases}$$

Now, for x in $(-\infty, 0)$ or $(0, \infty)$, $f(x)$ does not differ from $1/x$ by an arbitrary constant even though $f(x)$ and $1/x$ have their derivatives equal to $-1/x^2$ for x in $(-\infty, 0)$ or $(0, \infty)$. Hence, $-1/x^2$ has *two* antiderivatives that do not differ by a constant. The reason for this is that the set $(-\infty, 0)$ plus $(0, \infty)$ is *not* an interval (the point 0 is excluded!). However, for $x > 0$, $f(x) - 1/x = 5$, a constant—similarly for $x < 0$,

$$f(x) - 1/x = -8,$$

a constant. Thus, considering intervals is crucial!

You should also be aware of the fact that there are functions that have *no* antiderivatives. An example is the function:

$$f(x) = \begin{cases} -1, & -1 \le x < 0 \\ 0, & x = 0 \\ +1, & 0 < x \le 1. \end{cases}$$

We discuss this topic more fully in an appendix to this chapter.

At this point, it would be convenient to have a notation for antiderivatives.

DEFINITION. *If F is an antiderivative of f, we write*

$$\int f(x)\, dx = F(x) + C.$$

For example:

$$\int x^2\, dx = \frac{x^3}{3} + C,$$

$$\int t\, dt = \frac{t^2}{2} + C,$$

$$\int e^x\, dx = e^x + C.$$

Sometimes the symbol $\int f(x)\, dx$ is called an *indefinite integral* of f. The symbols dx and dt contribute little, if anything, in these examples. However, they become useful when the expression for a function involves unspecified constants. We refer to $f(x)$ as the *integrand*.

Hence,

$$\int x^2 t\, dx = \frac{x^3 t}{3} + C,$$

since dx singles out x as the variable, while

$$\int x^2 t\, dt = \frac{x^2 t^2}{2} + C,$$

since dt singles out t as the variable. Thus, dx tells us what the variable is in each case. In the above, we assume that t is *not* a function of x and x is *not* a function of t.

We now construct a table similar to Table 3.3.1 giving us antiderivatives of several well-known functions.

Table   5.1.1

$f(x)$	$\int f(x)\, dx$
1. K, where K is a constant	$Kx + C$
2. $x^n, n \neq -1$	$\dfrac{x^{n+1}}{n+1} + C$
3. $\dfrac{1}{x}$	$\log |x| + C$
4. e^x	$e^x + C$

Note that each entry can be checked by verifying that the derivative of the right-hand side yields the function on the left.

Exercises

1. Determine whether the following functions have antiderivatives on the intervals indicated. If so, find them.

 (a) $f(x) = \dfrac{1}{x^2}$ on the interval $[-1, 1]$.

 (b) $f(x) = \dfrac{1}{x^2}$ on the interval $(0, 1)]$.

 (c) $f(x) = x^{-1/2}$ on the interval $[0, 5]$.

 (d) $f(x) = x^{-1/3}$ on the interval $[-5, 5]$.

2. Calculate each of the indicated antiderivatives.

 (a) $\int x^{3/2}\, dx$

 (b) $\int x^{-3/2}\, dx$

 (c) $\int (t^2 + t)\, dt$

 (d) $\int \left(x^2 + \dfrac{1}{x} + e^x \right) dx$

 (e) $\int (1 - x)x^{1/2}\, dx$

 (f) $\int e^{\log x}\, dx$

 (g) $\int (x^2 + 1)^2\, dx$

 (h) $\int \log (e^x)\, dx$

 (i) $\int \left(12x^5 - \dfrac{5}{x^2} + 3 \right) dx$

 (j) $\int (3x - 5e^x)\, dx$

3. Let $h(x) = \sqrt{1 - x^2}$.

   (a) Find $h'(x)$.

   (b) What is $\int -x(1 - x^2)^{-1/2} \, dx$?

4. Let $g(x) = e^{4x + 9x^2}$.

   (a) Find $g'(x)$.

   (b) Evaluate $\int (4 + 18x)e^{4x + 9x^2} \, dx$.

5. Evaluate the following:

   (a) $\int \dfrac{x^3 + 1}{x^2} \, dx$ 　　　　　　(d) $\int \left(\sqrt{t} + \dfrac{1}{\sqrt{t}} \right) dt$

   (b) $\int \dfrac{dt}{t\sqrt{t}}$ 　　　　　　(e) $\int \left(t^{5/2} - \dfrac{1}{t^{2/5}} \right) dt$

   (c) $\int (x^2 - \sqrt{x}) \, dx$ 　　　　　　(f) $\int at^{9000} \, dt$, where a is a constant.

6. Let $g(x) = \frac{1}{3}(2x + 1)^{3/2}$.

   (a) Find $g'(x)$.

   (b) Evaluate: $\int (2x + 1)^{1/2} \, dx$.

7*. Let

$$f(x) = \frac{1 + |x|}{x} \quad \text{and} \quad g(x) = \frac{1}{x}.$$

   (a) Show that f and g are antiderivatives of $-1/x^2$, $x \neq 0$.

   (b) Show that there does not exist a constant k such that $f(x) = g(x) + k$ whenever $x \neq 1$.

   (c) Does this contradict the theorem about antiderivatives stated in this section? Why?

5.2 Properties of Antiderivatives

In general the problem of finding antiderivatives of functions is more difficult than differentiation. It is convenient to introduce, at this time, some techniques for finding antiderivatives.

First, we discuss properties of antidifferentiation analogous to those given in Sections 3.4–3.6 for derivatives.

PROPERTY 1. *If F is an antiderivative of f, then kF is an antiderivative of kf. Equivalently,*

$$\int kf(x) \, dx = k \int f(x) \, dx.$$

Proof. Simply observe that:

$$\frac{d}{dx}(kF) = k\frac{dF}{dx} = kf,$$

since $F'(x) = f(x)$. Hence kF is an antiderivative of kf.

Example 1. Find $\int 3x^2\, dx$.

Solution.

$$\int 3x^2\, dx = 3\int x^2\, dx = 3\cdot\frac{x^3}{3} + C = x^3 + C,$$

where C is an arbitrary constant. This is certainly easier than searching for an antiderivative of $3x^2$.

PROPERTY 2. *If F and G are antiderivatives of f and g, respectively, then $F \pm G$ is an antiderivative of $f \pm g$. Equivalently,*

$$\int (f \pm g)(x)\, dx = \int f(x)\, dx \pm \int g(x)\, dx.$$

Proof. We must show that

$$\frac{d}{dx}[F(x) \pm G(x)] = f(x) \pm g(x).$$

However,

$$\frac{d}{dx}[F(x) \pm G(x)] = \frac{d}{dx}F(x) \pm \frac{d}{dx}G(x) = f(x) \pm g(x),$$

since $F' - f$ and $G' = g$.

Example 2. Find $\int (3x^2 + 2x^3 + x)\, dx$.

Solution. By Property 2,

$$\int (3x^2 + 2x^3 + x)\, dx = \int 3x^2\, dx + \int 2x^3\, dx + \int x\, dx$$

$$= 3\int x^2\, dx + 2\int x^3\, dx + \int x\, dx,$$

by Property 1,

$$= 3\cdot\frac{x^3}{3} + 2\cdot\frac{x^4}{4} + \frac{x^2}{2} + C$$

$$= x^3 + \frac{x^4}{2} + \frac{x^2}{2} + C,$$

where C is an arbitrary constant.

Example 3. Find a function $f(x)$ satisfying the conditions that $f'(x) = x^2 + x + 1$ and $f(0) = 1$.

Solution. In order to find f, we must first find an antiderivative of $x^2 + x + 1$. Now,

$$\int (x^2 + x + 1)\, dx = \int x^2\, dx + \int x\, dx + \int dx = \frac{x^3}{3} + \frac{x^2}{2} + x + C.$$

Thus, $f(x) = x^3/3 + x^2/2 + x + C$. In order to find the function satisfying $f(0) = 1$, we observe after substituting 0 for x that $f(0) = C$, and hence $C = 1$. The function f satisfying the two given conditions is then given by

$$f(x) = \frac{x^3}{3} + \frac{x^2}{2} + x + 1.$$

There is really no direct analogue to the theorems concerning differentiation of products and quotients for antiderivatives. Note, however, that if F is an antiderivative of f and G is an antiderivative of g, then FG is not necessarily an antiderivative of $f \cdot g$. The following example demonstrates this.

Example 4. Let $f(x) = x^2$ and $g(x) = x$. Then $F(x) = x^3/3$ and $G(x) = x^2/2$. Hence, $F(x)G(x) = x^5/6$. Now, $f(x) \cdot g(x) = x^3$ and an antiderivative of x^3 is $x^4/4$. Clearly, $x^5/6$ and $x^4/4$ do not differ by only a constant. Thus, $F \cdot G$ is certainly not an antiderivative of $f \cdot g$.

We end this section with one more example—typical of the use of antiderivatives for applications.

Example 5. Find the equation of the curve $y = f(x)$ through the point $(2, -3)$ with slope at the point (x, y) given by $2x - 3$.

Solution. Since the slope at any point (x, y) is given by $2x - 3$, we must have $dy/dx = f'(x) = 2x - 3$. Thus, we must find an antiderivative of $2x - 3$ such that the curve goes through the point $(2, -3)$. Now,

$$y = \int (2x - 3)\, dx = x^2 - 3x + C.$$

To find C, we use the fact that when $y = -3$, $x = 2$. Hence, $-3 = 4 - 6 + C$ or $C = -1$ and the equation of the desired curve is given by

$$y = x^2 - 3x - 1.$$

Exercises

In Exercises 1–11, evaluate the following

1. $\int (x^2 - 3x + 9)\, dx.$

2. $\int (5 - 4x)\, dx.$

3. $\int (3x^{1/2} + 5x^{5/2} + x^{1/19})\, dx.$

4. $\int \left(17x^6 - x^5 + x^2 - \frac{1}{x}\right) dx.$

5. $\int (-x^{-1/3} + x^{-1/6} - x^{-1/9}) \, dx.$

6. $\int (x^{-1} - 3e^x) \, dx.$

7. $\int (2e^{-x} + e^x) \, dx.$

8. $\int \left(4e^{-x} - \frac{5}{x}\right) dx.$

9. $\int \left(\frac{1}{x} + \frac{3}{x^2} + x^{1/3} - x^{-2/3}\right) dx.$

10. $\int_3^2 [(x + 2)(x^2 - 1)] \, dx.$

11. $\int x^2(x^2 - 4) \, dx.$

12. Find a function $f(x)$ satisfying the conditions that

    (a) $f'(x) = x^2 - x, \quad f(0) = 1$

    (b) $f'(x) = x(x - 1)(x - 2), \quad f(2) = 4$

    (c) $f'(x) = e^x + 1/x, \quad f(1) = 1.$

13. Find the equation of the curve $y = f(x)$ through the given point, with slope at a typical point (x, y) as given.

    (a) point $(0, 0)$; slope $2x - 6x^3$.

    (b) point $(2, 1)$; slope $(3x - 1)(x - 2)$

    (c) point $(2, 8)$; slope $8x^3 - 2x.$

14. Suppose that the marginal cost of an item is given by $(x^2 - 1)/x^2$ (see Exercise 8, Section 3.1.), and the total cost is known to be 1 when unit 1 has been produced. Find the total cost of production.

15. Suppose that $f''(x) = x^{1/2} + x^{-1/2}$. Find $f(x)$, given that $f'(1) = 2$ and $f(0) = 0$. [HINT: Remember that f' is an antiderivative of f''. Thus, you must first find an antiderivative of f'' and then an antiderivative of f'.]

5.3 Antiderivatives by Substitution

At present we really have no powerful techniques for finding antiderivatives (only very simple rules). In this section we introduce such a tool, namely substitution. Another technique appears in Section 5.4.

It is convenient, at this stage, to introduce the formalism of differentials. In Leibnitz's notation for a derivative of a function f, df/dx, we never attempted to assign a meaning to the symbols df and dx separately. Nevertheless, the chain rule suggests that we may treat these much like numbers. In particular, if $y = f(x)$, where f is differentiable, we write $dy = f'(x)\,dx$. We say that dy is the "differential of f"—often $f'(x)\,dx$ is referred to as a first order differential.

Thus,

(i)   If $y = (4x^2 + 9)$, then $dy = 8x\,dx$.

(ii)   If $y = \log x + e^{2x}$, then $dy = (1/x + 2e^{2x})\,dx$.

(iii)   If $y = \sqrt{1 + x^3}$, then $dy = \frac{3}{2}x^2(1 + x^2)^{-1/2}\,dx$.

As a mnemonic aid write $dy/dx = f'(x)$ and then move the dx to obtain $dy = f'(x)\,dx$. The best way to think of differentials is as just a game. The rules of the game are clear enough and it is easy to play. To assuage any trepidations you might have, we do mention that:

(i)   Differentials can be presented in a more rigorous way and are really a disguised form of the chain rule.

(ii)   Differentials are very useful for finding antiderivatives.

We are now ready to study substitution. This topic is best introduced by an example.

Example 1.   Find $\int 5x^4(x^5 + 1)^9\,dx$.

Solution.   We try the substitution $y = x^5 + 1$. (Hints on what to try are given later.) This yields $\int 5x^4 y^9\,dx$. However, in substitution, we must replace every x including dx. To this end let us find dy. Clearly, $dy = 5x^4\,dx$. We can solve for dx, that is,

$$dx = \frac{1}{5x^4}\,dy.$$

Substituting this, we get

$$\int 5x^4 y^9 \frac{1}{5x^4}\,dy = \int y^9\,dy = \frac{y^{10}}{10} + C.$$

(We could have saved a step by substituting $5x^4\,dx$ for dy.)

Now, if we replace y by $(x^5 + 1)$, we obtain

$$\int 5x^4(x^5 + 1)^9\,dx = \frac{(x^5 + 1)^{10}}{10} + C.$$

If we differentiate the right-hand side of the preceding, we find that

$$\frac{d}{dx}\left[\frac{(x^5 + 1)^{10}}{10} + C\right] = 5x^4(x^5 + 1)^9,$$

that is, we have obtained the right answer. Is this a stroke of luck, a miracle, or is there a procedure here that works every time? Before answering that question let us consider one more example.

Example 2. Find $\int 3x(x+2)^{1/2}\,dx$.

Solution. Let $y = x + 2$. When we substitute, the original integral becomes $\int 3xy^{1/2}\,dx$. We must replace all the x's including dx. It is easy to replace the term $3x$, since $x = y - 2$, and so $3x = 3(y - 2)$. The integral now becomes $\int 3(y-2)y^{1/2}\,dx$. Since $y = x + 2$, we see that $dy = 1 \cdot dx$. Thus, our integral now has the form $\int 3(y-2)y^{1/2}\,dy$. This is easy to handle. Indeed,

$$\int 3(y-2)y^{1/2}\,dy = 3\int (y^{3/2} - 2y^{1/2})\,dy = 3\frac{y^{5/2}}{5/2} - 6\cdot\frac{y^{3/2}}{3/2} + C.$$

If we substitute $y = x + 2$, we find

$$\int 3x(x+2)^{1/2}\,dx = \tfrac{6}{5}(x+2)^{5/2} - 4(x+2)^{3/2} + C.$$

Again, if we differentiate the right-hand side of the equation we discover we have the correct answer!!

By now the reader should be persuaded that substitution is a method that really works. You need only differentiate your result in order to see this. We outline the basic rules of the procedure and illustrate them by examples. Later we explain why the method works. However, substitution is easier to use than to understand.

We are given an indefinite integral $\int f(x)\,dx$:

  (1) Choose a substitution $y = u(x)$. This choice is best made by experience. However, as a rule try differentiating the most complicated term in the integrand $f(x)$. If you get some other term in the integrand this way, you probably have a good choice for $u(x)$.
  (2) Replace all x's and dx's by y's and dy's. Note that $dx = (1/u'(x))\,dy$.
  (3) Find $\int g(y)\,dy = G(y)$.
  (4) Replace $G(y)$ by $G(u(x))$; that is, replace y by $u(x)$.
  (5) Check your answer by differentiating.

Let us now consider some examples.

Example 3.   Find $\int x^2 e^{(x^3+2)}\,dx$.

Solution.   First we observe that since x^2 is closely related to the derivative of $x^3 + 2$, we use the substitution $y = x^3 + 2$. Then $dy = 3x^2\,dx$ and hence,

$$\int x^2 e^{(x^3+2)}\,dx = \frac{1}{3}\int e^y\,dy = \frac{1}{3}e^y + C.$$

Substituting $y = x^3 + 2$, we obtain

$$\int x^2 e^{(x^3+2)}\,dx = \frac{1}{3}e^{(x^3+2)} + C.$$

Example 4.   Find

$$\int \frac{\log x}{x}\,dx.$$

Solution. We must first note that $(d/dx) \log x = 1/x$. Hence, the substitution we use is $y = \log x$. Then $dy = 1/x \, dx$ and

$$\int \frac{\log x}{x} \, dx = \int y \, dy = \frac{y^2}{2} + C.$$

Substituting $y = \log x$ yields

$$\int \frac{\log x}{x} \, dx = \frac{(\log x)^2}{2} + C.$$

Example 5. Find

$$\int \frac{x}{\sqrt{(1 - x^2)}} \, e^{\sqrt{(1 - x^2)}} \, dx.$$

Solution. This looks complicated, but if we recall that

$$\frac{d}{dx} \left[\sqrt{(1 - x^2)} \right] = \frac{-x}{\sqrt{(1 - x^2)}},$$

we see that the substitution to make is $y = \sqrt{(1 - x^2)}$. Then

$$dy = - \frac{x}{\sqrt{(1 - x^2)}} \, dx$$

and hence,

$$\int \frac{x}{\sqrt{(1 - x^2)}} \, e^{\sqrt{(1 - x^2)}} \, dx = - \int e^y \, dy = - e^y + C.$$

Substituting for y, we obtain

$$\int \frac{x}{\sqrt{(1 - x^2)}} \, e^{\sqrt{(1 - x^2)}} \, dx = - e^{\sqrt{(1 - x^2)}} + C.$$

Let us note again what we have done in each of the preceding examples. In the evaluation of each antiderivative of the form above, we first looked for a quantity whose derivative appeared in the integrand. This enabled us, with the formal use of differentials, to reduce the integration to evaluating antiderivatives of the form given in Table 5.1. It takes a lot of practice to become acquainted with the proper substitution to make in each case.

We have established certain conventions for handling symbols [dy, dx, etc.] that are as yet meaningless. However, we have still been able to apply this formalism to establish meaningful and correct results. We could have circumvented this by going through arguments like Examples 6 and 7 below. It should be clear to you, now, that differentials are a handy device. They allow us to make calculations with a minimum of effort. In order to give a proper treatment of differentials, we would have to introduce functions of two variables and base our discussion on them. We choose to omit this at the present stage of our development.

For those who wish to understand why substitution and the tricks with differentials work, we include two illustrative examples below. Note that the method you have just learned is considerably shorter and simpler.

Example 6.   Find $\int x\sqrt{1 + x^2}\, dx$.

Solution. In order to work this problem, we use the chain rule in reverse. Note that apart from a constant multiple (namely 2), x is the derivative of $1 + x^2$. This suggests that we make a substitution of the form $u(x) = 1 + x^2$. Then $du/dx = 2x$ so that $x\sqrt{1 + x^2} = \frac{1}{2}\sqrt{u}\, du/dx$. Observe that

$$\frac{1}{2}\int \sqrt{u}\, du = \frac{1}{2} \cdot \frac{u^{3/2}}{\frac{3}{2}} + C = \frac{1}{3} u^{3/2} + C.$$

Now,

$$\frac{d}{dx}\left(\frac{1}{3} u^{3/2} + C\right) = \frac{1}{2}\sqrt{u}\, \frac{du}{dx},$$

by the chain rule. Hence,

$$\int x\sqrt{(1 + x^2)}\, dx = \frac{1}{3} u^{3/2} + C.$$

At this stage we substitute back for u and obtain

$$\int x\sqrt{(1 + x^2)} = \frac{1}{3}(1 + x^2)^{3/2} + C.$$

Example 7.   Find $\int x^2 e^{x^3}\, dx$.

Solution. Again, apart from a constant multiple, x^2 is essentially the derivative x^3. This suggests a substitution of the form $u(x) = x^3$. Then $du/dx = 3x^2$ so that $x^2 e^{x^3} = \frac{1}{3} e^u\, du/dx$. Observe that

$$\frac{1}{3}\int e^u\, du = \frac{1}{3} e^u + C \qquad \text{and} \qquad \frac{d}{dx}\left[\frac{1}{3} e^u + C\right] = \frac{1}{3} e^u \frac{du}{dx}$$

by the chain rule. Hence

$$x^2 e^{x^3} = \frac{d}{dx}\left[\frac{1}{3} e^u + C\right]$$

or equivalently,

$$\int x^2 e^{x^3}\, dx = \frac{1}{3} e^{x^3} + C.$$

It would be rather awkward to go through such calculations as the above for each complicated problem. Thus, we introduced the formalism of differentials.

Exercises

1. Find the differentials of the following functions.

    (a) $y = x^3 + e^{2x}$ (e) $u = \sqrt{x^2 + 1}$

    (b) $u = \dfrac{4}{x} - 1$ (f) $u = \sqrt{x^3 + 4}$

    (c) $y = e^{(x^2 + 1)}$ (g) $y = x^4 + 1$

    (d) $y = x \log x - x$ (h) $u = e^x$

                          (i) $y = \log x$

Evaluate the following.

2. $\displaystyle\int \left(x^2 + x^3 + e^x + \frac{1}{x} \right) dx.$

3. $\displaystyle\int (x^{5/3} + x^{1/6} + x^{-2})\, dx.$

4. $\displaystyle\int (2x + 1)^{1/2}\, dx.$

5. $\displaystyle\int \frac{(x + 1)\, dx}{(x^2 + 2x + 3)^{1/3}}.$

6. $\displaystyle\int \frac{x^3\, dx}{\sqrt{(1 - x^4)}}$

7. $\displaystyle\int x(x + 1)^{1/2}\, dx.$

8. $\displaystyle\int (x + 1)(x + 3)^{3/2}\, dx.$

9. $\displaystyle\int x(x - 9)^{-1/2}\, dx.$

10. $\displaystyle\int (x - 1)(x + 1)^{100}\, dx.$

11. $\displaystyle\int \frac{\log x}{x}\, dx.$

12. $\displaystyle\int \frac{x}{x^2 + 1} \log(x^2 + 1)\, dx.$

13. $\displaystyle\int \frac{1}{x \log x}\, dx.$

14. $\displaystyle\int (x^2 + 1)e^{(x^3 + 3x)}\, dx.$

15. $\displaystyle\int \frac{\log x^2}{x}\, dx.$

16. $\displaystyle\int \frac{1}{x (\log x)^2}\, dx.$

17. $\displaystyle\int \frac{e^{\sqrt{x}}}{\sqrt{x}}\, dx.$

18. $\displaystyle\int x \sqrt{(1 - x^2)}\, dx.$

19. $\displaystyle\int \frac{x}{1 - x^2} \log \sqrt{(1 - x^2)}\, dx.$

20. $\displaystyle\int \frac{e^x}{1 + e^x}\, dx.$

21. $\displaystyle\int [x^2 e^{x^3} + 3x(x^2 + 1)^{1/2}]\, dx.$

22. $\displaystyle\int e^{(e^x)} e^x\, dx.$

23. Find a function $f(x)$ satisfying the conditions that $f'(x) = (x^2 + 1) e^{x^3 + 3x}$ with $f(0) = 1$.

24. Find a function $f(x)$ such that $f'(x) = x\sqrt{(1 + x^2)} + x^2$ and $f(1) = 0$.

25. Let $y = f(x)$ be a curve with slope equal to 3 at every point. Let $f(1) = -2$. Find $f(x)$.

26. If marginal revenue is known to be $x/\sqrt{9-x}$, find the revenue function $R(x)$ if we are told that $R(0) = 1$. [See Exercise 8, Section 3.1.]

27. The slope of a curve at (x, y) is $Ax(x^2 - 1)$, where A is some constant. The curve crosses the x axis at $x = 3$ and it crosses the y axis at $y = 2$. Find the equation of the curve. [HINT: Recall that the slope of a curve at a point (x, y) is given by dy/dx.]

5.4 The Newton Integral

Until now we have only discussed the question of indefinite integrals. However, in practical applications we usually must consider antiderivatives of functions with certain subsidiary conditions on the function. (See Example 3 of Section 5.2.) Thus, it is convenient to introduce at this time the concept of a definite integral. We begin by defining the Newton integral (a term introduced by J. W. Kitchen in his text *Calculus of One Variable*, Addison-Wesley, Reading, Mass., 1968).

DEFINITION. *If f has an antiderivative F on the interval $[a, b]$, we say that f is Newton integrable on $[a, b]$ and the Newton integral is defined to be the number*

$$\int_a^b f(x)\, dx = F(b) - F(a).$$

We refer to $\int_a^b f(x)\, dx$ as the definite integral of f from a to b. The function f shall be referred to as the integrand; a and b are called the *limits of integration*. The definition given above is legitimate since we can easily show that the integral is independent of the particular choice of F. It depends only on a, b, and f. In order to see this, we observe that if F and G are any two antiderivatives of f on $[a, b]$, then $F(b) - F(a) = G(b) - G(a)$. The proof of this is simple. If F and G are any two antiderivatives of f, then $F(x) = G(x) + C$, where C is an arbitrary constant. Now, $F(b) = G(b) + C$ and $F(a) = G(a) + C$; thus, $F(b) - F(a) = G(b) - G(a)$.

Let us now look at some examples.

Example 1.   Find $\int_{-1}^2 x^2\, dx$.

Solution.   Since $x^3/3$ is an antiderivative of x^2, we see that

$$\int_{-1}^2 x^2\, dx = \frac{8}{3} - \frac{(-1)}{3} = \frac{9}{3} = 3.$$

Example 2.   Find $\int_a^b x^n\, dx$, where n is any number except -1.

Solution.   Since $x^{n+1}/(n + 1)$ is an antiderivative of x^n, we see that

$$\int_a^b x^n\, dx = \frac{b^{n+1}}{n+1} - \frac{a^{n+1}}{n+1}.$$

Sometimes it will be convenient to use the following notation. If f is Newton integrable on $[a, b]$ and F is an antiderivative of f, then

$$\int_a^b f(x)\, dx = F(x)\Big|_a^b = F(b) - F(a).$$

Thus,

$$\int_a^b x^n \, dx = \frac{x^{n+1}}{n+1} \Big|_a^b = \frac{b^{n+1}}{n+1} - \frac{a^{n+1}}{n+1}.$$

Example 3.   Find $\int_0^1 e^x \, dx$.

Solution.   Since e^x is an antiderivative of e^x, we see that

$$\int_0^1 e^x \, dx = e^x \Big|_0^1 = e^1 - e^0 = e - 1.$$

Example 4.   Find $\int_1^2 1/x \, dx$.

Solution.   Since $\log |x|$ is an antiderivative of $1/x$, then

$$\int_1^2 \frac{1}{x} \, dx = \log |x| \Big|_1^2 = \log 2 - \log 1 = \log 2.$$

Definite integrals have a number of useful properties, which we list below.

PROPERTY 1.   *If f is Newton integrable on* [a, b] *and on* [b, c], *then f is Newton integrable on* [a, c] *and*

$$\int_a^c f(x) \, dx = \int_a^b f(x) \, dx + \int_b^c f(x) \, dx.$$

Proof of Property 1.   It is a simple matter to verify this result. Let F be an antiderivative of f on $[a, b]$ and G be an antiderivative of f on $[b, c]$. Let

$$H(x) = \begin{cases} F(x), & x \in [a, b] \\ G(x) + F(b) - G(b), & x \in [b, c]. \end{cases}$$

Clearly H is an antiderivative of f on $[a, c]$, since $H'(x) = F'(x) = f(x)$, $x \in [a, b]$, and $H'(x) = G'(x) = f(x)$, $x \in [b, c]$. Thus $H'(x) = f(x)$, $x \in [a, c]$. Hence, $\int_a^c f(x) \, dx = H(c) - H(a)$. Now

$$\int_a^b f(x) \, dx + \int_b^c f(x) \, dx = H(b) - H(a) + H(c) - H(b) = H(c) - H(a),$$

yielding the desired result.

PROPERTY 2.   *If f and g are Newton integrable on* [a, b], *then so is f + g and c · f, where c is any constant. In particular,*

$$\int_a^b [f(x) + g(x)] \, dx = \int_a^b f(x) \, dx + \int_a^b g(x) \, dx$$

and

$$\int_a^b cf(x) \, dx = c \int_a^b f(x) \, dx.$$

Proof of Property 2.   This is also simple to verify. Let F and G be the antiderivatives of f and g on $[a, b]$ respectively. Then from Property 2, Section 5.2, $F + G$ is an anti-derivative of $f + g$. Thus,

$$\int_a^b [f(x) + g(x)] \, dx = F(b) + G(b) - F(a) - G(a)$$

$$= [F(b) - F(a)] + [G(b) - G(a)] = \int_a^b f(x) \, dx + \int_a^b g(x) \, dx.$$

The second part follows directly from Property 1 of Section 5.2.

We now look at some examples using these properties.

Example 5.   Find $\int_0^4 |x(x - 2)| \, dx$.

Solution.   Since $|x(x - 2)| = x(x - 2) = x^2 - 2x$ for $x \geq 2$, the function $|x(x - 2)|$ is Newton integrable on $[2, 4]$. Since $|x(x - 2)| = -x(x - 2) = -x^2 + 2x$ for $0 \leq x \leq 2$, the function $|x(x - 2)|$ is Newton integrable on $[0, 2]$.

$$\int_0^4 |x(x - 2)| \, dx = \int_0^2 |x(x - 2)| \, dx + \int_2^4 |x(x - 2)| \, dx$$

$$= -\int_0^2 x(x - 2) \, dx + \int_2^4 x(x - 2) \, dx.$$

By Property 2

$$= -\int_0^2 x^2 \, dx + 2\int_0^2 x \, dx + \int_2^4 x^2 \, dx - 2\int_2^4 x \, dx$$

$$= -\frac{x^3}{3}\Big|_0^2 + x^2\Big|_0^2 + \frac{x^3}{3}\Big|_2^4 - x^2\Big|_2^4$$

$$= -\frac{8}{3} + 4 + \frac{64}{3} - \frac{8}{3} - 16 + 4 = \frac{24}{3}.$$

Without Property 1, we could not have evaluated this integral.

PROPERTY 3.   *Let f be Newton integrable on $[a, b]$, and suppose that $f(x) \geq 0$ on $[a, b]$. Then, $\int_a^b f(x) \, dx \geq 0$.*

Proof of Property 3.   This result is also easily obtained. Let F be an antiderivative of f on $[a, b]$. Then, $F'(x) = f(x) \geq 0$, and hence F is *increasing* on $[a, b]$. (Recall Section 4.2!) Thus if $b \geq a$, then $F(b) \geq F(a)$, and hence $\int_a^b f(x) \, dx = F(b) - F(a) \geq 0$.

The next property, which is a direct consequence of Property 3, is useful in applications.

PROPERTY 4.   *Let f and g be Newton integrable on $[a, b]$, and assume that for every $x \in [a, b]$, $f(x) \leq g(x)$. Then,*

$$\int_a^b f(x) \, dx \leq \int_a^b g(x) \, dx.$$

Proof of Property 4.   We need only realize that the function $g(x) - f(x) \geq 0$ on $[a, b]$ and apply Property 3 to obtain the desired result.

Now, let us investigate how substitutions can be performed to evaluate Newton integrals. Consider the following example.

Example 6.   Find

$$\int_0^1 \frac{x\,dx}{\sqrt{1 + x^2}}$$

Solution.   First we must find an antiderivative of $x/\sqrt{1 + x^2}$ which requires evaluating

$$\int \frac{x\,dx}{\sqrt{1 + x^2}}$$

Let $u = 1 + x^2$; then $du = 2x\,dx$ and

$$\int \frac{x\,dx}{\sqrt{1 + x^2}} = \frac{1}{2} \int u^{-1/2}\,du = u^{1/2} + C = \sqrt{1 + x^2} + C.$$

Hence, an antiderivative of $x/\sqrt{1 + x^2}$ is $\sqrt{1 + x^2}$. Thus,

$$\int_0^1 \frac{x\,dx}{\sqrt{1 + x^2}} = \sqrt{1 + x^2}\,\Big|_0^1 = \sqrt{2} - 1.$$

Now we can actually evaluate this definite integral without first finding an antiderivative of $x/\sqrt{1 + x^2}$ as follows: Let $u(x) = 1 + x^2$. Then when $x = 0$, $u(0) = 1$ and when $x = 1$, $u(1) = 2$. We assert that we can then write

$$\int_0^1 \frac{x\,dx}{\sqrt{1 + x^2}} = \frac{1}{2} \int_{u(0)}^{u(1)} u^{-1/2}\,du = \frac{1}{2} \int_1^2 u^{-1/2}\,du = u^{1/2}\,\Big|_1^2 = \sqrt{2} - 1.$$

This method avoids resubstituting $1 + x^2$ for u. The essential idea is that when we substitute u for $1 + x^2$, we also change the limits of integration. We will see that this can be done for most integrals. The result is stated as follows:

THEOREM 1.   *Let $u(x)$ be a differentiable function on $[a, b]$ and assume that for $x_0 \in [a, b]$ $c \leq u(x_0) \leq d$. If f is Newton integrable on $[c, d]$, then*

$$\int_a^b f[u(x)]u'(x)\,dx = \int_{u(a)}^{u(b)} f(u)\,du.$$

Proof.   The verification of the above theorem is straightforward. Let F be an antiderivative of f on $[u(a), u(b)]$. Then

$$\frac{d}{dx}\{F[u(x)]\} = F'[u(x)]u'(x) = f[u(x)]u'(x)$$

for all $x \in [a, b]$. Therefore, $F[u(x)]$ is an antiderivative of $f[u(x)]u'(x)$ and hence

$$\int_a^b f[u(x)]u'(x)\,dx = F[u(x)]\,\Big|_a^b = F[u(b)] - F[u(a)]$$

$$= \int_{u(a)}^{u(b)} f(u)\,du.$$

We again repeat that this theorem enables us to evaluate many Newton integrals without explicitly producing antiderivatives. Let us look at more examples.

Example 7.   Find

$$\int_0^1 \frac{e^x}{1+e^x}\, dx.$$

Solution.   Let $u = 1 + e^x$, then $u(0) = 2$ and $u(1) = 1 + e$; $du = e^x\, dx$. Hence,

$$\int_0^1 \frac{e^x\, dx}{1+e^x} = \int_{u(0)=2}^{u(1)=1+e} \frac{du}{u} = \log u \Big|_2^{1+e}$$

$$= \log(1 + e) - \log 2 = \log \frac{1+e}{2}.$$

In this example, $f[u(x)] = 1/u(x) = 1/(1 + e^x)$ and $u'(x) = e^x$. However, it is more convenient to use differentials than to go through this substitution each time.

Example 8.   Find $\int_0^2 (x + 1)(x^2 + 2x + 1)\, dx$.

Solution.   Let $u = x^2 + 2x + 1$; then $du = 2(x + 1)\, dx$. $u(0) = 1$ and $u(2) = 9$. Therefore,

$$\int_0^2 (x + 1)(x^2 + 2x + 1)\, dx = \frac{1}{2}\int_{u(0)=1}^{u(2)=9} u\, du = \frac{u^2}{4}\Big|_1^9 = \frac{81}{4} - \frac{1}{4} = 20.$$

Note that we could integrate $(x + 1)(x^2 + 2x + 1)$ by simply multiplying it out.

Until now, we have always evaluated $\int_a^b f(x)\, dx$ for $b > a$. Suppose that $b < a$. In this case we define $\int_a^b f(x)\, dx = F(b) - F(a)$, where F is any antiderivative of f on $[b, a]$. (Note that in this case we find any antiderivative on $[b, a]$ rather than on $[a, b]$.) Observe that since $a > b$, $\int_b^a f(x)\, dx = F(a) - F(b)$, where F is any antiderivative of f on $[b, a]$. Hence, $\int_a^b f(x)\, dx = -\int_b^a f(x)\, dx$. It is also true that $\int_a^a f(x)\, dx = 0$ since

$$\int_a^a f(x)\, dx = F(a) - F(a) = 0.$$

We end with another illustration of how the various properties of the Newton integral can be used to evaluate them.

Example 9.*   Find $\int_{-1}^1 x^2\, dx$.

Solution.   First we write $\int_{-1}^1 x^2\, dx = \int_{-1}^0 x^2\, dx + \int_0^1 x^2\, dx$. Now in the first integral let $u(x) = -x$. Then $du = -dx$, $u(-1) = 1$, $u(0) = 0$, and $\int_{-1}^0 x^2\, dx = -\int_1^0 u^2\, du = \int_0^1 u^2\, du$. Since the variable u is really a "dummy;" that is, $\int_0^1 u^2\, du = \int_0^1 x^2\, dx$ (the integral *only* depends on the function in the integral and the limits; thus the variable appearing in the integrand is incidental)—we see that

$$\int_{-1}^1 x^2\, dx = 2\int_0^1 x^2\, dx = 2 \cdot \frac{x^3}{3}\Big|_0^1 = \frac{2}{3}.$$

Example 9 has a useful generalization. Note that $f(x) = x^2$ is an even function—that is, $f(-x) = f(x)$. We claim that if f is an *even function*, then $\int_{-a}^{a} f(x)\,dx = 2\int_{0}^{a} f(x)\,dx$. This follows in exactly the same way as before. First write

$$\int_{a-}^{a} f(x)\,dx = \int_{-a}^{0} (f(x)\,dx + \int_{0}^{a} f(x)\,dx.$$

In the first integral let $u = -x$, $du = -dx$, $u(-a) = a$, and $u(0) = 0$. Thus,

$$\int_{a-}^{0} f(x)\,dx = -\int_{a}^{0} f(-u)\,du = \int_{0}^{a} f(-u)\,du = \int_{0}^{a} f(u)\,du = \int_{0}^{a} f(x)\,dx.$$

Hence, $\int_{-a}^{a} f(x)\,dx = 2\int_{0}^{a} f(x)\,dx$.

In exactly the same way, we can show that if f is an *odd function*, that is, $f(-u) = -f(u)$, then $\int_{-a}^{a}(f(x)\,dx = 0$. (See Exercise 4(b).)

Observe that the Newton integral is only *defined* for functions that have antiderivatives on closed intervals. It is quite useful to extend this concept of integration to functions that may not have antiderivatives. The example given in Section 5.1, namely,

$$f(x) = \begin{cases} -1, & x \in [-1, 0) \\ 0, & x = 0 \\ 1 & x \in (0, 1] \end{cases},$$

is such a function. This leads us to the Riemann integral, which we discuss in an appendix to this chapter. There are other functions that are Newton integrable but *not* Riemann integrable. We discuss this at greater length in the appendix.

An important theoretical question we have not settled in this chapter is: Does every continuous function have an antiderivative? The answer is yes, but we omit the proof. A proof can be found in Chapter 8 of J. W. Kitchen's text, *Calculus of One Variable*, published by Addison-Wesley, Reading, Mass., 1968.

Exercises

1. Evaluate the following definite integrals.

(a) $\displaystyle\int_{-1}^{3} x^2\,dx$

(e) $\displaystyle\int_{-2}^{3} e^{-1/2x}\,dx$

(b) $\displaystyle\int_{-1}^{1} (2x^2 - x^3)\,dx$

(f) $\displaystyle\int_{1}^{8} (1 + x^{1/3})\,dx$

(c) $\displaystyle\int_{1}^{4} \frac{dx}{2\sqrt{x}}$

(g) $\displaystyle\int_{1}^{2} \left(\frac{1}{x} + \frac{1}{x^2}\right) dx$

(d) $\displaystyle\int_{-3}^{-1} \left(\frac{1}{x^2} - \frac{1}{x^3}\right) dx$

(h) $\displaystyle\int_{0}^{3} (x^{1/2} + x^{3/2})\,dx$

2. Evaluate the following definite integrals.

(a) $\displaystyle\int_{-1}^{1} |x(x-1)|\,dx$

(c) $\displaystyle\int_{-1}^{2} |x^2(x-1)|\,dx$

(b) $\displaystyle\int_{-1}^{1} |x^2(x-1)|\,dx$

(d) $\displaystyle\int_{-1}^{2} \frac{dx}{|2x-5|}$

3. Evaluate the following.

(a) $\displaystyle\int_0^1 x^2(x^3 + 2)^2 \, dx$

(f) $\displaystyle\int_{-1}^0 e^x(e^x + 1)^3 \, dx$

(b) $\displaystyle\int_0^1 \frac{8x^2}{(x^3 + 2)^2} \, dx$

(g) $\displaystyle\int_1^2 \frac{(1 + x)^2}{\sqrt{x}} \, dx$

(c) $\displaystyle\int_0^1 \frac{x^2 \, dx}{\sqrt[4]{x^3 + 2}}$

(h) $\displaystyle\int_{50}^{100} e^{\log x} \frac{1}{x} \, dx$

(d) $\displaystyle\int_{-1}^2 \sqrt[3]{(1 - x^2)} x \, dx$

(i) $\displaystyle\int_{-4}^{-3} x(x + 6)^{1/2} \, dx$

(e) $\int_0^{1/\sqrt{2}} \sqrt{x^2 - 2x^4} \, dx$

(j) $\displaystyle\int_1^2 \frac{x}{x^2 + 1} \log(x^2 + 1) \, dx$

4. (a) Using a method similar to Example 9 of this section, show that

$$\int_{-1}^1 x^3 \, dx = 0.$$

(b) Show that if f is Newton integrable on $[-a, a]$ and odd on that interval [that is, $f(-x) = -f(x)$], then

$$\int_{-a}^a f(x) \, dx = 0.$$

5. Is the following function Newton integrable on $[-1, 1]$? $f(x) = |x|$. If your answer is yes, evaluate $\int_{-1}^1 f(x) \, dx$.
    [HINT: Solve the problem on $[-1, 0]$ and then $[0, 1]$ and see if you can match these at $x = 0$.]

6. Is the function $f(x) = 1/x^2$ Newton integrable on $[-1, 1]$? If not, why not? If so, evaluate $\int_{-1}^1 1/x^2 \, dx$. (Be careful!)

7*. $\int_0^t f(x) \, dx = t^2$. Find $f(x)$ when we assume f to be Newton integrable. [HINT: $\int_0^t f(x) \, dx = F(t) - F(0)$ when $F'(x) = f(x)$.]

5.5 Integration by Parts

In Section 5.2 we remarked that there was no rule analogous to the rule for the differentiation of the product of two functions for antiderivatives, and hence Newton integrals. The integration by parts formula comes closest to filling this gap. It also allows us to integrate a wide variety of rather complicated looking integrals. In Chapter 7, Section 7.2, we shall show how one could use tables of integrals for other problems.

This formula is based upon the rule for differentiating products,

$$\frac{d}{dx} [f(x)g(x)] = f(x) \frac{dg(x)}{dx} + g(x) \frac{df(x)}{dx}.$$

We take antiderivatives of both sides, yielding (apart from the arbitrary constant),

$$\int \frac{d}{dx} [f(x)g(x)] \, dx = \int f(x) \frac{dg(x)}{dx} \, dx + \int g(x) \frac{df(x)}{dx} \, dx$$

and obtain

$$f(x)g(x) = \int f(x) \frac{dg(x)}{dx} \, dx + \int g(x) \frac{df(x)}{dx} \, dx$$

or equivalently,

$$\int f(x) \frac{dg(x)}{dx} \, dx = f(x)g(x) - \int g(x) \frac{df(x)}{dx} \, dx.$$

If we let $u = f(x)$, $v = g(x)$, $du = f'(x) \, dx$, and $dv = g'(x) \, dx$, then simply rewriting the last equation yields

$$\int u \, dv = uv - \int v \, du.$$

It is a good idea to commit this formula to memory. We observe that for definite integrals, the formula reads

$$\int_a^b u \, dv = u(x)v(x) \Big|_a^b - \int_a^b v \, du.$$

When using this technique, one tries to choose u and v in such a way that the integral appearing on the right is easier to evaluate than that on the left. It takes a good deal of practice to learn to choose the u and v properly.

Let us look at some examples.

Example 1. Find $\int xe^x \, dx$.

Solution. In this case we choose $u = x$ and $dv = e^x \, dx$. Then, $du = dx$ and $v = e^x$. (To find v, of course, we must also find an antiderivative.) Thus,

$$\int xe^x \, dx = \int u \, dv = u(x)v(x) - \int v \, du = xe^x - \int e^x \, dx = xe^x - e^x + C = (x - 1)e^x + C.$$

Example 2. Find $\int \log x \, dx$.

Solution. In this case we are not left much choice. Choose $u = \log x$, and $dv = dx$. Obviously $v = x$ and $du = (1/x) \, dx$. Thus,

$$\int \log x \, dx = \int u \, dv = uv - \int v \, du$$

$$= x \log x - \int x \cdot \frac{1}{x} \, dx = x \log x - \int dx = x \log x - x + C.$$

Note that we really use integration by parts to evaluate integrals of form $\int f(x)g(x) \, dx$.

We choose $u = f(x)$ and $dv = g(x)\,dx$ if (i) $\int v\,du$ is easier to compute than $\int u\,dv$, and (ii) it is a simple matter to find an antiderivative of $g(x)$. If this is not the case, then we choose $u = g(x)$ and $dv = f(x)\,dx$.

The next examples show that we must be careful how we choose u and v.

Example 3.   Let us attempt to evaluate $\int x^{-1}\,dx$ by parts.

Solution.   We would probably set $u = x^{-1}$, $du = -x^2\,dx$, $dv = dx$, and $v = x$. This would yield

$$\int x^{-1}\,dx = 1 + \int x^{-2}x\,dx = 1 + \int x^{-1}\,dx.$$

Obviously this is false! (Otherwise $0 = 1$). It is clear in this case that an *arbitrary* constant *must* be included—that is, $\int x^{-1}\,dx = 1 + \int x^{-1}\,dx + C$—which implies $C = -1$. (For integration by parts with a definite integral this sort of thing can not happen; see Exercise 7.) However, we were *not* able to evaluate $\int x^{-1}\,dx$ in this manner. We already know that $\int x^{-1}\,dx = \log|x| + C$. The use of integration by parts was not relevant.

Example 4.   Evaluate $\int x^2 e^x\,dx$.

Solution.   Suppose we begin by letting $u = e^x$, $du = e^x\,dx$, $dv = x^2\,dx$, and $v = x^3/3$. Then,

$$\int x^2 e^x\,dx = u(x)v(x) - \int v\,du = \frac{x^3 e^x}{3} - \frac{1}{3}\int x^3 e^x\,dx.$$

Clearly we are worse off than when we started—$\int x^3 e^x\,dx$ is more difficult to calculate than $\int x^2 e^x\,dx$. Now try $u = x^2$, $dv = e^x\,dx$; then $du = 2x\,dx$ and $v = e^x$. Thus $\int x^2 e^x\,dx = x^2 e^x - 2\int x e^x\,dx$. By Example 1 (that is, another application of integration by parts on $\int x e^x\,dx$); $\int x e^x\,dx = (x - 1)e^x + C$. Thus,

$$\int x^2 e^x\,dx = x^2 e^x - 2(x - 1)e^x + C.$$

This shows that one must use some insight to choose u and dv appropriately.

Example 5.   Evaluate $\int_0^1 x(1 + x)^{1/2}\,dx$.

Solution.   It is best in this example if we choose $u = x$ and $dv = (1 + x)^{1/2}\,dx$. (Otherwise, $\int v\,du$ becomes extremely complicated—try it!) In this case, $du = dx$ and $v = \int (1 + x)^{1/2}\,dx = \frac{2}{3}(x + 1)^{3/2}$. Then

$$\int_0^1 x(1 + x)^{1/2}\,dx = u(x)v(x)\Big|_0^1 - \int_0^1 v\,du$$

$$= \frac{2}{3}x(1 + x)^{3/2}\Big|_0^1 - \frac{2}{3}\int_0^1 (x + 1)^{3/2}\,dx.$$

We must evaluate $\int_0^1 (x + 1)^{3/2} dx$. Let us make the substitution $w = x + 1$; then $w(0) = 1$, $w(1) = 2$, and $dw = dx$. Thus,

$$\int_0^1 (x + 1)^{3/2} dx = \int_1^2 w^{3/2} dw = \frac{2}{5} w^{5/2} \Big|_1^2 = \frac{2}{5} \cdot 2^{5/2} - \frac{2}{5},$$

and

$$\int_0^1 x(1 + x)^{1/2} dx = \frac{2}{3} \cdot 2^{3/2} - \frac{4}{15} \cdot 2^{5/2} + \frac{4}{15}$$

$$= \frac{4\sqrt{2}}{3} - \frac{16}{15}\sqrt{2} + \frac{4}{15} = \frac{4(\sqrt{2} + 1)}{15}.$$

Exercises

1. Find, using integration by parts, the following antiderivatives.

   (a) $\int x \log x \, dx$ (d) $\int x^2 \log x \, dx$

   (b) $\int x^2 \sqrt{1 + x} \, dx$ (e) $\int x^{1/2} \log x \, dx$

   (c) $\int x^3 e^x \, dx$ (f) $\int \log x \, dx$

2. Find the following Newton integrals, using integration by parts.

   (a) $\int_0^1 xe^{-x} \, dx$ (c) $\int_0^4 x^2\sqrt{4 - x} \, dx$

   (b) $\int_1^e x \log x \, dx$ (d) $\int_{-1}^0 x^2\sqrt{x^2 + 1} \, dx$

   [HINT: Let $u = x$, $dv = x(x^2 + 1)^{1/2} \, dx$.]

3. Show that the integrals $\int_a^b f(x)g'(x) \, dx$ and $\int_a^b f'(x)g(x) \, dx$ have as their sum $f(b)g(b) - f(a)g(a)$. [HINT: Integrate one of them by parts.]

4. Use integration by parts to show that

$$\int \sqrt{1 - x^2} \, dx = x\sqrt{1 - x^2} + \int \frac{x^2 \, dx}{\sqrt{1 - x^2}}.$$

Write $x^2 = (x^2 - 1) + 1$ in the second integral and deduce the formula

$$\int \sqrt{1 - x^2} \, dx = \frac{1}{2} x\sqrt{1 - x^2} + \frac{1}{2} \int \frac{dx}{\sqrt{1 - x^2}}.$$

5. Show that the formula

$$\int_a^b xg'(x) \, dx = xg(x) \Big|_a^b - \int_a^b g(x) \, dx$$

is a special case of the integration by parts formula for definite integrals.

6.   Evaluate $\int_0^1 x(1+x)^{1/2}\, dx$ by making the substitution $w = x + 1$ initially.

7.   Evaluate $\int_a^b x^{-1}\, dx$ as in Example 3 by integration by parts. (Assume $a > 0$, $b > 0$.) Note that the right- and left-hand sides are equal. Why?

5.6 Area and the Integral

Most of you probably know how to find the areas of rectangles, triangles, and certain other polygons from geometry. Thus far you have not had the techniques for determining the area of a region bounded by an arbitrary curve. In this section, we carefully investigate this problem.

Let f be a continuous function on $[a, b]$ and suppose that $f(x) \geq 0$ on $[a, b]$. (See Figure 5.6.1.) Consider the following questions.

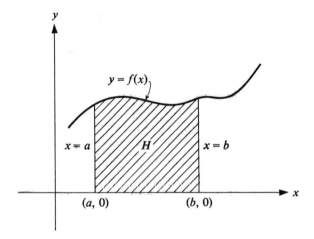

FIGURE 5.6.1

(1) What is the area of the shaded region H? (That is, the area bounded above by the curve $y = f(x)$ and to the left by $x = a$ and to the right by $x = b$.)

(2) Does there exist a differentiable function F on $[a, b]$ such that $F'(x) = f(x)$?

Off hand the two questions do not seem to be related. The surprising thing is that they are. So the reader may share our joy and astonishment at this turn of events, we state the answers in rough form now. The answers to (1) and (2) are:

(1) Area $(H) = \int_a^b f(x)\, dx$.

(2) Let $A(x_0) = $ area under the curve bounded above by $y = f(x)$, to the left by $x = a$, and to the right by $x = x_0$. Then, $A'(x_0) = f(x_0)$. (See Figure 5.6.2.)

We now turn to the problem of determining the area of a region bounded above by a curve $y = f(x)$, to the left by $x = a$, and to the right by $x = b$. We assume first that f is continuous on $[a, b]$ and $f(x) \geq 0$ for $x \in [a, b]$. (See Figure 5.6.1.) Let us call $H = \{(x, y) \mid 0 \leq y \leq f(x)$ and $a \leq x \leq b\}$. Our problem is to determine the area of H, which we denote by Area (H). Observe that the area is really a function that assigns to each set

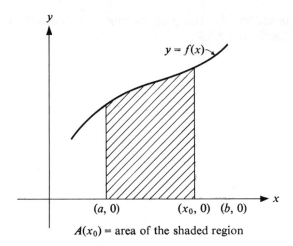

$A(x_0)$ = area of the shaded region

FIGURE 5.6.2

H a nonnegative number, which we call Area (H). (We assume that there is a well-defined quantity called area.) H is often referred to as an *ordinate set* for f.

Before proceeding further, let us list three of the fundamental properties that we expect area to have. (These are intuitively plausible from geometrical considerations.)

PROPERTY 1.   *If H and K are two ordinate sets with $H \subset K$, then Area $(H) \leq Area (K)$.* (See Figure 5.6.3.)

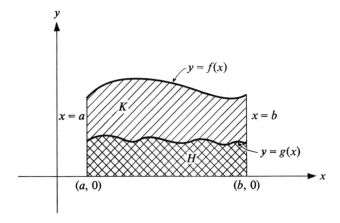

FIGURE 5.6.3

Suppose on $[a, b]$, $g(x) \leq f(x)$ (as in Figure 5.6.3); then $H = \{(x, y) \mid 0 \leq y \leq g(x)$ and $a \leq x \leq b\}$ and $K = \{(x, y) \mid 0 \leq y \leq f(x)$ and $a \leq x \leq b\}$. Clearly $H \subset K$. Property 1 simply states that Area$(H) \leq$ Area(K).

PROPERTY 2.   *If S is an ordinate set and the line $x = c$ divides S into two ordinate sets H and K, then Area$(S) = Area(H) + Area(K)$.*

(Note that $S = H$ joined with K. This property simply states that the whole is equal to the sum of its parts! See Figure 5.6.4.)

In Figure 5.6.4,

$$S = \{(x, y) \mid 0 \leq y \leq f(x), a \leq x \leq b\},$$

$$H = \{(x, y) \mid 0 \leq y \leq f(x), a \leq x \leq c\},$$

and

$$K = \{(x, y) \mid 0 \leq y \leq f(x), c \leq x \leq b\}.$$

Property 2 simply states that Area(S) = Area(H) + Area(K).

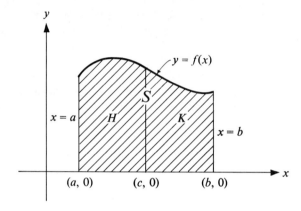

FIGURE 5.6.4

PROPERTY 3.   *If S is the set* $\{(x, y) \mid a \leq x \leq b$ *and* $c \leq y \leq d\}$ *(that is, a rectangle), then* *Area*(S) = $(d - c)(b - a)$. (That is, the area of a rectangle is its length times its width— see Figure 5.6.5.)

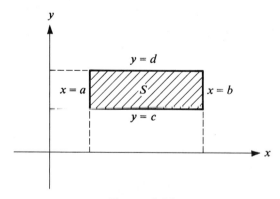

FIGURE 5.6.5

First, observe that $\int_a^b f(x)\,dx$ possesses Properties 1, 2, and 3, since:

(1) If $f(x) \le g(x)$, then by Property 4, Section 5.3,

$$\int_a^b f(x)\,dx \le \int_a^b g(x)\,dx.$$

(2) By Property 1, Section 5.3,

$$\int_a^b f(x)\,dx = \int_a^c f(x)\,dx + \int_c^b f(x)\,dx.$$

(3) If $f(x) = c$, where c is a constant then

$$\int_a^b c\,dx = c(b - a).$$

In like manner, $\int_a^b d\,dx = d(b - a)$ and thus

$$\int_a^b d\,dx - \int_a^b c\,dx = (d - c)(b - a).$$

Before starting the proofs, let us clarify our notation and hypothesis. We designate the Area(K) by $A(x_0)$; K is the set of points bounded above by $y = f(x)$, to the left by $x = a$ and to the right by $x = x_0$. (See Figure 5.6.6.)

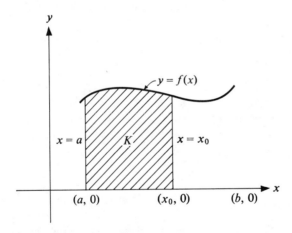

FIGURE 5.6.6

We assume the following:

(a)   There exists a well-defined quantity called the area under the curve f from a to x_0, for $a \le x_0 \le b$. (We do not give a definition of area, but assume that such a definition exists.)

(b)   The area has the properties stated above: Properties 1, 2, and 3.

In more advanced courses, it is shown that there is a definition of area with the stated properties.

If you think about it for a moment, it is clear that $A(x)$ is a function. What is more, it is continuous. In fact, we will show that it has a derivative.

We now investigate the derivative of $A(x)$. Let $x_0 \geq 0$ and x be any number " near " x_0 (see Figure 5.6.7) such that $x, x_0 \in (a, b)$ with $x > x_0$. We see that $A(x) - A(x_0)$ is the area of the shaded region in the above figure. Now, since f is continuous on $[a, b]$ and hence on $[x_0, x]$, we know that f assumes a local maximum and a local minimum value

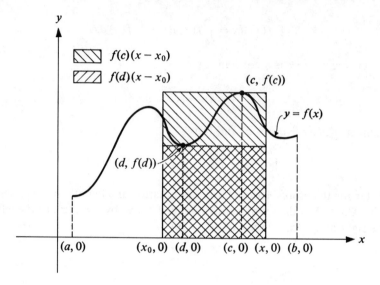

FIGURE 5.6.7

on this interval at points $(c, f(c))$ and $(d, f(d))$, respectively. (This was discussed in Section 4.2.) It is clear from Figure 5.6.7 that $f(d)(x - x_0) \leq A(x) - A(x_0) \leq f(c)(x - x_0)$, where $f(d)(x - x_0)$ is the area of the rectangle whose height is $f(d)$ and width $x - x_0$ and $f(c)(x - x_0)$ is the area of a rectangle whose height is $f(c)$ and width is $x - x_0$. Thus,

$$f(d) \leq \frac{A(x) - A(x_0)}{x - x_0} \leq f(c).$$

Since f is continuous,

$$\lim_{x \to x_0^+} f(d) = f(x_0) \qquad \text{and} \qquad \lim_{x \to x_0^+} f(c) = f(x_0).$$

Hence, by the squeezing principle of limits,

$$\lim_{x \to x_0^+} \frac{A(x) - A(x_0)}{x - x_0} = f(x_0).$$

By selecting $x < x_0$, we can easily show that

$$\lim_{x \to x_0^-} \frac{A(x) - A(x_0)}{x - x_0} = f(x_0)$$

and hence for each $x_0 \in (a, b)$,

$$A'(x_0) = \lim_{x \to x_0} \frac{A(x) - A(x_0)}{x - x_0} = f(x_0).$$

We have shown then that $A(x)$ is a differentiable function on $[a, b]$ and $A'(x) = f(x)$ for $x \in (a, b)$. This answers Question (2). Moreover, $\int_a^b f(x)\, dx = A(x) \mid_a^b = A(b) - A(a) = A(b)$, since $A(a) = 0$ by definition of the area. Since $A(b) = \text{Area}(H)$ we have proved that $\text{Area}(H) = \int_a^b f(x)\, dx$. [$H$ refers to the region in Fig. 5.6.1.]

Note that we answered Questions (1) and (2) simultaneously. We showed that the area function $A(x)$ satisfied the condition $A'(x) = f(x)$. Then we used this fact to show that Area $(H) = A(b) = \int_a^b f(x)\, dx$. The result we have just obtained are two forms of the *fundamental theorem of calculus*, and its importance, both practical and theoretical, cannot be stressed too greatly. For that reason we repeat them here in the form of two theorems.

THEOREM 1. *Let $f(x)$ be continuous on $[a, b]$ and suppose that $f(x) \geq 0$ on $[a, b]$. If Area(H) denotes the area of the set of points bounded above by the curve $y = f(x)$ and on the sides by $x = a$ and $x = b$, then $\int_a^b f(x)\, dx$ exists (the Newton integral of f exists) and Area $(H) = \int_a^b f(x)\, dx$.*

THEOREM 2. *Let $f(x)$ be continuous on $[a, b]$. Set $F(t) = \int_a^t f(x)\, dx$ for every t in $[a, b]$. Then F is differentiable on $[a, b]$, and $F'(t_0) = f(t_0)$, where $t_0 \in [a, b]$.*

One further comment is in order. We did not define the area $A(b)$ of the region bounded by the curve $y = f(x)$, $x = a$, and $x = b$, but assumed there was a definition of area that satisfied Properties 1, 2, and 3. We then proved that with this definition of area, $A(b) = \text{Area}(H) = \int_a^b f(x)\, dx$. In other words, we proved that no matter what definition you choose for area, you always end up with the same number, namely $\int_a^b f(x)\, dx$ for Area(H) if your area has Properties 1, 2, and 3. This result is quite remarkable. It can be interpreted as saying there is really only one way to define area, and all definitions lead to the same answer.

Let us now consider some examples of the above discussion.

Example 1. Find the area under the curve $y = x^2$ from 0 to 1.

Solution. First we graph the function. (See Figure 5.6.8.) We see that $f(x) = x^2$ is greater than or equal to zero of the interval $[0, 1]$. Hence,

$$\text{Area} = \int_0^1 x^2\, dx = \frac{x^3}{3} \Big|_0^1 = \frac{1}{3}.$$

We have thus found the area under part of a parabola. This problem was solved by Archimedes, a military consultant, about 250 B.C.

Example 2. Find the area bounded by the curve $y = 2x - x^2$ and the x axis.

Solution. First, graph the function $2x - x^2$. Note that $f'(x) = 2(1 - x)$ and $f''(x) = -2$. Thus, $f'(x) = 0$ for $x = 1$ and since $f''(1) < 0$, f has a maximum at $x = 1$, $y = 1$.

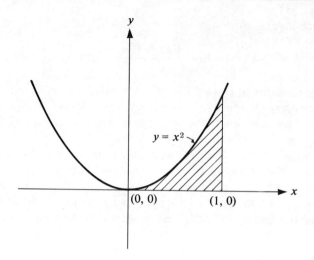

FIGURE 5.6.8

The graph is increasing for $x < 1$ and decreasing for $x > 1$, and crosses the x axis at $x = 0$ and $x = 2$ only. Thus, the area we want is that of the shaded region in Figure 5.6.9. The area is given by

$$\int_0^2 (2x - x^2)\, dx.$$

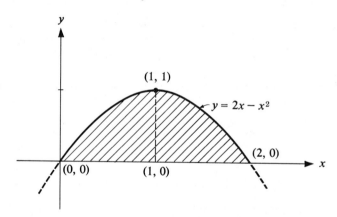

FIGURE 5.6.9

Now,

$$\int_0^2 (2x - x^2)\, dx = x^2 - \frac{x^3}{3}\Big|_0^2 - 4 - \frac{8}{3} = \frac{4}{3}.$$

In solving *any* area problem, it is first necessary to sketch the region in question. This is an aid in setting up the proper limits of integration.

Thus far we have restricted ourselves only to functions f such that $f(x) \geq 0$ on $[a, b]$. We now show how to extend our discussions to functions which are negative over part or all of $[a, b]$.

In Figure 5.6.10(a), we wish to find Area(B), and in Figure 5.6.10(b), Area(B plus C) = Area(B) + Area(C). In the first case (Figure 5.6.10(a)), we take Area(B) = $-\int_a^b f(x)\,dx$, while in the second case, Area(B plus C) = $\int_a^c f(x)\,dx - \int_c^b f(x)\,dx$. (Note that this will

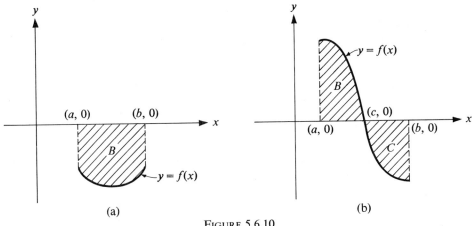

(a)   (b)

FIGURE 5.6.10

always guarantee that the area is a positive number.) We proceed to illustrate this with some examples.

Example 3.   Find the area bounded by $y = x^2 - 1$ and $x = 0$ and $x = 1$.

Solution.   The graph of $x^2 - 1$, $x \in [0, 1]$ is shown in Figure 5.6.11.

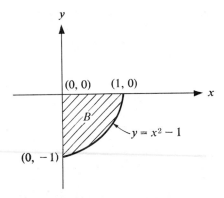

FIGURE 5.6.11

Hence,

$$\text{Area}(B) = -\int_0^1 (x^2 - 1)\,dx = -\left(\frac{x^3}{3} - x\right)\Big|_0^1 = -\left(\frac{1}{3} - 1\right) = \frac{2}{3}.$$

Example 4.   Find the area bounded by the curve $y = x^3$ and the lines $x = -1$ and $x = 1$.

Solution.   The graph of $y = x^3$, $x \in (-1, 1)$ is shown in Figure 5.6.12.

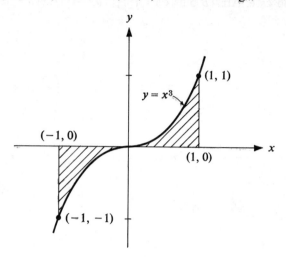

FIGURE 5.6.12

The area in question is the area of the shaded region. The desired area is given by

$$- \int_{-1}^{0} x^3 \, dx + \int_{0}^{1} x^3 \, dx = - \frac{x^4}{4} \Big|_{-1}^{0} + \frac{x^4}{4} \Big|_{0}^{1} = \frac{1}{4} + \frac{1}{4} = \frac{1}{2}.$$

Recall from Exercise 4, Section 5.4 that $\int_{-1}^{1} x^3 \, dx = 0$. But the area under the curve $y = x^3$ between -1 and $+1$ is obviously not 0. It is the introduction of the minus sign for functions lying below the x axis that prevents us from obtaining 0 as the area.

Next we consider the problem of determining the area of a region bounded by the graphs of two functions f and g, both continuous on $[a, b]$ and the vertical lines $x = a$ and $x = b$. Suppose, as in Figure 5.6.13, $f(x) \geq g(x)$ for $x \in [a, b]$. We wish to determine

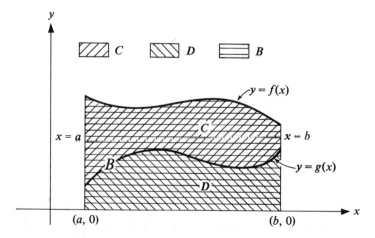

FIGURE 5.6.13

the area of the shaded region *C*. Now it should be clear to you that if we denote the ordinate set of *f* by *B* and the ordinate set of *g* by *D*, then

$$\text{Area}(C) = \text{Area}(B) - \text{Area}(D) = \int_a^b f(x)\,dx - \int_a^b g(x)\,dx$$

$$= \int_a^b [f(x) - g(x)]\,dx.$$

The above result holds *even* if *f* and *g* both take on negative values on [*a*, *b*].

Example 5.   Find the area bounded by the curves $y = x(x - 2)$ and $y = x/2$ over the interval [0, 2].

Solution.   First we draw the graphs of the two functions. We must determine the area of the region *B* in Figure 5.6.14. Clearly on [0, 2], $x/2 > x(x - 2)$ (look at the graphs).

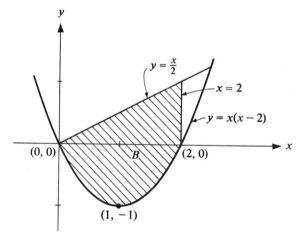

FIGURE 5.6.14

Hence,

$$\text{Area}(B) = \int_0^2 \left[\frac{x}{2} - x(x - 2) \right] dx = \int_0^2 \left[\frac{5x}{2} - x^2 \right] dx$$

$$= \frac{5}{4}x^2 - \frac{x^3}{3} \Big|_0^2 = 5 - \frac{8}{3} = \frac{7}{3}.$$

Example 6.   Find the area bounded by the curves $y = x$ and $y = x^3$.

Solution.   First we must draw the graphs and find their points of intersection. Obviously the two curves intersect at points for which $x = x^3$. This occurs when $x = 0$, $x = -1$, and $x = 1$. Their graphs are shown in Figure 5.6.15. We wish to find

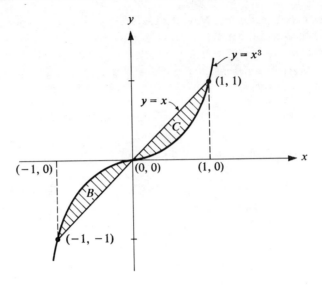

FIGURE 5.6.15

Area(B plus C) = Area(B) + Area(C). Note that for $x \in [0, 1]$, $x \geq x^3$. Thus,

$$\text{Area}(B \text{ plus } C) = \int_{-1}^{0} (x^3 - x)\, dx + \int_{0}^{1} (x - x^3)\, dx$$

$$- \left(\frac{x^4}{4} - \frac{x^2}{2}\right)\Big|_{-1}^{0} + \left(\frac{x^2}{2} - \frac{x^4}{4}\right)\Big|_{0}^{1} = -\frac{1}{4} + \frac{1}{2} + \frac{1}{2} - \frac{1}{4} = \frac{1}{2}.$$

Exercises

In Exercises 1–10, first sketch the graphs of the functions and indicate the area of the region you are calculating by shading.

1. Find the area of the region lying above the x axis and under the parabola $y = 4x - x^2$.
2. Find the area of the region bounded by $y = 1/x$, $x = 1$, and $x = 3$.
3. Find the area of the region bounded by $y = e^x$, $x = 0$, and $x = 1$.
4. Find the area of the region bounded by $y = -x^2 - 2x + 15$ and the x axis between $x = 0$ and $x = 5$.
5. Find the area of the region bounded by the parabolas $y = 6x - x^2$ and $y = x^2 - 2x$.
6. Find the area of the region bounded by the curve $y = e^x$ and the lines $y = 2x + 1$ and $x = -1/2$.
7. Find the area of the region bounded by the curve $y = e^{-x}$ and the lines $y = -3x + 1$, $x = \frac{1}{3}$. *Note that* $e^0 = 1$.
8. (a) Find the area of the region bounded by the curves $y = x$ and $y = x^3/4$ over the interval $[-1, 2]$.
    (b) Find the area of the region bounded by the curves $y = x$ and $y = x^3/4$.
9. Find the area of the region inside the closed curve $y^2 = x^2 - x^4$.

10. Find the area of the region bounded by $y = |x|$ and $y = x^2 - 1$ over the interval $[-1, 1]$.

11. Let $F(x) = x - x^2$ and $g(x) = bx$. Determine b so that the region above the graph of g and below the graph of f has area $\frac{9}{2}$.

12. Show that the area of the region which lies between the parabolas $y^2 = 2ax$ and $x^2 = 2by(a, b > 0)$ is $4(ab/3)$.

13. The graph of $y = \sqrt{a^2 - x^2}$ over $-a \leq x \leq a$ is a semicircle of radius a.

   (a) Using this fact, explain why it is true that

   $$\int_{-a}^{a} \sqrt{a^2 - x^2}\, dx = \frac{1}{2}\pi a^2.$$

   (b) Evaluate

   $$\int_{0}^{a} \sqrt{a^2 - x^2}\, dx.$$

14. The area bounded by the curve $y = x^2$ and the line $y = 4$ is divided into two equal portions by the line $y = c$. Find c.

15. (a) If $F(x) = \int_0^x \sqrt{1 - t^2}\, dt$, $0 < x < 1$, what is $F'(x)$? (Use Theorem 2 of this section.)

   (b) Suppose $F(x) = \int_0^{x^2} \sqrt{1 - t^2}\, dt$, $0 < x < 1$. How could you find $F'(x)$? Find it.

16.* The area bounded by the x axis, the curve $y = f(x)$, and the lines $x = a$, $x = b$ is equal to $\sqrt{b^2 - a^2}$ for all $b > a$. Find $f(x)$.

5.7 Other Applications of Integration

In this section, we discuss some applications of integration to the social and biological sciences.

Example 1. The Indiana University Commons is considering the purchase of an automatic doughnut making machine. The cost of the machine is \$25,000. The management believes that after a short period of installation and adjustment, cost savings will be realized through increased production. The rate of cost savings over 6 years is thought to be $c(x) = 2000x$, where x represents years and $c(x)$ represents dollars per year savings at any given time. Would the machine pay for itself in 4 years? If not, in how many years of operation would the machine pay for itself?

Solution. Since the rate of cost savings is given by $c(x) = 2000x$, we see that the actual savings after n years, $S(n)$, must be given by

$$S(n) = \int_0^n 2000x\, dx.$$

Thus the total savings during the first four years is given by

$$S(4) = \int_0^4 2000x\, dx = 2000 \cdot \frac{x^2}{2}\Big|_0^4 = 16,000.$$

Clearly in four years the machine would not pay for itself since the savings are only $16,000. To determine in how many years the machine would pay for itself, we must find a number n so that

$$\int_0^n 2000x \, dx = \frac{2000x^2}{2} \bigg|_0^n = \frac{2000n^2}{2} = 25{,}000.$$

Thus $n^2 = 25$ or $n = 5$. At the indicated rate of saving, it would take 5 years for this new doughnut machine to pay for itself.

Example 2. (*Consumer and Producer Surplus*)   In economics a demand function for a specified commodity indicates the relationship between the quantity of the commodity demanded and other variables such as the price of the commodity. We consider an ideal situation (which turns out in many cases to be a reasonable assumption!), where the demand and the price are the only variables, assuming that " all other things are equal." Let x be the quantity demanded and p the price for each unit of x (in some monetary unit). The relationship between x and p may be expressed in three ways: (1) the price p may be given directly in terms of x, that is, $p = f(x)$; or (2) x may be given directly in terms of p, that is, $x = g(p)$; or (3) a relationship exists between x and p such as $x^2 + p^2 - 9 = 0$, or in general, a relationship of the form $h(x, p) = 0$. Any of these three cases is called the demand law. We remark at this point that the demand is assumed to be a *monotonically decreasing function* of the price—that is, the lower the price the greater the demand; the higher the price the lower the demand. (Lipstick is a famous example to the contrary.)

A supply function for a specified commodity represents a relation between the quantity offered on the market by the producers and the price of the commodity. Suppose x is the quantity supplied and p the price of one unit of x. Then the relationship between x and p can be expressed as (i) $p = F(x)$, or (ii) $x = G(p)$, or (iii) $H(p, x) = 0$. We assume that the *price* is a monotonically increasing function of the quantity supplied—that is, the higher the price, the greater the amount offered on the market, since more and more producers are attracted to manufacture that item.

Market equilibrium is said to occur under pure competition if the demand is equal to the supply. The equilibrium price and equilibrium supply are defined to be the coordinates of the point of intersection of the supply and demand curves. The equilibrium amount and price can be found algebraically by solving the two equations simultaneously. (See Figure 5.7.1.)

If a demand curve is given and the market demand x_0 and the corresponding price p_0 are determined in some way, then consumers who would have been willing to pay more than the market price p_0 have gained by the setting of the price at p_0 rather than at the maximum price they would have been willing to pay. If a supply curve were given and x_0, p_0 were determined by pure competition, then x_0, p_0 would be the equilibrium supply and price. Under certain economic assumptions the total consumer gain, as a result of this fixed price p_0, called the consumer's surplus, is represented (Figure 5.7.2) by the area above the line $p = p_0$ and below the *demand* curve. In Figure 5.7.2, we consider the case of pure competition and let $p = f(x)$ be the demand law and $p = F(x)$, the supply law. Then, $p = p_0$ is the market equilibrium price.

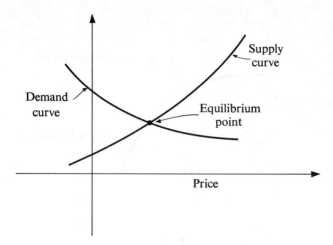

FIGURE 5.7.1

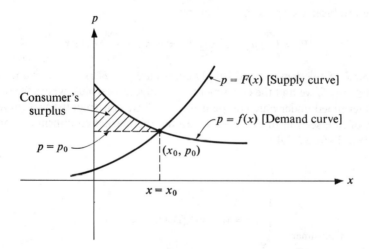

FIGURE 5.7.2

It is clear from Figure 5.7.2 that the consumer's surplus C_s is given by

$$C_s = \int_0^{x_0} [f(x) - p_0]\, dx = \int_0^{x_0} f(x)\, dx - p_0 x_0.$$

If a supply curve is given and if the amount supplied x_0 and the corresponding price p_0 are determined in some way, then producers who would have been willing to supply the commodity below the price p_0 have gained by the setting of the price at p_0. The total producer's gain, called the producer's surplus, is represented by the area below the line $p = p_0$ and above the supply curve. (See Figure 5.7.3.)

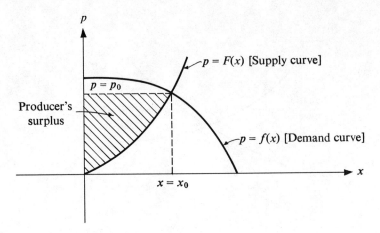

FIGURE 5.7.3

In Figure 5.7.3, x_0 and p_0 represent the market equilibrium amount and price, respectively. The producer's surplus P_s is given by

$$P_s = \int_0^{x_0} [p_0 - F(x)]\, dx = p_0 x_0 - \int_0^{x_0} F(x)\, dx.$$

For example, if the demand and supply curves are $p = F(x) = 14 + x$ and $p = f(x) = (6 - x)^2$, $0 \le x \le 6$, we find the consumer and producer's surplus, where the demand and price are determined under pure competition. First, we construct the two curves and find their point of intersection in order to determine the market equilibrium supply x_0 and price p_0. (See Figure 5.7.4.)

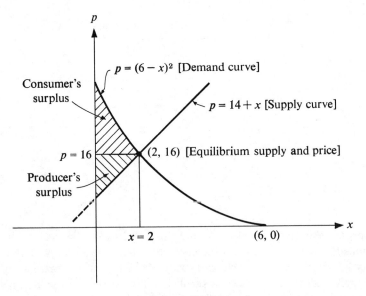

FIGURE 5.7.4

Algebraically, the equilibrium supply is determined by setting the supply equal to the demand (since we assume pure competition). Thus,

$$(6 - x)^2 = 14 + x \quad or \quad 36 - 12x + x^2 = 14 + x \quad or \quad x^2 - 13x + 22 = 0$$

yielding $(x - 11)(x - 2) = 0$. Since $0 \le x \le 6, x \neq 11$, and $x = 2$. Thus, $x_0 = 2$ and $p_0 = 14 + x_0 = 16$. The consumer's surplus is then

$$C_s = \int_0^2 [6 - x]^2 \, dx - 2 \cdot 16 = -\frac{(6 - x)^3}{3} \Big|_0^2 - 32$$

$$= -\frac{64}{3} + \frac{216}{3} - 32 = \frac{56}{3}$$

and the producer's surplus is

$$P_s = 32 - \int_0^2 (14 + x) \, dx = 32 - \left(14x + \frac{x^2}{2}\right) \Big|_0^2$$

$$= 32 - 28 + 2 = 6.$$

Hence, the consumer's surplus is $18.67 and the producer's surplus is $6.00.

Example 3.   The process of finding the average value of a finite number of data is familiar to most students. For example, if $a_1, a_2, \ldots, a_n$ were the grades of a class of students on a certain hour exam, then the class average of the test is

$$y_{avg} = \frac{a_1 + a_2 + \cdots + a_n}{n} \tag{5.7.1}$$

When the number of the data is not finite, the above equation is not feasible. For example, if the data y are given by a continuous function $y = f(x)$, $a \le x \le b$, it is not possible to use Equation (5.7.1). In this case, we define the average value of y, with respect to x by

$$[y_{avg}]_x = \frac{1}{b - a} \int_a^b f(x) \, dx. \tag{5.7.2}$$

Thus the average value of the function $y = x^2$ for $x \in [1, 3]$ is given by

$$[y_{avg}]_x = \frac{1}{2} \int_1^3 x^2 \, dx = \frac{1}{2} \left[\frac{27}{3} - \frac{1}{3}\right] = \frac{13}{3}.$$

If we multiply both sides of Equation (5.7.2) by $(b - a)$, we see that

$$(b - a)[y_{avg}]_x = \int_a^b f(x) \, dx.$$

The left-hand side is simply the area of a rectangle of height $[y_{avg}]_x$ and width $b - a$. (See Figure 5.7.5.) Thus, $[y_{avg}]_x$ is that ordinate of the curve $y = f(x)$ that should be used as the altitude if one wishes to contruct a rectangle whose base is the interval $[a, b]$ and whose area is the area under the region bounded by $y = f(x)$, the x axis, $x = a$, and $x = b$.

We now consider an example of some biological phenomena.

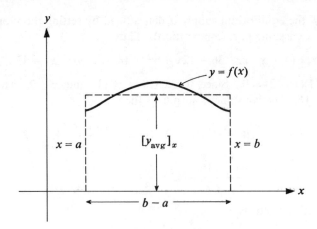

FIGURE 5.7.5

Example 4. A growing culture of bacteria weighs $5e^{.3t}$ grams at time t. What is its average rate of growth during the first three hours—that is, from time $t = 0$ to time $t = 3$?

Solution. Letting $y = 5e^{.3t}$, we see that the rate of growth of the culture at any time t is given by $dy/dt = 1.5e^{.3t}$ grams per hour. (Let $y = f(t)$ be the amount of a substance at time t. Then the rate of growth of the substance at time t is $dy/dt = f'(t)$. We take up this topic at considerable length in Chapter 6.) The average during the first three hours of the growth rate, considered as a function of time t, is defined as

$$\frac{1}{3}\int_0^3 (1.5)e^{.3t}\, dt = \left[\left(\frac{1.5}{3}\right)\frac{1}{.3}e^{.3t}\right]\Big|_0^3 = \frac{5}{3}(e^{.9} - 1).$$

Thus the average growth rate in question is $\frac{5}{3}(e^{.9} - 1)$, or about 2.43 grams per hour.

In the preceding example we asked for the average growth rate with respect to time. Suppose we ask for the average growth rate with respect to the amount of bacteria present. In this case, the growth rate at any time t is still $dy/dt = 1.5e^{.3t}$, where $y = 5e^{.3t}$. As a function of weight y, this growth rate is $dy/dt = .3y$. Thus, we wish the average of $.3y$ the first three hours. This is the average of $.3y$ in the interval $5e^{.3(0)} = 5$ to $5e^{.3(3)} = 5e^{.9}$. The average growth rate with respect to weight y is

$$\frac{\int_5^{5e^{.9}} .3y\, dy}{5(e^{.9} - 1)} = \frac{0.3}{10(e^{.9} - 1)} \cdot y^2\Big|_5^{5e^{.9}} = \frac{0.15[(5e^{.9})^2 - 5^2]}{5e^{.9} - 5}.$$

This works out to about 2.59 grams per hour.

Example 5. (*Interest Rates*) Banks compute interest over varying lengths of time: yearly, semi-annually, monthly, and even daily. (The latter practice sometimes stems from an effort to avoid certain regulations of the Federal Reserve Board.) It is possible to compute interest instantaneously. If P_0 is the initial amount of money and if interest is computed instantaneously at the rate r, then the accrued interest I over a period of time T is given by the formula

$$I = \int_0^T P_0\, re^{rt}\, dt.$$

Thus if $P(T)$ represents the principal (amount of money) at time T, then

$$P(T) = P_0 + \int_0^T P_0\, re^{rt}\, dt.$$

(It is not terribly difficult to derive these formulas, but we will not do so here. See Chapter 8 for the derivation.) Let us now consider some specific applications of the above.

Problem 1.   How much interest does \$1000 draw at a rate of $5\% = 5/100$ compounded instantaneously for 1 year?

Solution.   Here $T = 1$, $P_0 = 1000$, and $r = 0.05$. Thus,

$$I = \int_0^1 P_0\, re^{rt}\, dt = P_0\, e^{rt}\Big|_0^1 = P_0\, e^r - P_0 = P_0(e^r - 1).$$

We now substitute the known values (it is easier to integrate before substituting) to obtain $I = 1000(e^{.05} - 1)$. By checking Table A.1, we find

$$I = 1000(1.0513 - 1) \doteq 51.30.$$

Hence the interest accrued is \$51.30. The interest rate in this problem is thus equivalent to 5.13% computed annually.

Problem 2.   How long will it take an amount of money P to double if interest is computed instantaneously at a rate of 10%?

Solution.   The unknown in this problem is T. Our equation becomes

$$P(T) = 2P = P + \int_0^T Pe^{rt}\, dt$$

or

$$2P = P + Pe^{rt}\Big|_0^T = Pe^{rT} \qquad \text{or} \qquad 2 = e^{rT}.$$

We now substitute $r = 1/10$ to obtain $2 = e^{T/10}$. To determine T we first take logarithms of both sides of the equation and find that

$$\log 2 = \log e^{T/10} = T/10 \qquad \text{or} \qquad T = 10 \log 2.$$

Checking Table A.2 in the back of the text, we find

$$T = 10(0.6931) = 6.931.$$

Thus the money will double in slightly less than seven years.

Exercises

1. Indiana University is considering the purchase of a new computer. The cost of the machine is $50,000. The director of the computing center believes that after a short period of adjustment, savings in efficiency of various departments utilizing the computer will offset the cost. It is thought that the rate of cost savings will be $C(x) = 6000x + 2000$ over 10 years, where x represents years and $C(x)$ is dollars per year savings at any given time. Will the machine pay for itself during its 10-year life? At what time would the break-even point come?

2. In Exercise 1, suppose another machine costing $100,000 is being considered with an expected life of 20 years. Assume that the cost savings function $C(x)$ is the same as in Exercise 1. Would the machine pay for itself in 10 years? Where would the breakeven point be? Suppose that the University would consider the $100,000 machine against two consecutive $50,000 machines (with installation and procedural change costs for each.) Which would be less expensive for a 20-year period?

3. If the demand law is $p = 39 - 3x^2$, find the consumer's surplus (a) if $x_0 = \frac{5}{2}$ and (b) if the commodity is free—that is, $p_0 = 0$. Draw the appropriate diagram.

4. The quantity demanded and the corresponding price, under pure competition, are determined by the demand and supply laws

$$p = 36 - x^2 \quad \text{and} \quad p = 6 + \frac{t^2}{4},$$

where x is the supply and t is the demand. Determine the corresponding consumer's surplus and producer's surplus. Draw the appropriate diagram.

5. If the supply curve is $p = 4e^{x/3}$, and $x_0 = 3$, find the producer's surplus.

6. (a) Find the average value of $f(x) = x^3$ for $x \in [1, 3]$.
   (b) How does this average compare to $f(2)$? To the average of $f(1)$ and $f(3)$?

7. The amount of a quantity y present is a function of time t. Show that the average of the growth rate dy/dt, considered as a function of time, is simply "the change in y divided by the length of time considered."

8. A growing culture of bacteria weights 10^t grams at time t. What is its average weight of growth during the first three hours (a) with respect to time t, and (b) with respect to the amount of bacteria present?

9. Find the average value of y with respect to x for that part of the curve $y = \sqrt{ax}$ between $x = a$ and $x = 3a$.

10. If $200 draws interest at a rate of 4% compounded instantaneously, how much interest will accrue in 3 years?

11. How long will it take for an amount P to triple at a rate of 20% compounded instantaneously?

12. An amount of money P draws interest at the rate r compounded instantaneously. If the amount doubles in 15 years, what is r?

13. An amount of money P draws interest at 5% compounded instantaneously. At the end of 10 years the total amount is $2000. What was the initial amount P?

14. If interest is compounded instantaneously at the rate of 10%, how much money must you start with to obtain $1000 in interest at the end of one year?

5.8 Summary of Chapter 5

(1) A function F, differentiable on an interval I (open or closed), is said to be an *antiderivative* of f if

$$F'(x) = f(x) \qquad \text{for} \qquad x \in I.$$

We also write $F(x) = \int f(x)\, dx + C$.

(2) Antiderivatives of common functions. (In all parts, C is an arbitrary constant.):

   (a) $\int K\, dx = Kx + C$, where K is a constant
   (b) $\int x^n\, dx = x^{n+1}/(n+1) + C$ for $x \neq -1$
   (c) $\int x^{-1}\, dx = \log|x| + C$ for $x \neq 0$
   (d) $\int e^x\, dx = e^x + C$.

(3) *Properties of antiderivatives*

   (a) $\int [f(x) + g(x)]\, dx = \int f(x)\, dx + \int g(x)\, dx$
   (b) $\int kf(x)\, dx = k \int f(x)\, dx$, where k is a constant.

(4) A function f is said to be *Newton integrable* on $[a, b]$ if f has an antiderivative F on $[a, b]$. We define the Newton integral as

$$\int_a^b f(x)\, dx = F(x) \Big|_a^b = F(b) - F(a).$$

(5) *Substitution*

Let $u(x)$ be a differentiable function of $[a, b]$. Then

$$\int_a^b f[u(x)]u'(x)\, dx = \int_{u(a)}^{u(b)} f(u)\, du.$$

For antiderivatives,

$$\int f[u(x)]u'(x)\, dx = \int f(u)\, du + C,$$

where C is an arbitrary constant.

(6) *Integration by parts*

$$\int_a^b f(x)g'(x)\, dx = f(x)g(x) \Big|_a^b - \int_a^b g(x)f'(x)\, dx$$

or for antiderivatives,

$$\int f(x)g'(x)\, dx = f(x)g(x) - \int g(x)f'(x)\, dx + C,$$

where C is an arbitrary constant.

(7) The area of the region bounded above by the curve $y = f(x)$, below by the x axis, on the left by $x = a$, and on the right by $x = b$ (see Figure 5.8.1) is given by

$$\text{Area}(H) = \int_a^b f(x)\,dx.$$

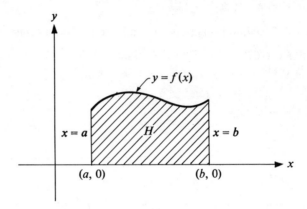

FIGURE 5.8.1

(a)   if the curve $y = f(x)$ lies *below* the x axis, then

$$\text{Area}(H) = -\int_a^b f(x)\,dx.$$

(b)   In Figure 5.8.2, $\text{Area}(H) + \text{Area}(I) = \int_a^c f(x)\,dx - \int_c^b f(x)\,dx$.

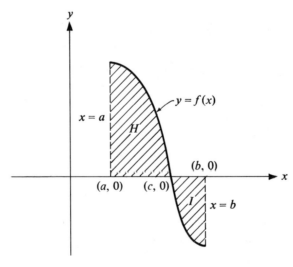

FIGURE 5.8.2

(c) In Figure 5.8.3, $f(x) \le g(x)$, $x \in [a, b]$. The area between the two curves, Area(I), is given by

$$\text{Area}(I) = \text{Area}(B) - \text{Area}(A) = \int_a^b g(x)\, dx - \int_a^b f(x)\, dx$$

$$= \int_a^b [g(x) - f(x)]\, dx.$$

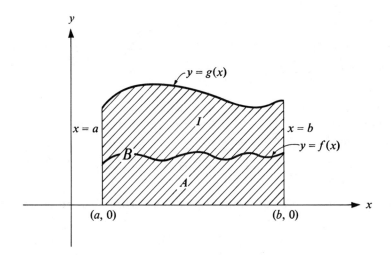

FIGURE 5.8.3

REVIEW EXERCISES

1. Evaluate the following.

(a) $\int (5x^2 - 9x^{-10} + 10x^{-5/3})\, dx$

(b) $\int \left(104e^x - \frac{1}{x} + 3\right) dx$

(c) $\int_{-1}^1 e^{10x+1}\, dx$

(d) $\int \frac{\log(2x+4)}{x}\, dx$

(e) $\int \frac{x}{\sqrt{1-x^2}}\, dx$

(f) $\int_0^1 \frac{\log(x+1)}{x+1}\, dx$

(g) $\int x(4x^2 - 9)^5\, dx$

(h) $\int_{-1}^0 x^2 e^{(x^3+1)}\, dx$

(i) $\int x(x-3)^{1/2}\, dx$

(j) $\int x^2 e^x\, dx$

(k) $\int_1^2 x \log x\, dx$

(l) $\int \frac{(\log x)^5}{x}\, dx$

(m) $\int \frac{\log x}{x^4}\, dx$

(n) $\displaystyle\int_1^2 \frac{\log(x+1)}{x^2}\,dx$ (p) $\displaystyle\int_1^{10} \left(\frac{1}{x}+x\right) dx$

$\left[\text{HINT:}\quad \dfrac{1}{x(x+1)} = \dfrac{1}{x} - \dfrac{1}{x+1}\right]$ (q) $\displaystyle\int_1^2 \frac{x}{\sqrt{3x^2-1}}\,dx$

(o) $\displaystyle\int_3^5 xe^{(x+1)}\,dx$ (r) $\displaystyle\int_0^1 x(2x+1)^{1/3}\,dx$

2. Find the area enclosed by the curves $y = x^2$ and $y = x^5$.

3. Find the area enclosed by the curves $y = x^2$ and $y = x^4$. (Draw a picture.)

4. Find the area bounded by the curve $f(x) = e^x - 2$, the x axis, and the lines $x = 0$ and $x = 1$.

5. The amount of money in the Monopoly Bank over a two-year period is given by the function $f(x) = 2x^2 - 3x + 9$ for $0 \le x \le 2$. What was the average amount of money in the bank over this period?

6*. The area bounded by the x axis, the curve $y = f(x)$, and the lines $x = 1$ and $x = t$ is equal to $(t^2 + 1)^{1/2} - (2)^{1/2}$ for all $t > 1$. Find $f(x)$.

7. The rate of change of sales s with respect to advertising expenditure x is given by

$$\frac{ds}{dx} = 5 + \frac{10}{x} + x \log x \qquad \text{for} \qquad 1 \le x \le 5.$$

Determine the total change in sales as advertising expenditure is increased from 1 to 5.

8. The marginal cost of producing Supergoodie Cake Mix is given by

$$m(x) = 225 - 1200e^{-2x},$$

where x, the number of packages produced, is in dozens, and the cost is in cents. Find a formula for $C(x)$, the total cost of producing x dozen packages of the cake mix, given that 2 dozen can be produced at a cost of $8.00. How much will it cost to produce 4 dozen packages of the mix?

9. Evaluate the Newton integral, $\int_{-1}^1 (x^3 - x)\,dx$ and explain the result by drawing the region whose area the definite integral is supposed to represent.

10. Let $y = x^2 + 1$ be the supply curve for a certain product (y is the supply and x is the price). Let $y = -x + 7$ be the demand curve for the same product (y is the demand and x is the price). Assuming pure competition, find the equilibrium price and supply. Find the consumer surplus and the producer surplus.

11. If $200 draws interest at the rate of 5% compounded instantaneously, find the total amount at the end of 10 years.

12. How long will it take for an amount of money P to double if it draws 1% interest compounded instantaneously?

13.   A piece of real estate worth $20 billion in 1969 is alleged to have been worth $20 in 1639. What rate of interest, compounded instantaneously would yield this increase in the same time?

14.   Find all differentiable functions f which satisfy the equation

$$[f(x)]^2 = \int_0^x f(t)\, dt.$$

[HINT: First differentiate both sides of the equation with respect to x, the left-hand side by the chain rule, the right by the fundamental theorem of calculus. Then, perform the obvious cancellation. Now, integrate both sides of the equation and use the fact that $f(0) = 0$.]

APPENDIX 1

5.9   The Riemann Integral

In this section, we briefly discuss another approach to integration. This is based upon an idea of Georg Friedrich Riemann (1826–1866).

Let f be any function bounded on $[a, b]$. (A function f is bounded on $[a, b]$ if there exists a number M such that $|f(x)| \le M$, $x \in [a, b]$.) If n is a positive integer, we choose points $x_0, x_1, x_2, \ldots, x_n$ such that $a = x_0 \le x_1 \le x_2 \le \cdots \le x_n = b$. We then choose points $c_1, c_2, \ldots, c_n$ such that $c_i \in [x_{i-1}, x_i]$. Finally, we consider the sum

$$R_a^b(n) = f(c_1)(x_1 - x_0) + f(c_2)(x_2 - x_1) + \cdots + f(c_n)(x_n - x_{n-1}). \qquad (5.9.1)$$

This sum is called a Riemann sum. We regard the function f and the interval $[a, b]$ as fixed; but the integer n and the points x_i, c_i may be chosen in various ways. In most cases, the number $R_a^b(n)$ obtained in this manner will vary as we vary n and the choice of the x_i's and c_i's.

If we increase n and space the points $x_0, x_1, \ldots, x_n$ in such a way that the maximum of the distance between consecutive points (that is, $|x_i - x_{i-1}|$) approaches 0 as $n \to \infty$, then $R_a^b(n)$ may or may not approach a limit. If it does, we say the function f is Riemann integrable. Thus, if

$$\lim_{n \to \infty} R_a^b(n) = \lim_{\substack{\max|x_i - x_{i-1}| \to 0 \\ n \to \infty}} [f(c_1)(x_1 - x_0) + \cdots + f(c_n)(x_n - x_{n-1})]$$

exists, we say f is Riemann integrable.

We denote the limit by $I_a^b(f)$ to indicate that it depends only on the function f and the interval $[a, b]$. If f is continuous on $[a, b]$, then f is Riemann integrable on $[a, b]$, although this fact is not simple to prove. At this point, it is natural to ask about the relation between the Riemann and the Newton integral. More specifically, (i) if a function is Riemann integrable, is it Newton integrable and vice versa; (ii) if a function is both Riemann and Newton integrable, do the integrals have the same value?

For the first question, the answer is no in both directions. For example, the function

$$f(x) = \begin{cases} 0, & 0 \le x < 1 \\ 2, & 1 \le x \le 2 \end{cases}$$

is Riemann integrable, but is not Newton integrable since f does not have an anti-derivative on $[0, 2]$. This is not too hard to show. To produce a function that is Newton integrable but not Riemann integrable is more difficult. One must construct spiky functions on a Cantor set with positive measure, a reasonable problem for a first-year graduate student but too involved for us.

However, the answer to the second question is yes. The question is not as difficult as it seems, so let us write down the answer with a proof.

THEOREM.   *Let f be both Riemann and Newton integrable on $[a, b]$. Then*

$$I_a^b(f) = \int_a^b f(x)\, dx = F(b) - F(a),$$

where $F'(x) = f(x)$ for $x \in [a, b]$.

Proof.   Since f is Newton integrable, we know that a function $F(x)$ exists with the property that $F'(x) = f(x)$. Since f is Riemann integrable, we know that the

$$\lim_{\substack{\max|x_i - x_{i-1}| \to 0 \\ n \to \infty}} [f(c_1)(x_1 - x_0) + \cdots + f(c_n)(x_n - x_{n-1})]$$

exists for any choice of x_i and c_i as long as $c_i \in [x_{i-1}, x_i]$ and $\max|x_i - x_{i-1}| \to 0$. Since the limit exists for any choice of the c_i's, let us choose them to our advantage. Consider the $x_0, x_1, \ldots, x_n$ to be fixed. By the mean-value theorem, we know that

$$\frac{F(x_i) - F(x_{i-1})}{x_i - x_{i-1}} = F'(p_i)$$

for some $p_i \in (x_{i-1}, x_i)$. Thus,

$$F(x_i) - F(x_{i-1}) = F'(p_i)(x_i - x_{i-1}) = f(p_i)(x_i - x_{i-1}).$$

We will choose $c_i = p_i$. Then

$$
\begin{aligned}
R_a^b(n) &= [f(p_1)(x_1 - x_0) + \cdots + f(p_n)(x_n - x_{n-1})] \\
&= [(F(x_1) - F(x_0)) + (F(x_2) - F(x_1)) + \cdots + (F(x_n) - F(x_{n-1}))] \\
&= F(x_n) - F(x_0) = F(b) - F(a).
\end{aligned}
$$

(The second from the last line reduces to the last line if we perform all available cancellations.) Thus,

$$\lim_{n \to \infty} R_a^b(n) = I_a^b(f) = F(b) - F(a) = \int_a^b f(x)\, dx.$$

The first equality is just the definition of the Riemann integral. The last equality is just the definition of the Newton integral.

As a special case of this theorem, we note that if f is continuous on $[a, b]$, then f is both Newton and Riemann integrable (with the same value, of course). This case is intimately related to the Fundamental Theorem of Calculus.

It is not important for us to go into a detailed discussion as to the merits of the Riemann integral over the Newton integral. We only remark that for most applications to social and biological science, the Newton integral suffices.

For a more detailed discussion of the ideas presented in this section, the reader is referred to either J. W. Kitchen, *Calculus of One Variable*, Addison-Wesley, Reading, Mass., 1968, or T. Apostol, *Calculus, Volume I*, Blaisdell, Publishing Company, Waltham, Mass., 1961.

APPENDIX 2

5.10   Volumes

Integrals can also be used to calculate the volume of various solids. We first discuss the volume of a solid of known cross-sectional area. Many solids are conveniently described in terms of three-dimensional Cartesian coordinates. We introduce such a system by first selecting three mutually perpendicular lines—called the coordinate axes. Their common point of intersection is called the origin of the coordinate system. Using a common unit of length, we introduce coordinates onto each of the coordinate axes (called the x, y, and z axes). (See Figure 5.10.1.) We do this in such a way that if the index and middle

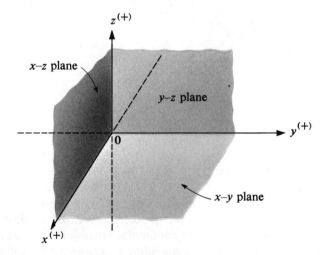

FIGURE 5.10.1

finger of the right hand is pointed in the direction of the positive x and y axes, then the right thumb is pointed in the direction of the positive z axis. The three planes determined by pairs of coordinate axes are called the *coordinate planes*. These planes are the x–y plane formed by the x and y axes, the y–z plane formed by the y and z axes, and the x–z plane formed by the x and z axes. Just as we were able to establish a one-to-one correspondence between points in the two-dimensional coordinate system and ordered pairs of numbers, we could do exactly the same between points in the three-dimensional system and ordered triples of numbers. Given any point P_0 in space, we let x_0, y_0, and z_0 be the perpendicular projections of P_0 onto the x, y, and z axes. We then write for P_0 the ordered triple (x_0, y_0, z_0). (See Figure 5.10.2.)

We now proceed to show that if B is any solid whose cross-sectional area is given by $A(x)$, $a \leq x \leq b$, then the volume of H is

$$\text{volume}(H) = \int_a^b A(x)\, dx.$$

Although this is true more generally, we will assume that $A(x)$ is continuous.

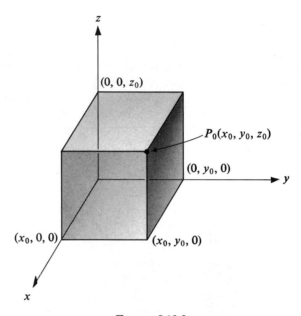

FIGURE 5.10.2

Let us consider the solid illustrated in Figure 5.10.3. We wish to determine the volume of the solid H, whose cross-sectional area is known for any $a \leq x \leq b$ to be $A(x)$. (*Caution:* Do not confuse the present notation with that of the previous section.) In order to show that $\text{volume}(H) = \int_a^b A(x)\, dx$, we proceed as we did in determining areas. Imagine that the solid is generated by the continuous expansion of the set $A(a)$, indicated in Figure 5.10.3 to the right. Thus, corresponding to x_0, we get the solid S_{x_0}, which is that part of H that lies between the planes $x = a$ and $x = x_0$, indicated in Figure 5.10.4) We denote the volume of S_x by $V(x)$ and the volume of S_{x_0} by $V(x_0)$. We study the rate at which the volume increases. In particular, we show that for any $x_0 \in [a, b]$, $V'(x_0) = A(x_0)$. First observe that the volume $[V(x) - V(x_0)]$ is the volume of the slice shown in Figure 5.10.5. Now if we consider the solid cylinder C_{x_0} whose cross-sectional area is $A(x_0)$ and height is $x - x_0$, (see Figure 5.10.6) and the solid cylinder C_x whose cross-sectional area is $A(x)$ and height is $(x - x_0)$, (see Figure 5.10.7) then the solid slice $K \subset C_x$ and $C_{x_0} \subset K$. (See Figure 5.10.8.) Thus, we see that geometrically

$$\text{volume}(C_{x_0}) \leq V(x) - V(x_0) \leq \text{volume}(C_x);$$

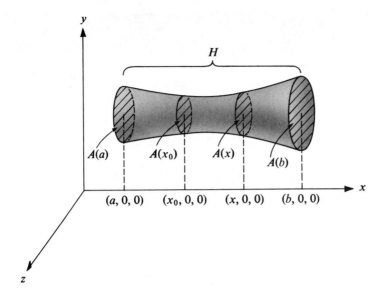

FIGURE 5.10.3

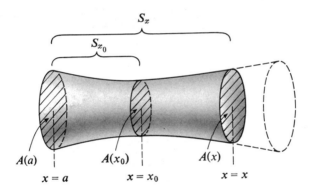

FIGURE 5.10.4

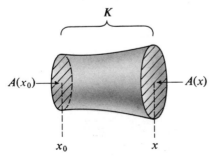

FIGURE 5.10.5

The cylinder C_{x_0}

FIGURE 5.10.6

The cylinder C_x

FIGURE 5.10.7

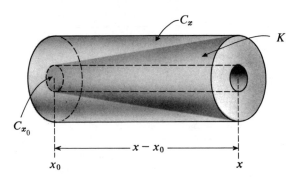

FIGURE 5.10.8

but the volume of a cylinder is simply the area of its base times its height. Hence, volume(C_x) $= A(x)(x - x_0)$ and volume(C_{x_0}) $= A(x_0)(x - x_0)$ and

$$A(x_0)(x - x_0) \leq V(x) - V(x_0) \leq A(x)(x - x_0)$$

or

$$A(x_0) \leq \frac{V(x) - V(x_0)}{x - x_0} \leq A(x).$$

Since A is continuous on $[a, b]$, $\lim_{x \to x_0^+} A(x) = A(x_0)$ and by the squeezing principle,

$$\lim_{x \to x_0^+} \frac{V(x) - V(x_0)}{x - x_0} = A(x_0).$$

If we take $x < x_0$, then we find

$$\lim_{x \to x_0^-} \frac{V(x) - V(x_0)}{x - x_0} = A(x_0),$$

thus yielding

$$V'(x_0) = \lim_{x \to x_0} \frac{V(x) - V(x_0)}{x - x_0} = A(x_0).$$

We have shown, then, that for any $x \in (a, b)$, $V'(x) = A(x)$ and hence,

$$\text{volume}(H) = V(b) - V(a) = \int_a^b A(x)\, dx,$$

where A is the cross-sectional area of the solid.

Example 1.   A solid has a circular base of radius 4. Find the volume of the solid if every plane section perpendicular to a fixed diameter is an equilateral triangle.

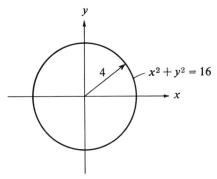

FIGURE 5.10.9

Solution.   Since the solid has a circular base of radius 4, we first sketch the base in the x–y plane. (See Figure 5.10.9) Next, we use the fact that every plane section perpendicular to a fixed diameter is an equilateral triangle. Hence, the solid must look like so:

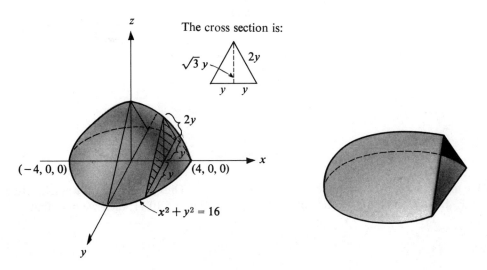

FIGURE 5.10.10

where $y = \sqrt{16 - x^2}$. Hence,

$$A(x) = \tfrac{1}{2}(2y)\sqrt{3}\, y = \sqrt{3}\, y^2 = \sqrt{3}(16 - x^2).$$

Thus, the volume V is given by

$$V = \sqrt{3} \int_{-4}^{4} (16 - x^2)\, dx.$$

Since $16 - x^2$ is an *even* function,

$$V = 2\sqrt{3} \int_{0}^{4} (16 - x^2)\, dx = 2\sqrt{3} \left[16x - \frac{x^3}{3} \right] \Big|_{0}^{4}$$

$$= 2\sqrt{3} \left(64 - \frac{64}{3} \right) = \sqrt{3} \cdot \frac{256}{3}.$$

Example 2. The base of a certain solid is an equilateral triangle of side s with one vertex at the origin and an altitude along the x axis. Each plane section perpendicular to the x axis is a square, one side of which lies in the base of the solid. Find the volume of the solid.

  Solution. The base is shown in Figure 5.10.11, and the solid is shown in Figure 5.10.12. The volume is given by $V = \int_{0}^{h} A(x)\, dx$. (We call h the length of the altitude.)

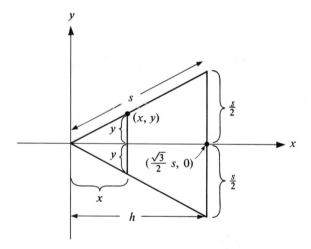

FIGURE 5.10.11

We must compute the cross-sectional area $A(x) = 4y^2$. Thus, we must find y as a function of x. If we let h be the altitude of the triangle in Figure 5.10.11, then we note that

$$h^2 = s^2 - \frac{s^2}{4} = \frac{3s^2}{4} \qquad \text{or} \qquad h = \frac{\sqrt{3}}{2} s.$$

Now, using similar triangles,

$$\frac{y}{x} = \frac{s/2}{h} = \frac{s/2}{\sqrt{3}\, s/2} = \frac{1}{\sqrt{3}}$$

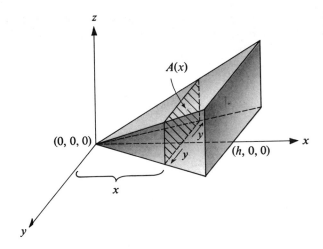

FIGURE 5.10.12

and thus, $A(x) = \frac{4}{3}x^2$.

Hence,

$$V = \int_0^h A(x)\,dx = \int_0^{\sqrt{3}s/2} \frac{4}{3}x^2\,dx = \frac{4}{9}x^3 \Big|_0^{\sqrt{3}s/2}$$

$$= \frac{4}{9} \cdot \frac{3\sqrt{3}}{8} \cdot s^3 = \frac{\sqrt{3}\,s^3}{6}.$$

Solids can also be generated by revolving ordinate sets, that is, a set

$$S = \{(x, y)\,|\,a \le x \le b \text{ and } 0 \le y \le f(x)\}$$

about the x axis. We call these solids, solids of revolution. Consider a function f that is continuous and nonnegative on $[a, b]$. (See Figure 5.10.13.) Let B be the solid generated

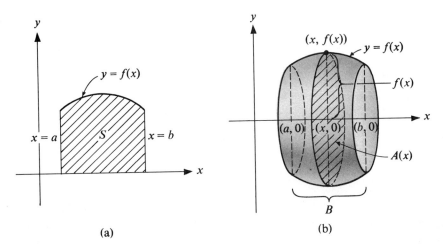

(a)

(b)

FIGURE 5.10.13

by revolving S about the x axis. The cross-sectional area of B at x, $A(x) = \pi[f(x)]^2$, since the cross section of B corresponding to x is a circle of radius $f(x)$. Hence, the volume of B, $V(B)$, is given by

$$V(B) = \pi \int_a^b [f(x)]^2 \, dx.$$

Example 3.   Find the volume of a sphere of radius r by rotating the ordinate set of the function

$$f(x) = \sqrt{r^2 - x^2} \qquad \text{for} \qquad -r \le x \le r.$$

Solution.   (See Figure 5.10.14.)

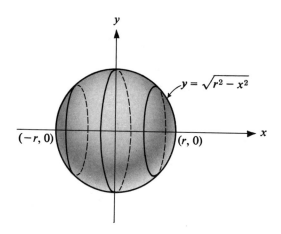

FIGURE 5.10.14

By the above result,

$$V = \pi \int_{-r}^{r} (r^2 - x^2) \, dx = 2\pi \int_0^r (r^2 - x^2) \, dx$$

$$= 2\pi \left(r^2 x - \frac{x^3}{3} \right) \Big|_0^r = \frac{4\pi r^3}{3}.$$

Example 4.   Find the volume generated by revolving about the x axis the area bounded by the curve $y = e^x$, $x = 1$, and $x = 2$.

Solution.   First, sketch the region in question. (See Figure 5.10.15).

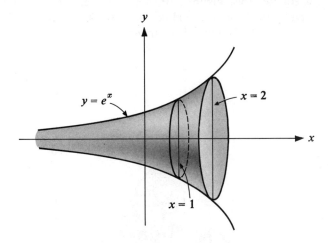

FIGURE 5.10.15

$$V = \pi \int_1^2 (e^x)^2 \, dx = \pi \int_1^2 e^{2x} \, dx.$$

Let $u = 2x$, $du = 2dx$; then $u(1) = 2$, $u(2) = 4$, and

$$\int_1^2 e^{2x} \, dx = \frac{1}{2} \int_2^4 e^u \, du = \frac{e^4 - e^2}{2}.$$

Thus,

$$V = \frac{\pi(e^4 - e^2)}{2}.$$

Exercises

In all exercises, first sketch the solid whose volume is to be found.

1. The base of a certain solid is the circle $x^2 + y^2 = a^2$. Each plane section of the solid cut out by a plane perpendicular to the x axis is a square with one edge of the square in the base of the solid. Find the volume of the solid.

2. The base of a solid is the figure bounded by the parabola $y^2 = x$ and the line $x = 1$. Every cross section perpendicular to the x axis is a square. Find the volume of the solid.

3. Find the volume in Exercise 2 if the cross sections perpendicular to the x axis are equilateral triangles.

4. A tower is 60 ft high and 30 ft square at the base, and every cross section parallel to the base is square. (See Figure 5.10.16.) The side, s feet, of any of these cross sections is given by the formula $s = (15x^2 - x^3/6)/600$, where x is the distance in feet from the peak. Find the volume of the tower.

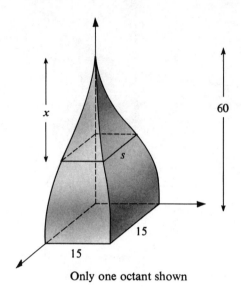

Only one octant shown

FIGURE 5.10.16

5. Find the volume of the solid generated by revolving about the x axis the region bounded by the graphs of:

(a) $f(x) = \sqrt{x}$, the x axis, $x = 0$, and $x = 1$.

(b) $f(x) = x + 1/x$, the x axis, $x = 1$, and $x = 4$.

(c) $f(x) = \sqrt{x}(x^2 + 1)^{1/4}$, the x axis, $x = 0$, and $x = 1$.

(d) $f(x) = xe^{\sqrt{x}}$, the x axis, $x = 1$, and $x = 2$.

(e) $f(x) = \log x$, the x axis, $x = 1$, and $x = 2$.

(f) $f(x) = x(x^3 + 5)^{20}$, the x axis, $x = 0$, and $x = 2$.

6. Find the volume of the solid generated by revolving about the x axis the region bounded by the parabola $y = 9 - x^2$ and the straight line $y = 8$. [HINT: Find two volumes and subtract.]

7. Find the volume of the solid generated by revolving about the x axis the region bounded by $y = x^2$ and $y = x^3$.

8. Suppose the region in Exercise 7 were revolved about the y axis. Find the volume of the solid so generated.

9. Find the volume generated by revolving the region bounded by $f(x) = 1 - x^2$ and the x axis about the line $y = -1$.

6

Miscellaneous Topics I. Differentiation

6.1 Implicit Differentiation

Consider the two functions $f(x) = \sqrt{1 - x^2}$ and $f(x) = -\sqrt{1 - x^2}$ whose graphs are shown in Figure 6.1.1. Their graphs are the top and bottom halves of a circle, respectively. It is convenient to combine both in one equation: $x^2 + y^2 - 1 = 0$, which is the equation of a circle.

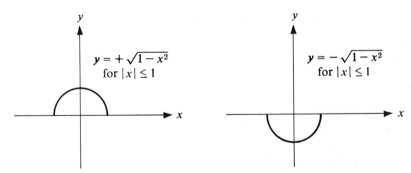

FIGURE 6.1.1

If we are given a "function" of the form $F(x, y) = 0$, is it possible to find dy/dx without first solving for y? If so, this would save considerable time and effort. (In fact, it may *not* be possible to solve for y in terms of x.)

Let us first rewrite the equation of a circle as

$$x^2 + [f(x)]^2 = 1.$$

Now formally differentiate the equation. Thus,

$$\frac{d}{dx}(x^2 + [f(x)]^2) = \frac{d}{dx}(1).$$

By our rule for sums this becomes

$$\frac{d}{dx}(x^2) + \frac{d}{dx}[f(x)]^2 = 0.$$

The first term is easy, $(d/dx)(x^2) = 2x$; but how about $(d/dx)[f(x)]^2$? At first glance, one might think we have no way of handling such expressions, but recall the chain rule. In particular,

$$\frac{d}{dx}[u(x)]^n = n[u(x)]^{n-1}\frac{du(x)}{dx}.$$

Thus, $(d/dx)[f(x)]^2 = 2f(x)\,df/dx$. Our formally differentiated equation thus becomes

$$2x + 2f(x)\frac{df}{dx} = 0 \quad \text{or} \quad \frac{df}{dx} = -\frac{x}{f(x)}.$$

We can, if we wish, write this as $dy/dx = -x/y$. Before going further a comment is in order. First, although we make no attempt to prove it, this formal differentiation is valid (that is, yields the correct value for the derivative). Second, since $x^2 + y^2 = 1$ represents not *one* but *two* functions (see Figure 6.1.1), how can the expression $dy/dx = -x/y$ give the correct derivative for both? The answer lies in the y or $f(x)$ term in $-x/y$ or $-x/f(x)$. It is this term that automatically makes the necessary adjustments in the derivative $dy/dx = -x/y$.

RULE. *To differentiate implicitly a function of the form $F(x, y) = 0$, simply differentiate formally, using the chain rule on terms containing y. (We assume, of course, that y is a function of x, say $y = f(x)$.)*

Let us consider another example.

Example 2. If $x^2y + e^y + y^5 + \log xy = 0$, find dy/dx.

Solution. This time we will not replace y and f. Differentiating both sides formally with respect to x, we obtain

$$\frac{d}{dx}(x^2y) + \frac{d}{dx}(e^y) + \frac{d}{dx}(y^5) + \frac{d}{dx}(\log xy) = 0$$

or

$$y\frac{d}{dx}(x^2) + x^2\frac{d}{dx}(y) + e^y\frac{dy}{dx} + 5y^4\frac{dy}{dx} + \frac{1}{xy}\frac{d}{dx}(xy) = 0.$$

(We have used the chain rule on the 2nd and 3rd terms.) Continuing on we obtain

$$2yx + x^2\frac{dy}{dx} + e^y\frac{dy}{dx} + 5y^4\frac{dy}{dx} + \frac{1}{xy}\left(y + x\frac{dy}{dx}\right) = 0.$$

If we wish, we can now solve for dy/dx. Thus,

$$\frac{dy}{dx}\left(x^2 + e^y + 5y^4 + \frac{x}{xy}\right) + 2yx + \frac{y}{xy} = 0$$

and so

$$\frac{dy}{dx} = \frac{-(2yx + 1/x)}{(x^2 + e^y + 5y^4 + 1/y)}.$$

We give one more example to illustrate the method.

Example 3. If $e^{5x^3y^2} = 0$, find dy/dx.

Solution. Differentiating both sides we obtain $(d/dx)(e^{5x^3y^2}) = 0$. Since $5x^3y^2$ can be considered a function of x, by the chain rule we have

$$e^{5x^3y^2}\frac{d}{dx}(5x^3y^2) = 0$$

or

$$e^{5x^3y^2}\left[y^2\frac{d}{dx}(5x^3) + 5x^3\frac{d}{dx}(y^2)\right] = 0$$

or

$$e^{5x^3y^2}\left[15y^2x^2 + 10x^3y\frac{dy}{dx}\right] = 0.$$

We will not solve for dy/dx. Note, though, that no matter how complicated the equation $F(x, y) = 0$, one can solve for dy/dx, and in fact, one need only solve an equation of the form $A\,dy/dx + B = 0$ to obtain dy/dx.

Actually implicit differentiation is a special case of a more general situation. Given any equation $F(x, y, u, v, \ldots) = 0$, where $y, u, v, \ldots$ are functions of x, we can differentiate it formally to obtain a relation between $x, y, u, v, \ldots$ and their derivatives. This last statement is imposingly abstract; let us consider an example. Let $e^u + w^2 + uv = 0$ when u, v, w are functions of x. We differentiate with respect to x and obtain

$$\frac{d}{dx}(e^u) + \frac{d}{dx}(w^2) + \frac{d}{dx}(uv) = 0.$$

By the chain rule, this yields

$$e^u\frac{du}{dx} + 2w\frac{dw}{dx} + u\frac{dv}{dx} + v\frac{du}{dx} = 0.$$

This is really all there is to it.

We consider another example.

Example 4. If $V = \frac{4}{3}\pi r^3$ where V and r are functions of t, find dV/dt.

Solution. Differentiating both sides with respect to t, we obtain

$$\frac{dV}{dt} = \frac{d}{dt}\left(\frac{4\pi}{3}r^3\right) = \frac{4\pi}{3}\frac{d}{dt}(r^3) = \frac{4\pi}{3}\cdot 3r^2\frac{dr}{dt}$$

by the chain rule.

As a final example, we look at

Example 5. Differentiate the expression $u + u^2 + \log uv^3$ with respect to x when u, v are functions of x.

Solution.

$$\frac{d}{dx}(u + u^2 + \log uv^3) = \frac{du}{dx} + \frac{d}{dx}(u^2) + \frac{d}{dx}(\log uv^3)$$

$$= \frac{du}{dx} + 2u\frac{du}{dx} + \frac{1}{uv^3}\frac{d}{dx}(uv^3) = \frac{du}{dx} + 2u\frac{du}{dx} + \frac{1}{uv^3}\left(v^3\frac{du}{dx} + 3v^2u\frac{dv}{dx}\right).$$

Exercises

1. Find dy/dx for each of the equations below.

   (a) $xy = 1$

   (b) $3x^2y + 4xy^2 + 5 = 0$

   (c) $e^x + \log y = 0$

   (d) $e^y \log x + 1 = 0$

   (e) $x^5 \log 2y + y^3 \log x = 1$

   (f) $xy^3 + 2x^2y^2 + x^3y = 9$

   (g) $e^{x^2y} = 1$

   (h) $\dfrac{x^2}{4} + \dfrac{y^2}{9} = 1.$

2. (a) If $x^2 + y^2 = 4$, find dy/dx at $(1, \sqrt{3})$.

   (b) If $e^{xy^2} = e$, find dy/dx at $(1, -1)$.

   (c) If $x^2/4 - y^2/9 = -3$, find dy/dx at $(2, 6)$.

   (d) If $x^2 + x + 2 - y^2 = 0$, find dy/dx at $(1, 2)$.

   (e) If $x + \log xy^3 = 8$, find dy/dx at $(8, \frac{1}{2})$.

3. If A, B, C are functions of x, formally differentiate the following equation with respect to x,

   $$A^2B + B^2C + C^2A = 0.$$

4. Given that U and V are functions of t, find dV/dt when $V^3 + 2\log U = 0$.

5. Find

   $$\frac{d}{dx}[e^{u(x)} + \log v(x)].$$

6. If $A = \pi r^2$ when A and r are functions of t, find dr/dt.
7. If u, v, w are functons of x, find

$$\frac{d}{dx}[2uv^2 + \log(u - 2w) + e^{(w^2 + u)}].$$

8. Find the equation of the line tangent to the curve

$$x^3 - 2x^2y^2 + 3y = x - 2$$

    at $(1, 2)$.
9. Find the equation of the line tangent to the curve

$$x = \frac{2a^3}{2a^2 - y^2}$$

    at $(2a, -a)$.
10. Find the value of d^2y/dx^2 at the points, where $dy/dx = 0$, of the following:

    (a) $2\sqrt{x} + 3\sqrt{y} = 8$.

    (b) $x^2 = y^4 - y^2$.

6.2 Inverse Functions and Their Derivatives

Recall from Chapter 1 the definition of a function: namely, a rule that assigns to elements in one set of numbers *one* and *only* one number in another set. That is, if a is any number in the domain of f, then there is precisely one number b such that $f(a) = b$. Geometrically, this means that the vertical line through a meets the graph of the function f at just one point $(a, f(a))$. (See Figure 6.2.1.)

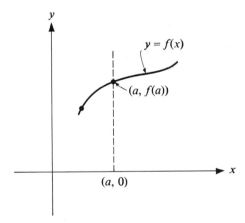

FIGURE 6.2.1

However, it is quite possible, given a point b in the range of values of f, to have two or more numbers $a_1, a_2, \ldots$, in the domain such that $f(a_1) = f(a_2) = f(a_3) = \cdots = b$. Geometrically, this means that the horizontal line through b may meet the graph of f at more than one point. (See Figure 6.2.2.)

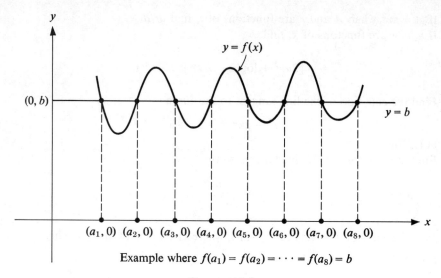

Example where $f(a_1) = f(a_2) = \cdots = f(a_8) = b$

FIGURE 6.2.2

There are, of course, some functions, such as $f(x) = 2x + 3$, $g(x) = x^3$, $h(x) = e^x$, etc., for which this does not happen; that is, any horizontal line meets the graph of the function in at most one point. Whenever this is the case, we can define a new function that is, roughly speaking, the reverse or inverse of the original function.

For the moment we concentrate our attention on functions that take on a given value *once at most*. That is, we are interested in the functions that take on every value in the range of the function exactly once. In this situation there is a simple way to define a new function g. If $f(2) = 6$ we set $g(6) = 2$. Thus if f takes P_1 to P_2, g takes P_2 back to P_1. This rule defines a function g whose domain is the range of f. More precisely, g takes every point in the range of f back into the domain of f. (See Figure 6.2.3.)

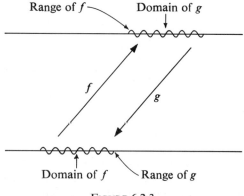

FIGURE 6.2.3

Example. Let f be the function that assigns to every father his eldest son. Then the function g assigns to the eldest son, his father.

We will now give a mathematically precise definition of the foregoing concepts.

DEFINITION. *A function f is 1-1 (read one-to-one) if $x_1 \neq x_2$ implies that $f(x_1) \neq f(x_2)$ for all x_1, x_2 in the domain of f. An equivalent formulation that is sometimes more convenient to check is: f is 1-1 if $f(x_1) = f(x_2)$ implies $x_1 = x_2$.*

DEFINITION. *Let the function f be 1-1. Then there exists a function g called the inverse function of f, which is defined as follows: $g[f(x)] = x$ for all x in the domain of f. That is, if $f(x) = y$, then $g(y) = x$. We shall denote the inverse function g by f^{-1}. Thus $f^{-1}[f(x)] = x$. Moreover, it is easy to see that $f[f^{-1}(x)] = x$.*

Often, given an equation defining a function f, it is possible to find an equation defining its inverse function. For example, let $f(x) = x^3$. Now, in order to find $f^{-1}(x)$, we must solve the equation $f[f^{-1}(x)] = x$ or $[f^{-1}(x)]^3 = x$ and thus $f^{-1}(x) = x^{1/3}$. It is easy to see that $f^{-1}[f(x)] = f^{-1}[x^3] = (x^3)^{1/3} = x$. Note that if $x_1 \neq x_2$, then $x_1^3 \neq x_2^3$; hence, the above is valid since an inverse exists.

Now, consider the function defined by $f(x) = x^2$, $-1 \leq x \leq 1$. (See Figure 6.2.4.) From Figure 6.2.4, we see that although $x_1 \neq x_2$, $f(x_1) = f(x_2)$. For example, if $x_1 = -\frac{1}{2}$ and $x_2 = \frac{1}{2}$, $x_1 \neq x_2$, but $f(x_1) = (-\frac{1}{2})^2 = (\frac{1}{2})^2 = f(x_2)$. Thus, the function

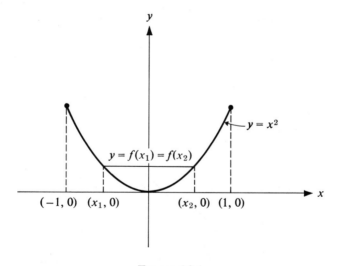

FIGURE 6.2.4

$f(x) = x^2$, $-1 \leq x \leq 1$ *does not* have an inverse. If we consider the function, $f(x) = x^2$, for $x \geq 0$, then f does have an inverse. It is clear from Figure 6.2.5 that if $x_1 \neq x_2$, then $f(x_1) \neq f(x_2)$. To find the inverse, we solve $f[f^{-1}(x)] = x$, $x \geq 0$ yielding $[f^{-1}(x)]^2 = x$, and thus $[f^{-1}(x)] = x^{1/2}$, $x \geq 0$.

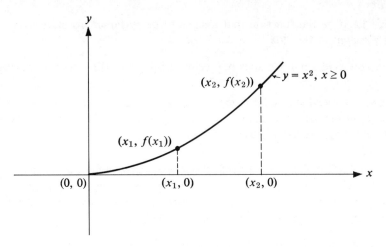

FIGURE 6.2.5

If we have the graph of a function f, it is quite easy to obtain the graph of f^{-1}. Indeed, the graph of f^{-1} is just the graph of f rotated around the line $y = x$. (See Figure 6.2.6.) Let us see why this is the case. If we rotate the point (x_1, y_1) about the line $y = x$, we obtain the point (y_1, x_1). Let (a, b) be a point in the graph of f, that is, $f(a) = b$. Then $f^{-1}(b) = a$ by definition of f^{-1}. If we rotate the point $(a, b) = (a, f(a))$ about the line $y = x$, we get the point $(b, a) = (b, f^{-1}(b))$. But the point $(b, f^{-1}(b))$ is a point in the graph of f^{-1}. This argument shows that if we rotate the graph of f about the line $y = x$, we end up in the graph of f^{-1}. It is not hard to see that we get *every* point in the graph of f^{-1} in this way. (Just reverse the roles of f and f^{-1}.)

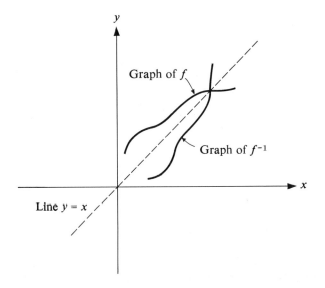

FIGURE 6.2.6

Let us consider some examples.

Example 1.   Let $f(x) = 2x + 1$. Does $f^{-1}(x)$ exist? If so, find f^{-1} and construct its graph.

Solution.   To show that f^{-1} exists, we must show that $f(x_1) = f(x_2)$ implies that $x_1 = x_2$. Suppose $f(x_1) = f(x_2)$, that is, $2x_1 + 1 = 2x_2 + 1$. Then clearly $x_1 = x_2$. To find f^{-1}, we must solve the equations

$$f[f^{-1}(x)] = x$$

or

$$2[f^{-1}(x)] + 1 = x$$

since $f(x) = 2x + 1$. Thus, $2f^{-1}(x) = x - 1$ and therefore

$$f^{-1}(x) = \frac{x - 1}{2}.$$

The graph is drawn in Figure 6.2.7. Note that geometrically the point $(-\frac{1}{2}, 0)$ becomes $(0, -\frac{1}{2})$ and $(0, 1)$ becomes $(1, 0)$. Thus, if the graph of f is a straight line, once we

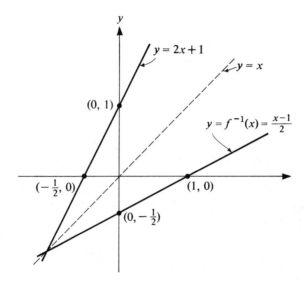

FIGURE 6.2.7

know two points on the line, it is easy to construct the graph and write the equation of the line representing the inverse function. (It is not difficult—see Exercise 11 of this section—to show that the inverse of any linear function is again a linear function.)

Example 2.   Let $f(x) = e^x$. We have seen in Section 1.5 of Chapter 1 that if $x_1 \neq x_2$, then $e^{x_1} \neq e^{x_2}$. Also, we know that the inverse function $f^{-1}(x) = \log x$. In like manner, if $f(x) = \log x$, its inverse function is e^x. Applying the results in this section to find the inverse of the exponential function, we must solve

$$f[f^{-1}(x)] = x \qquad \text{or} \qquad e^{f^{-1}(x)} = x,$$

which yields

$$f^{-1}(x) = \log x.$$

The graphs of e^x and its inverse $\log x$ are again displayed in Figure 6.2.8.

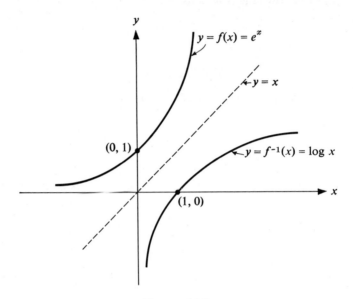

FIGURE 6.2.8

Let us now investigate the problem of finding derivatives of the inverse functions once we know the original function. It turns out that we can actually calculate the derivative of the inverse function with only a knowledge of the derivative of the original function.

For convenience we let g be the inverse function of f. Hence $g[f(x)] = x$. We differentiate this equation using the chain rule. Thus, $(d/dx)g[f(x)] = (d/dx)(x)$ and so $g'[f(x)]f'(x) = 1$. Hence we see that if $f(a) = b$, then

$$g'(b) = \frac{1}{f'(a)} \qquad \text{or} \qquad (f^{-1})'(b) = \frac{1}{f'(a)}.$$

Written in the Leibnitz notation, this relation becomes

$$\frac{d}{dx} f^{-1}(x)\bigg|_{x=b} = \frac{1}{\dfrac{d}{dx}[f(x)]\bigg|_{x=a}},$$

where $b = f(a)$. We now apply this rule to some examples.

Example 3.   Suppose $f(x) = x^3$. Find $(f^{-1})'$ (2).

Solution.   First we do the problem by means of the formula. Let $g = f^{-1}$. Then

$g'(2) = 1/f'(2^{1/3})$ since $f(2^{1/3}) = 2$. But since $f(x) = x^3, f'(x) = 3x^2$ and $f'(2^{1/3}) = 3 \cdot 2^{2/3}$. Thus, $g'(2) = 1/(3 \cdot 2^{2/3}) = (f^{-1})'(2)$. This problem can be done without recourse to the formula. We start with the equation $f[g(x)] = x$, where $g = f^{-1}$. This equation can be written as

$$[g(x)]^3 = x$$

since $f(x) = x^3$. Now differentiating by the chain rule, we find that

$$3[g(x)]^2 g'(x) = 1.$$

Hence,

$$g'(x) = \frac{1}{3[g(x)]^2}.$$

Finally,

$$g'(2) = \frac{1}{3 \cdot 2^{2/3}}$$

since $g(2) = 2^{1/3}$. Of course, in this case, since we know that $f^{-1}(x) = x^{1/3}$, it is just as easy to find f^{-1} by differentiating $x^{1/3}$. Let us look at a different sort of example.

Example 4.   Consider the function $g(x) = x^{1/n}$, for $x \geq 0$ and n a positive integer. Using the fact that $(d/dx)[x]^n = nx^{n-1}$, verify that

$$g'(x) = \frac{1}{n} x^{[(1/n)-1]}.$$

Solution.   Let $f(x) = x^n$, $x \geq 0$. It should be clear that an inverse exists and $f^{-1}(x) = g(x) = x^{1/n}$. Now, $f[g(x)] = x$ and thus $[g(x)]^n = x$. Using the chain rule, we see that $n[g(x)]^{n-1}g'(x) = 1$ and thus

$$g'(x) = \frac{1}{n[g(x)]^{n-1}} = \frac{1}{n[x^{1/n}]^{n-1}}$$

$$= \frac{1}{n} \frac{1}{x^{1-(1/n)}} = \frac{1}{n} x^{(1/n)-1}.$$

This provides justification for what we have assumed until now.

Example 5.   Use the fact that $(d/dx)(e^x) = e^x$ to show that $(d/dx)(\log x) = 1/x$.

Solution.   If $f(x) = e^x$, then $f^{-1}(x) = \log x$ is the inverse function. Hence,

$$f[f^{-1}(x)] = x \qquad \text{or} \qquad e^{f^{-1}(x)} = x.$$

Thus,

$$\frac{d}{dx}[e^{f^{-1}(x)}] = e^{f^{-1}(x)}\left[\frac{df^{-1}}{dx}(x)\right] = 1.$$

Since $e^{f^{-1}(x)} = e^{\log x} = x$, we have

$$\frac{df^{-1}}{dx}(x) = \frac{1}{x} \qquad \text{or} \qquad \frac{d}{dx}(\log x) = \frac{1}{x}.$$

Exercises

1. Find the inverses of the following if they exist and draw the graphs of both f and f^{-1}.

    (a) $f(x) = x + 2$ (d) $f(x) = x^5$

    (b) $f(x) = 3x + 1$ (e) $f(x) = \dfrac{x - 1}{x + 1}$

    (c) $f(x) = \sqrt[3]{x}$ (f) $f(x) = x^2 + 1$.

2. Justify, using intuitive arguments, the following statement: "If $f'(x) > 0$ for all x, then f^{-1} exists."

3. Show that the function f defined by $f(x) = \sqrt{1 - x^2}, 0 < x \le 1$, has an inverse function. Find where the graph of f has a vertical tangent.

4. Let f be a function defined by

    $$f(x) = \frac{ax + b}{cx - a},$$

    where a, b, c are fixed nonzero numbers. Show that f^{-1} exists and $f^{-1} = f$.

5. Let $g(x) = x^4 + 13$, h being the inverse of g. Compute $h'(14)$ by first finding a formula for h and then do it, by using the fact that

    $$\frac{d}{dx}\, [f^{-1}](x) = \frac{1}{f'[f^{-1}(x)]}.$$

6. Find $h'(19)$ if h is the inverse of the function $g(x) = x^7 + x^5 + 17$.

7. Let $f(x) = x^5 + x^3 + 1$. Does f^{-1} exist? (Use Exercise 2.) Find $(f^{-1})'(1)$, $(f^{-1})'(-1)$, and $(f^{-1})'(41)$. What must you do in order to find $(f^{-1})'(0)$?

8. Let F be the function defined by

    $$F(x) = \int_0^x \frac{dt}{1 + t^2}.$$

    Show that if $y = F^{-1}(x)$, then y satisfies the equation $y' = 1 + y^2$.

9. (a) If h is the inverse of a function g, where g'' exists, find h'' in terms of the derivatives for g.

    (b) Let $g(x) = (x - 1)/(x + 1)$. Apply (a) to finding h'', where h is the inverse of g.

10. Find dy/dx if y is implicitly defined as a function of x by

    $$ye^x - xe^y = \frac{xy}{\log(xy)}.$$

11. Show that the inverse of any function of the form $f(x) = ax + b$, where a and b are constants, always exists if $a \ne 0$. What is the graph of f^{-1}? Hence, conclude that the inverse of any linear function is again a linear function.

12.* If $g[f(x)] = 3x^2 + 1$, find g' in terms of f and f'.

6.3 Rate, Change, and Velocity

The world of experience offers no shortage of quantities that are a function of time. The gross national product, batting averages, the position of a rocket, the number of cars in Chicago, the Dow Jones industrial average, the price of copper, the water level in Lake Michigan, and the number of people in the United States are all functions of time. For at least some of these quantities there is considerable interest—emotional, financial, and otherwise—in the rate at which the quantity is changing. It is a remarkable and extremely useful fact that the derivative enables us to determine this rate of change for such a diverse range of phenomena. We have observed some of this in various examples in Chapter 3. Let us now consider another example in some detail.

A car moves along a straight road R. Its distance from a city C is given as a function of time by $S(t)$. We might determine the position of the car at time t_1 and its position a short time later at time t_2. (See Figure 6.3.1.) Then the quantity $S(t_2) - S(t_1)$ represents

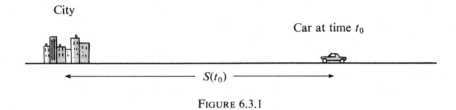

FIGURE 6.3.1

the distance traveled in the time interval between t_1 and t_2. The expression

$$[S(t_2) - S(t_1)]/(t_2 - t_1)$$

represents the average speed over this time period. (See Figure 6.3.2.)

Certainly this last calculation is familiar to everyone and is used in every day life to calculate the average speed of cars, boats, bicycles, etc. Of course, this average speed

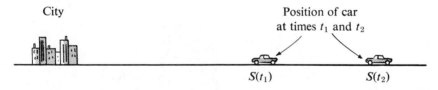

FIGURE 6.3.2

does not tell us how fast the car was going at t_1. To make a good estimate of the speed at t_1, one might choose t_2 close to t_1; that is, make the time interval for the average small. Let us do this by taking a limit. Consider

$$\lim_{t \to t_1} \frac{S(t) - S(t_1)}{t - t_1}.$$

On the one hand this is the limit of the average speed of the car over the time interval from t_1 to t; thus we would expect it to approach the speed of the car at $t = t_1$. On the other hand, the expression is just $S'(t_1)$ (if $S(t)$ is differentiable at t_1).

This discussion was only intended to make plausible our next statement.

DEFINITION.   *If $S(t)$ represents the position of a particle at time t, then the velocity of the particle at time t_0 is $S'(t_0)$, if S is differentiable at t_0.*

This definition only applies to a particle that travels in a straight line and we restrict ourselves to this case.

GENERAL DEFINITION.   *If $g(t)$ represents some quantity, then the **rate of change** of $g(t)$ at t_0 equals $g'(t_0)$, if the derivative exists.*

We should observe that we defined the *velocity*, not the *speed*, to be $S'(t_0)$. The number $S'(t_0)$ may be either plus or minus and the sign may be interpreted as saying the particle is either coming or going, moving up or down, etc. The speed, on the other hand, is *always* a positive quantity. The relation between them is simple; speed $= |$velocity$|$. This will be amply illustrated in the examples.

Example 1.   A particle moves along the x axis in such a way that its position (x coordinate) at time t is $5t^3 - 9t + 3$. Find the velocity of the particle when $t = 7$.

Solution.   Let $S(t) = 5t^3 - 9t + 3$ represent the position of the particle at time t. Then $S'(t) = 15t^2 - 9$. The velocity of the particle at $t = 7$ equals $S'(7) = 15 \cdot 7^2 - 9 = 726$.

Example 2.   A ball is dropped from the top of a building that is 1024 ft high. Its distance $S(t)$ from the top of the building is $16t^2$ at time t (assuming it was dropped at $t = 0$). What is the speed of the ball when it hits the ground?

Solution.   First we draw a picture to illustrate the problems. (See Figure 6.3.3.) To

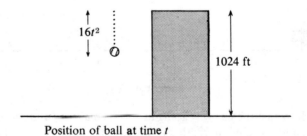

Position of ball at time t

FIGURE 6.3.3

determine when the ball will hit the ground, we set $16t^2 = 1024$. Thus, $t^2 = 64$ or $t = 8$. Since $S(t) = 16t^2$, $S'(t) = 32t$. The velocity at $t = 8$ is $S'(8) = 32 \cdot 8 = 256$. Thus the speed of the ball when it hits the ground is 256 ft/sec (about 180 miles per hour).

Example 3. A ball is thrown straight upward at time $t = 0$. Its distance (in feet), $S(t)$, above the ground is $288 - 16t^2$ at time t (in seconds). (See Figure 6.3.4.)

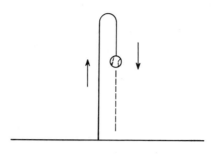

FIGURE 6.3.4

(a) What is its velocity at the moment it leaves the ground?

(b) How high does it get before it comes down (the high point is just when the velocity is zero)?

(c) When does it reach the ground?

(d) What is its velocity when it reaches the ground?

Solution. As before we set $S(t) = 288t - 16t^2$, where $S(t)$ represents the distance from the ground to the ball. To answer (a), we find $S'(t) = 288 - 32t$. Thus the velocity when the ball leaves the ground is just $S'(0) = 288$ ft/sec. To find the high point, we set $0 = S'(t) = 288 - 32t$. Thus, $t = 288/32 = 9$. At $t = 9$, $S(9) = 288 \cdot 9 - 16 \cdot 9^2 = 1296$ ft. To find when the ball reaches the ground, we set $S(t) = 0 = 288t - 16t^2$. Thus, $16t(18 - t) = 0$, and so the ball returns to the ground after 18 sec. (The other root of the equation $t = 0$ simply represents the time the ball left the ground.) The velocity of the ball when it strikes the ground is

$$S'(18) = 288 - 32 \cdot 18 = -288.$$

This is just the velocity with which it left the ground, except for the sign. This is to be expected if there is no air resistance. Note that $S'(0) = 288$ while $S'(18) = -288$. The velocity is positive when the ball is going up and negative when it comes down, since it is going in opposite directions.

We will now consider another type of "rate" problem.

Example 4.* A rock is thrown into a pond and a ripple moves out at a constant velocity of 5 ft/sec. What is the rate of change of the surface area encompassed by the ripple when the ripple is 10 ft from the place where the rock struck? (See Figure 6.3.5.)

Solution. Note that the ripple moves out in a straight line from the place where the rock strikes as shown by the arrow. At first this may seem like a very different sort of rate problem and one beyond the scope of present techniques. As we shall see, it is not so difficult. Let r be the distance of the ripple from the center. The area surrounded by the ripple is $A = \pi r^2$. Certainly A and r are functions of time. What we do now is simply differentiate the equation $A = \pi r^2$ with respect to t, obtaining $dA/dt = 2\pi r \, dr/dt$.

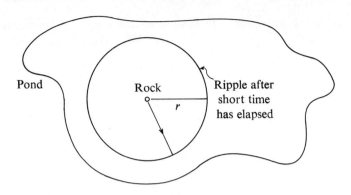

FIGURE 6.3.5

Now dA/dt is the rate of change of the surface area, which is what we are looking for. The quantity r is familiar and easy to handle. Finally, dr/dt is just the rate of change of the radius, or the velocity of the ripple and that was given. More precisely, $dr/dt = 5$. Thus the rate of change of the surface area where the ripple is 10 ft from the center is

$$\frac{dA}{dt} = 2 \cdot \pi \cdot 10 \cdot 5 = 100\pi \text{ (sq ft/sec)}.$$

Example 5.   A water tank in the shape of an inverted cone is being filled with water at a rate of 5 cubic ft/min. (See Figure 6.3.6 for the dimensions of the cone.) How fast is

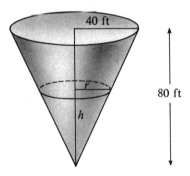

FIGURE 6.3.6

the surface of the water rising when the water is 10 ft deep? (Use the fact that in Figure 6.3.6, the volume of the cone is $\frac{1}{3}\pi r^2 h$.)

Solution.   If the water is h ft deep, the volume of the water is $V = (\pi/3)r^2h$. First, observe that $r/h = 40/80$ so that $r = h/2$. Thus, $V = (\pi/12)h^3$. Now both the volume and the height of the water are functions of time. Differentiating both sides with respect to t, we obtain

$$\frac{dV}{dt} = \frac{\pi}{12} 3h^2 \frac{dh}{dt}.$$

But $dV/dt = 5$ since the tank is being filled at a rate of 5 cubic ft/min. Thus

$$5 = (\pi/4)h^2 \, dh/dt \quad \text{or} \quad dh/dt = 20/\pi h^2.$$

Thus when $h = 10$, $dh/dt = 20/\pi 10^2 = 1/5\pi$ ft/sec. Note that dh/dt, the "velocity" of the surface level, decreases as h increases. Why is this obvious on physical grounds?

Rather than do another rate problem, we give a general schema for handling such problems.

1. Draw a big picture of the situation and label it clearly.
2. Write an equation relating the quantities known and unknown in the problem (it may be necessary to use auxiliary equations to do this).
3. Differentiate the equation with respect to time.
4. Substitute the known quantities to obtain an answer.

It is also useful to discuss the converse to the problems treated earlier in this section, that is, given the rate of change of a function g, find the function. From Chapter 5, we know that this is simply a problem in integration. Let us just consider a few examples.

Example 6.   The velocity, at time t, of a moving body is given by gt, where g is a constant. If the body's position at time $t = 0$ is given by s_0, find the distance s as a function of time t.

Solution.   We know that the velocity is given by ds/dt. Hence, $ds/dt = gt$ and $s(t) = \frac{1}{2}gt^2 + c_1$, where c_1 is an arbitrary constant. However, using the fact that $s(0) = s_0$, and $c_1 = s_0$; $s(t)$ is given by

$$s(t) = \frac{1}{2}gt^2 + s_0.$$

Example 7.   Acceleration a is simply the rate of change of velocity v. That is, if s is the distance as a function of time, then $v = ds/dt$ and $a = dv/dt = d^2s/dt^2$. If a stone were thrown straight down from a stationary balloon 10,000 ft above the ground with a speed of 48 ft/sec, assuming that the acceleration is 32 ft/sec^2, locate the stone and find its speed 20 seconds later.

Solution.   Let us assume that positive distance and velocity are directed upward. (See Figure 6.3.7.) When the stone leaves the balloon, it is subject to the law $a = dv/dt = -32$ ft/sec^2 and hence, $v = -32t + c_1$, where c_1 is a constant. However, we are told that when $t = 0$, $v = -48$ ft/sec. Hence, $c_1 = -48$ and $v(t) = ds/dt = -32t - 48$. (The minus sign—since we said that positive distance and acceleration is upward, and the stone is thrown downward.) Thus,

$$s(t) = -32\frac{t^2}{2} - 48t + c_2,$$

where c_2 is a constant. However, at time $t = 0$, $s(0) = 10,000$. Thus, $c_2 = 10,000$ and

$$s(t) = -16t^2 - 48t + 10,000.$$

The stone's position 20 seconds after being dropped is $s(20) = -16(20)^2 - 48(20) + 10,000 = 2640$ feet above the ground and its speed, since $v(20) = -32(20) - 48 = -688$, is 688 ft/sec.

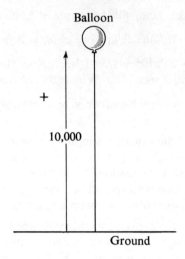

FIGURE 6.3.7

Exercises

1. A particle moves along the x axis in such a way that its x coordinate is equal to $(t^3 - e^{2t})$ at time t.
    (a) What is the velocity of the particle at $t = 2$?
    (b) When, if ever, is the velocity 0?
2. A particle moves along the positive x axis in such a way that its distance from the origin is $[4t^2 + \log(1 + t^2)]$ at time t.
    (a) What is the velocity of the particle when $t = 0$?
    (b) What is the velocity of the particle when $t = 1$?
3. Let the expression below represent the x coordinate of a particle at time t. Find the velocity at the time indicated.

    (a) $t^5 - 6t^2 + 9, \quad t = -2$

    (b) $e^{4t^2} - t, \quad t = 3$

    (c) $t^6 - 4t^4 + t^2, \quad t = 1$

    (d) $2t + 4 \log t$ for $t > 0, \quad t = 9$

    (e) $t\sqrt{1 + t^2}, \quad t = -1$

    (f) $(1 - t^2)/\sqrt{1 + t}, \quad t = 3$

4. In Exercise 3(f) above discuss what happens at $t = -1$.
5. A rocket is fired in a vertical direction at $t = 0$. The equation of its distance from the ground is $h(t) = 4800 - 16t^2$.
    (a) What is its velocity when it leaves the ground?
    (b) What is its velocity 10 seconds after it is launched?
    (c) What is its maximum altitude?
    (d) At what time does it return to the ground?
    (e) What is its velocity when it reaches the ground?

6. A bowling ball is dropped from the top of a building 512 ft high at time $t = 0$. Its distance from the ground is $512 - 16t^2$ at time t.
    (a) What is its initial velocity (that is, at $t = 0$)?
    (b) When does it strike the ground?
    (c) What is its velocity when it strikes the ground?

7. A tank in the shape of a cone is filled with water. (See Figure 6.3.8.) The water flows out at a rate of 12 cubic ft/min. How fast is the surface level dropping when the release valve is first opened? When the water is 30 ft deep? [HINT: Recall that the volume of a cone is $\frac{1}{3}\pi r^2 h$.]

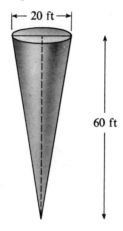

FIGURE 6.3.8

8. A ladder 40 ft long is standing against a building. The base of the ladder is moved away from the building at the rate of 2 ft/sec. How fast is the top of the ladder moving down the wall when the base of the ladder is 0 ft, 20 ft, 30 ft, from the foot of the building? (See Figure 6.3.9.)

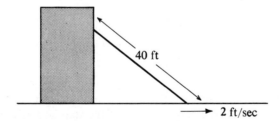

FIGURE 6.3.9

9. A rubber balloon is being inflated with air at a constant rate of 6 cubic ft/sec. The volume V and area A are related by the equation $A^3 = 36\pi V^2$. What is the rate of change of the area when the volume is 20 cubic ft?

10. As in Exercise 9, a balloon is being inflated at a constant rate of 6 cubic ft/sec. The volume is $V = (4\pi/3)r^2$, where r is the radius. How fast is the radius changing when the radius is 10 ft, 20 ft? What is the rate of change of the radius when the radius is 0? Does this seem reasonable?

11. A ball is rolled over a level lawn with an initial velocity of 25 ft/sec. Due to friction, the velocity decreases at the rate of 6 ft/sec^2. How far will the ball roll?

12. The acceleration at time t seconds is known to be $(12t + 2)$ ft/sec^2. At time $t = 3$ sec, the moving object is at $s = 5$ ft and its velocity is -20 ft/sec. Find its position at $t = 5$ sec.

13. The acceleration at time t of an object moving on a straight line is $12t^2$ ft/sec^2. Its position at time $t = 2$ sec is known to be 28 ft, and at time $t = 3$ sec, it is 98 ft. Find its position at time t.

14. A ball was dropped from a balloon 640 ft above the ground. If the balloon were rising at the rate of 48 ft/sec, find

    (a) the greatest distance above the ground attained by the ball.

    (b) the time the ball was in the air.

    (c) the speed of the ball when it struck the ground.

    [HINT: Assume the acceleration a is governed by $a(t) = -32$ ft/sec^2.]

6.4 L'Hôpital's Rule

Suppose we wish to find $\lim_{x \to x_0} f(x)/g(x)$ when f, g are continuous at x_0. In most cases we can apply Property 6 of Section 2.4 to obtain

$$\lim_{x \to x_0} \frac{f(x)}{g(x)} = \frac{\lim_{x \to x_0} f(x)}{\lim_{x \to x_0} g(x)} = \frac{f(x_0)}{g(x_0)}.$$

But what if $f(x_0) = g(x_0) = 0$? Although the above method does not apply, we now present a technique for handling this case.

THEOREM (L'HÔPITAL'S RULE).   *Let f and g be differentiable at x_0. Let $f(x_0) = g(x_0) = 0$. Let $g(x) \neq 0$ near x_0. Then*

$$\lim_{x \to x_0} \frac{f(x)}{g(x)} = \lim_{x \to x_0} \frac{f'(x)}{g'(x)}$$

if $g'(x_0) \neq 0$.

The proof of this theorem uses an extended version of the mean-value theorem. We omit the proof here and refer the reader to any of the standard advanced calculus texts. In spite of the many hypotheses, this theorem is quite easy to apply in practice.

Example 1.   Find $\lim_{x \to 0} (e^x - 1)/2x$.

Solution.   Note that both the numerator and denominator are 0 at $x = 0$. Since $f(x) = e^x - 1, f'(x) = e^x$. Since $g(x) = 2x, g'(x) = 2$. Thus,

$$\lim_{x \to 0} \frac{e^x - 1}{2x} = \lim_{x \to 0} \frac{e^x}{2} = \frac{1}{2}.$$

Example 2. Find $\lim_{x \to 0} (1 + x - e^x)/x^2$.

Solution. If $f(x) = 1 + x - e^x$ and $g(x) = x^2$, then $f(0) = g(0) = 0$. By L'Hôpital's rule,

$$\lim_{x \to 0} \frac{1 + x - e^x}{x^2} = \lim_{x \to 0} \frac{1 - e^x}{2x}.$$

But again both the numerator and denominator vanish at $x = 0$, so that it is not clear what $\lim_{x \to 0} (1 - e^x)/2x$ is. However, we can simply apply L'Hôpital's rule again! Thus,

$$\lim_{x \to 0} \frac{1 - e^x}{2x} = \lim_{x \to 0} \frac{-e^x}{2} = -\frac{1}{2}.$$

Example 3. Find $\lim_{x \to 1} \log x/(1 - x)$.

Solution. By L'Hôpital's rule,

$$\lim_{x \to 1} \frac{\log x}{1 - x} = \lim_{x \to 1} \frac{1/x}{-1} = -1.$$

Remark. In Example 2, we applied L'Hôpital's rule twice to obtain the answer. One might think that it is permissible to apply the rule as many times as one wishes to a quotient: that is,

$$\lim_{x \to x_0} \frac{f(x)}{g(x)} = \lim_{x \to x_0} \frac{f'(x)}{g'(x)} = \lim_{x \to x_0} \frac{f''(x)}{g''(x)} = \cdots.$$

This is true as long as both numerator and denominator vanish at x_0. But note in Example 3, if we had continued another step we would have obtained

$$\lim_{x \to 1} \frac{\log x}{1 - x} = \lim_{x \to 1} \frac{1/x}{-1} \overset{?}{=} \lim_{x \to 1} \frac{-1/x^2}{0} = \frac{-1}{0} = -\infty.$$

Since we know the term before the question mark equals -1, this is clearly not valid.

Exercises

1. Find:

   (a) $\displaystyle \lim_{x \to 2} \frac{x - 2}{e^{(x-2)} - 1}$

   (b) $\displaystyle \lim_{x \to 1} \frac{1 - e^{(1-x)}}{\log x}$

   (c) $\displaystyle \lim_{x \to 1} \frac{x \log x}{\sqrt{1 - x^2}}$

   (d) $\displaystyle \lim_{x \to 0} \frac{xe^x - x}{\log(1 - x)}$

   (e) $\displaystyle \lim_{x \to 2} \frac{\sqrt{x^2 - 4}}{x - 2}$

2. There is a version of L'Hôpital's rule that is valid for the case when $\lim_{x \to x_0} f(x) = \pm \infty = \lim_{x \to x_0} g(x)$. (*Note:* In this situation, the expression $\lim_{x \to x_0} f(x)/\lim_{x \to x_0} g(x)$ is ambiguous.) As before,

$$\lim_{x \to x_0} \frac{f(x)}{g(x)} = \lim_{x \to x_0} \frac{f'(x)}{g'(x)}$$

if this limit exists. For example, let us find $\lim_{x \to 0} \log x/x^{-2}$. Note that $\lim_{x \to 0} \log x = -\infty$ and $\lim_{x \to 0} x^{-2} = \infty$. By the second version of L'Hôpital's rule,

$$\lim_{x \to 0} \frac{\log x}{x^{-2}} = \lim_{x \to 0} \frac{1/x}{-2x^{-3}} = \lim_{x \to 0} \frac{x^3}{-2x} = \lim_{x \to 0} \frac{3x^2}{-2} = 0.$$

(Why is the penultimate step valid?)

3. Find $\lim_{x \to 0} x \log x$. [HINT: Rewrite the initial expression as either $\log x/(1/x)$ or $x/(1/\log x)$ and apply Exercise 2. *Caution:* One rewrite works but the other does not.]
4. Find $\lim_{x \to 0} \sqrt{x} \log x$.
5. Find $\lim_{x \to 0} x^x$ for $x > 0$. [HINT: First consider $\log x^x = x \log x$.]

6.5 * Trigonometric Functions

In this section we introduce the derivatives of the well-known trigonometric functions. We discussed the functions sine, cosine, and tangent in Section 1.7. As a reminder, we will present their graphs below in Figures 6.5.1, 6.5.2, and 6.5.3. Also, recall that $\tan x = \sin x/\cos x$. We introduce the derivatives of these functions in table 6.5.1.

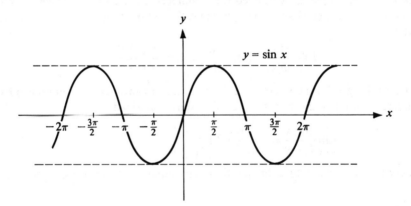

FIGURE 6.5.1

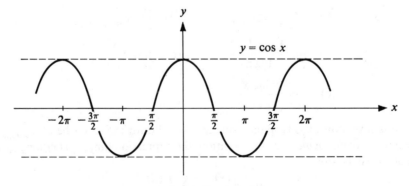

FIGURE 6.5.2

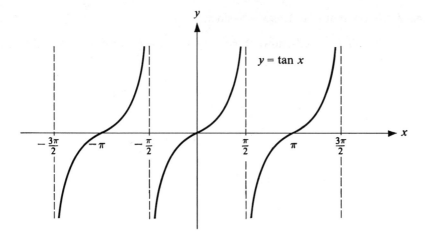

FIGURE 6.5.3

Table 6.5.1

$f(x)$	$f'(x)$
$\sin x$	$\cos x$
$\cos x$	$-\sin x$
$\tan x$	$\dfrac{1}{(\cos x)^2} = (\cos x)^{-2}$

Using the chain rule for composite functions, one can easily obtain the formulas in Table 6.5.2.

Table 6.5.2

1. $\dfrac{d}{dx} \sin u(x) = (\cos u(x)) \dfrac{du}{dx}$

2. $\dfrac{d}{dx} \cos u(x) = (-\sin u(x)) \dfrac{du}{dx}$

3. $\dfrac{d}{dx} \tan u(x) = \dfrac{1}{(\cos u(x))^2} \dfrac{du}{dx}$

We will not derive the formulas in Tables 6.5.1 and 6.5.2. However, we do point out that once you know that $(d/dx)\sin x = \cos x$, the rest can be obtained without much trouble. This is illustrated both in the exercises at the end of the section and in the next example.

Example 1.   Show that $(d/dx)\cos x = -\sin x$.

   Solution.   Recall the well-known formula $(\sin x)^2 + (\cos x)^2 = 1$. We differentiate this implicitly:

$$0 = \frac{d}{dx}(1) = \frac{d}{dx}[(\sin x)^2 + (\cos x)^2]$$

$$= \frac{d}{dx}(\sin x)^2 + \frac{d}{dx}(\cos x)^2$$

$$= 2\sin x \frac{d}{dx}\sin x + 2\cos x \frac{d}{dx}\cos x$$

(this last step is just the chain rule)

$$= 2\sin x \cos x + 2\cos x \frac{d}{dx}(\cos x).$$

Of course, we don't know what $(d/dx)\cos x$ is—that is what we are trying to determine! Solving this equation, we obtain

$$2\cos x \frac{d}{dx}(\cos x) = -2\cos x \sin x.$$

Thus,

$$\frac{d}{dx}(\cos x) = -\sin x,$$

at least when $\cos x \neq 0$. (Why?) We stop at this point.

Example 2.   If $f(x) = \sin(\cos x^2)$, find $f'(x)$.

   Solution.

$$\frac{d}{dx}[\sin(\cos x^2)] = \cos[\cos x^2]\frac{d}{dx}(\cos x^2) \qquad \text{(by Table 6.5.2)}$$

$$= \cos[\cos x^2](-\sin x^2)\frac{d}{dx}x^2 \qquad \text{(by Table 6.5.2)}$$

$$= -2x \sin x^2 \cos[\cos x^2].$$

   We now introduce the inverse trigonometric functions. Observe from Figures 6.5.1, 6.5.2, and 6.5.3 that since the trigonometric functions are periodic, they do not have the property that if $x_1 \neq x_2$ then $f(x_1) \neq f(x_2)$. In fact, just draw any line parallel to the x axis through the graphs and it is easy to see that there are many points $a_1, a_2, \ldots$ such that $f(a_1) = f(a_2) = \cdots$ etc. However, these functions may be restricted to intervals on which they are increasing, and in this manner partial inverses may be defined.

Let us begin with the function $f(x) = \sin x$, where $-\pi/2 \le x \le \pi/2$. (See Figure 6.5.4.) Observe that for $-\pi/2 \le x \le \pi/2, f'(x) = \cos x$ and $\cos x \ge 0$ on this interval. Thus, by Exercise 2 in Section 6.2, we see that on $[-\pi/2, \pi/2]$ the sine function has an inverse,

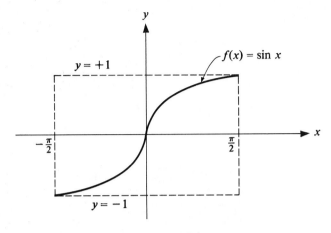

FIGURE 6.5.4

$f^{-1}(x)$. We call $f^{-1}(x)$ the arc sine function and write $f^{-1}(x) = \text{arc sin } x$. Thus, if $y = \text{arc sin } x$, then $x = \sin y$ by definition of inverse. The domain of arc sin x must be $[-1, 1]$. Its graph is shown in Figure 6.5.5.

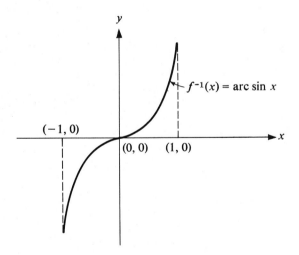

FIGURE 6.5.5

Partial inverses can be defined for all of the trigonometric functions in exactly the same way as for the sine function. We consider only one more case, that of the tangent function. It is clear from the graph in Figure 6.5.3 that the tangent does not have an inverse function. However, consider the function $g(x) = \tan x$, $-\pi/2 < x < \pi/2$. Its graph is shown in Figure 6.5.6. It is clear that in the interval $(-\pi/2, \pi/2)$ the function is increasing

—that is, $f'(x) > 0$, and hence an inverse $g^{-1}(x)$ exists. We call this inverse the *arc tangent function* and write $g^{-1}(x) = \text{arc tan } x$. The domain must be the set of *all* real numbers. Its graph, obtained by reflecting the tangent function about the line $y = x$ is shown in Figure 6.5.7.

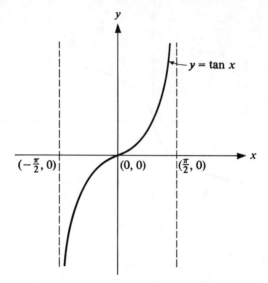

FIGURE 6.5.6

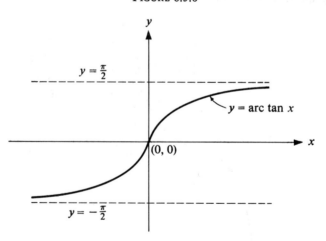

FIGURE 6.5.7

It is a simple matter to obtain the derivatives of the arc sin and arc tan functions. Surprisingly, these derivatives are simple algebraic functions. We begin with the arc sin function. Let $y = \text{arc sin } x$, $-1 \le x \le 1$. It follows that $x = \sin y$, and hence by implicit differentiation (and Table 6.5.1), $1 = \cos y \, dy/dx$. Thus,

$$\frac{dy}{dx} = \frac{d}{dx} (\text{arc sin } x) = \frac{1}{\cos y} = \frac{1}{\sqrt{1 - x^2}}$$

(since $\cos y = \sqrt{1 - \sin^2 y} = \sqrt{1 - x^2}$). In like manner, if $y = \arctan x$, then $\tan y = x$ and

$$\sec^2 y \cdot \frac{dy}{dx} = 1 \qquad \text{or} \qquad \frac{dy}{dx} = \frac{1}{\sec^2 y} = \frac{1}{1 + \tan^2 y} = \frac{1}{1 + x^2}.$$

We summarize our results in Tables 6.5.3 and 6.5.4.

Table 6.5.3

$f(x)$	$f'(x)$
arc sin x	$\dfrac{1}{\sqrt{1 - x^2}}$
arc tan x	$\dfrac{1}{1 + x^2}$

Table 6.5.4

$f(x)$	$f'(x)$
arc sin $u(x)$	$\dfrac{1}{\sqrt{1 - [u(x)]^2}} \dfrac{du}{dx}$
arc tan $u(x)$	$\dfrac{1}{1 + [u(x)]^2} \dfrac{du}{dx}$

The results in Table 6.5.4 are to be verified in Exercise 8 of this section.

Exercises

1. Find the derivative of each of the functions below.

(a) $\cos 5x$

(b) $\sin\sqrt{1 - x}$

(c) $\sqrt{\sin(1 - x)}$

(d) $e^{\sin 3x}$

(e) $\log(\cos 2x)$

(f) $\tan(x + 5)^3$

(g) $\tan e^x$

(h) $\log\left[\tan\dfrac{1 + x}{1 - x}\right]$

(i) $\log[\sin(\cos x)]$

(j) $\dfrac{e^x}{(\sin x)^2}$

(k) $\tan(\log x)$

(l) $\sin[\sin(\sin x)]$

(m) $\dfrac{e^{\tan x}}{\sqrt{1 + \sin x}}$

2. Find the following limits (use L'Hôpital's rule).

(a) $\lim\limits_{x\to 0} \dfrac{\sin x}{x}$

(d) $\lim\limits_{x\to 1} \dfrac{\log x}{\cos(\pi/2)x}$

(b) $\lim\limits_{x\to 0} \dfrac{1 - \cos x}{x}$

(e) $\lim\limits_{x\to 0} \dfrac{\tan x}{x}$

(c) $\lim\limits_{x\to 0} \dfrac{1 - \cos x}{x^2}$

3. Using the fact that (i) $\cos x = \sin(x + \pi/2)$ and (ii) $\cos(x + \pi/2) = -\sin x$, show that $(d/dx)\cos x = -\sin x$. [You may use the fact that $(d/dx)\sin x = \cos x$.]
4. Show that $(d/dx)\tan x = 1/(\cos x)^2$. [HINT: $\tan x = \sin x/\cos x$.]
5. Derive the expressions in Table 6.5.2.
6. How would you define the inverse function, arc cos x, viewing the cosine only on the interval $[0, \pi]$?
7. What is the derivative of arc cos x?
8. Differentiate the following.

(a) arc $\sin(x^2 - 1)$

(d) x arc sin x

(b) arc $\cos(2x + 5)$

(e) arc sin x + arc tan x

(c) arc tan $\sqrt{x}$

(f) arc $\tan(\sin x^2)$

9. Derive the formulas in Table 6.5.4.
10. A man whose eyes are 6 ft above the ground is walking toward a lamp post 10 ft tall at a rate of 5 ft/sec. (See Figure 6.5.8.) He keeps his eyes fixed on the light.

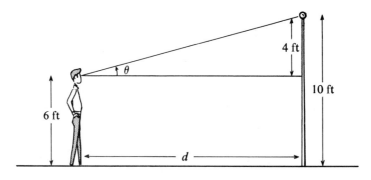

FIGURE 6.5.8

How fast is his head rotating when he is 24 ft from the lamp post? (*Remark:* If θ is the angle indicated, then $d\theta/dt$ is the rate at which his head is rotating—called angular velocity by some.) [HINT: Obtain an equation relating θ and d. Differentiate it with respect to time.]

11. A searchlight turns at the constant rate of 2 revolutions per minute. (It throws a narrow beam of light that we consider to be a straight line.) The beam is parallel to the ground. There is a wall 100 ft from the light. (See Figure 6.5.9.) How fast is

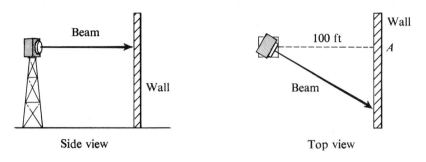

Side view   Top view

FIGURE 6.5.9

the beam moving along the wall when it is 100 ft from point A, the point nearest the light? [HINT: 2 revolutions per minute means the light has an angular velocity of 4π radius per minute.]

6.6 Summary of Chapter 6

(1) If we are given a function $y = g(x)$ in the form $F(x, y) = 0$, it is possible to find dy/dx without first solving for y. Simply differentiate formally using the chain rule on terms containing y.

(2) (a) If f is a function such that if $x_1, x_2 \in$ domain of f with $x_1 \neq x_2$ implying $f(x_1) \neq f(x_2)$, then f has associated with it an inverse function f^{-1} defined by

$$f[f^{-1}(x)] = x \qquad \text{and} \qquad f^{-1}[f(x)] = x.$$

(b) If $f'(x) > 0$ for $x \in$ domain of f, then f has an inverse function f^{-1} defined as above.

(c) If f and f^{-1} are differentiable, then implicit differentiation yields that

$$(f^{-1})'(x) = \frac{1}{f'[f^{-1}(x)]}.$$

(3) If $s(t)$ represents the position of a particle at time t, the *velocity* of the particle at time $t = t_0$ is given by $s'(t_0)$, if s is differentiable at t_0 and the acceleration at time $t = t_0$ is given by $s''(t_0)$ (provided that s'' exists). Given the velocity and acceleration with certain initial conditions, it is possible to find the distance function s.

(4) L'HÔPITAL'S RULE. *Let f and g be differentiable at x_0. If $f(x_0) = g(x_0) = 0$ and $g(x) \neq 0$ near x_0, then*

$$\lim_{x \to x_0} \frac{f(x)}{g(x)} = \lim_{x \to x_0} \frac{f'(x)}{g'(x)}$$

provided that near x_0, $g'(x) \neq 0$.

REVIEW EXERCISES

1. Find dy/dx if y is implicitly defined as a function of x by:

   (a) $2xy + y^{1/2} = x + y$

   (b) $x^2 y^2 = x^2 + y^2$

   (c) $(x + y)^4 + (x - y)^4 = x^5 + y^5$

   (d) $e^{x^2 y} + x \log xy = x^3$

   (e) $x^3 - xy + y^3 = 10$

   (f) $\dfrac{1}{y} + \dfrac{1}{x} = 1$

   (g) $e^y \log xy = y^2$

2. Find the equation of the line tangent to the following curves at the indicated points.

   (a) $x^2 + y^2 = 25$ at $(3, -4)$

   (b) $x^2 y^2 = 9$ at $(-1, 3)$

   (c) $(y - x)^2 = 2x + 4$ at $(6, 2)$

3. The total surface area of a right circular cone of height h and radius of base r is $S = \pi(r^2 + r\sqrt{r^2 + h^2})$. If S is constant, find dr/dh when $r = 3$ and $h = 4$.

4. A ladder 15 ft long rests against a house. It slides down, the lower end slipping along the level ground at the rate of 2 ft/sec. How fast is the upper end of the ladder sliding down the wall when it is 12 ft from the ground?

5. Two airplanes fly eastward on parallel courses 12 miles apart. One flies at 240 miles per hour, the other at 300 miles per hour. How fast is the distance between the planes changing when the slower plane is 5 miles farther east than the faster plane?

6. Suppose that a particle moves along a straight line in such a way that its distance from some fixed point on the line at any time t is given by $s(t) = t^2 - 2\log(t + 1)$, where t is measured in seconds and s in feet.

   (a) Compute the velocity and acceleration at the end of 2 and at the end of 3 sec.

   (b) What happens to the velocity and acceleration as t becomes very large?

7. A ball rolling up a certain incline is slowed down at the rate of 9 ft/sec². If the ball is moving 12 ft/sec when it passes at a certain point, how far does it roll before it stops and begins to roll down? [HINT: Use the fact that acceleration

$$a = \frac{dv}{dt} = \frac{d^2 s}{dt^2} = \frac{dv}{ds} \cdot \frac{ds}{dt} = v\frac{dv}{ds}.$$

8. Evaluate the following limits

    (a) $\displaystyle\lim_{h\to 0}\frac{\sqrt{4+h}-2}{h}$
           (d) $\displaystyle\lim_{x\to 0^+} x\log x$

    (b) $\displaystyle\lim_{x\to 0}\frac{e^x-(1+x)}{x^2}$
           (e) $\displaystyle\lim_{x\to 0^+} x^x$ [HINT: $x^x = e^{x\log x}$]

    (c) $\displaystyle\lim_{x\to 1^+}\left(\frac{1}{x-1}-\frac{1}{\sqrt{x-1}}\right)$
        (f) $\displaystyle\lim_{x\to 1} x^{1/(1-x)}$ [HINT: $x^{1/(1-x)} = e^{\log x/(1-x)}$]

9. Determine whether or not an inverse exists for each of the following functions. If so, explicitly find the inverse and graph both the inverse and the original function.

    (a) $f(x) = 1/x, \quad x \neq 0$

    (b) $f(x) = 3x - 2$

    (c) $f(x) = x^4$

    (d) $f(x) = x^4, \quad x > 0$

    (e) $f(x) = |x|$.

10. An object moves along the curve $y = x^3$. At what points on the curve will the x and y coordinates be changing at the same rate?

11. Assume that the brakes on an automobile produce a constant deceleration of p ft/sec^2.

    (a) Determine what p must be to bring an automobile traveling 60 mi/hr (88 ft/sec) to rest in a distance of 100 ft from the point where the brakes are applied.

    (b) With the same p, how far would a car traveling 30 mi/hr travel before being brought to a stop?

7

Miscellaneous Topics II. Integration

In this section we first discuss finding antiderivatives of the basic trigonometric functions, sin, cos, and tan. We recall from Chapter 4 that

$$\frac{d}{dx}\sin x = \cos x \qquad \text{and} \qquad \frac{d}{dx}\cos x = -\sin x.$$

Hence, we can immediately conclude that

$$\int \sin x \, dx = -\cos x + C \qquad (7.1.1)$$

and

$$\int \cos x \, dx = \sin x + C, \qquad (7.1.2)$$

where C is any arbitrary constant.

In order to find an antiderivative of the tangent function, we first note that

$$\int \tan x \, dx = \int \frac{\sin x}{\cos x} \, dx.$$

Let us try a substitution. If in the second integral we let $u = \cos x$ and $du = -\sin x \, dx$, then

$$\int \frac{\sin x}{\cos x} \, dx = -\int \frac{du}{u} = -\log|u| + C = -\log|\cos x| + C$$

$$= \log \left| \frac{1}{\cos x} \right| + C = \log|\sec x| + C,$$

where C is an arbitrary constant. We also note that

$$\frac{d}{dx} \tan x = \frac{d}{dx} \left(\frac{\sin x}{\cos x} \right) = \frac{\cos^2 x + \sin^2 x}{\cos^2 x} = \frac{1}{\cos^2 x} = \sec^2 x.$$

Hence,

$$\int \sec^2 x \, dx = \tan x + C,$$

where C is an arbitrary constant. We summarize the results in Table 7.1.1.

Table 7.1.1

$f(x)$	$\int f(x)\, dx$
1. $\sin x$	$-\cos x + C$
2. $\cos x$	$\sin x + C$
3. $\tan x$	$\log|\sec x| + C$
4. $\sec^2 x$	$\tan x + C$

Let us now consider the following examples.

Example 1.   Evaluate the following definite integral

$$\int_0^{\pi/2} 2 \cos 2x \, dx.$$

Solution.   Let $u = 2x$; $du = 2\, dx$; and

$$\int_0^{\pi/2} 2 \cos 2x \, dx = \int_0^{\pi} \cos u \, du = \sin u \Big|_0^{\pi} = 0.$$

Example 2.   Evaluate

$$\int \cos x \, e^{\sin x} \, dx.$$

Solution.   If we try a substitution, $u = \sin x$, then $du = \cos x \, dx$ and

$$\int \cos x \, e^{\sin x} \, dx = \int e^u \, du = e^u + C = e^{\sin x} + C.$$

Example 3.   Evaluate $\int_0^\pi x \cos x \, dx$.

Solution.   It is clear that substitution will not work in this case. Let us try integration by parts with $u = x$ and $dv = \cos x \, dx$, yielding $v = \sin x$ and $du = dx$. Then,

$$\int_0^\pi x \cos x \, dx = x \sin x \bigg|_0^\pi - \int_0^\pi \sin x \, dx = 0 + \cos x \bigg|_0^\pi = -2.$$

Example 4.   Find the area bounded by the curves $y = \sin x$, $y = \cos x$, the y axis and the first point where these curves intersect for $x > 0$.

Solution.   We first draw a graph of the two curves and indicate the region in question. (See Figure 7.1.1.) The area of the region in question is given by

$$\int_0^{\pi/4} (\cos x - \sin x) \, dx = \sin x + \cos x \bigg|_0^{\pi/4} = \sqrt{2} - 1.$$

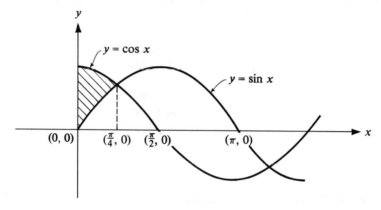

FIGURE 7.1.1

Example 5.   Find the volume of the solid generated by revolving the curve $y = \cos x$ between $x = 0$ and $x = \pi/4$ about the x axis.

Solution.   The volume of the solid given in Figure 7.1.2 is

$$V = \pi \int_0^{\pi/4} \cos^2 x \, dx.$$

Now, in order to find an antiderivative of $\cos^2 x$, it is necessary for us to use the result that

$$\cos^2 x = \frac{1 + \cos 2x}{2}$$

since we observe from Table 7.1.1 that there is apparently no simple way of calculating an antiderivative of $\cos^2 x$. [*Note:* If the integrand were $\sin^2 x$, then we would use the

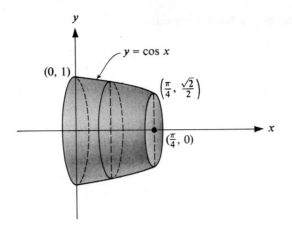

FIGURE 7.1.2

result $\sin^2 x = (1 - \cos 2x)/2$.] Thus,

$$\int_0^{\pi/4} \cos^2 x \, dx = \int_0^{\pi/4} \frac{1 + \cos 2x}{2}$$

$$= \frac{1}{2}\int_0^{\pi/4} dx + \frac{1}{2}\int_0^{\pi/4} \cos 2x \, dx.$$

In the second integral we let $u = 2x$ obtaining $du = 2 \, dx$ and

$$\int_0^{\pi/4} \cos 2x \, dx = \frac{1}{2}\int_0^{\pi/2} \cos u \, du = \frac{1}{2}\sin u \Big|_0^{\pi/2} = \frac{1}{2},$$

$$\int_0^{\pi/4} \cos^2 x \, dx = \frac{\pi}{8} + \frac{1}{4}$$

and

$$V = \frac{\pi^2}{8} + \frac{\pi}{4}.$$

Exercises

1. Evaluate the following.

(a) $\int \cos 5x \, dx$

(b) $\int x \sin x^2 \, dx$

(c) $\int \tan 2x \, dx$

(d) $\int \sec^2 x \, e^{\tan x} \, dx$

(e) $\int \sin x \log|\cos x| \, dx$

(f) $\int x \sin 2x \, dx$

2. Evaluate the following definite integrals.

(a) $\displaystyle\int_{\pi/2}^{3\pi/4} \sin 2x \, dx$ (d) $\displaystyle\int_{0}^{\pi/2} \sqrt{1 - \sin x} \, \cos x \, dx$

(b) $\displaystyle\int_{0}^{\pi/2} \sin^2 x \cos x \, dx$ (e) $\displaystyle\int_{0}^{\pi/2} \frac{\cos x}{1 + \sin x} \, dx$

(c) $\displaystyle\int_{\pi/6}^{\pi/3} \tan x \sec^2 x \, dx$

3. Find the area bounded by the curve $y = \sin x$, the y axis, and the line $y = 1$.
4. Find the area bounded by $y = \sin x$, $y = \cos x$, and the x axis, for $0 \le x \le \pi/2$.
5. Find the volume of the solid generated by revolving the curve $y = \sin x$, $0 \le x \le \pi/2$ about the x axis.
6. Find the volume of the solid generated by revolving the curve $y = \sec x$, $0 \le x \le \pi/4$ about the x axis.
7. Find the area bounded by the curve $y = \tan x \sec^2 x$, the x axis, and, the line $x = \pi/3$.

7.2   The Use of Tables of Integrals

The integration techniques introduced in Chapter 5—integration by parts and substitution—are adequate for many of the functions needed in applications. However, there still are many important types of functions for which neither of the above techniques will apply. For example, consider the evaluation of the integral

$$\int \frac{(2x + 1)}{(3x^2 + 2x + 1)^3} \, dx \,.$$

In order to evaluate such an integral, we would be required to use techniques involving trigonometric functions and an algebraic device called partial fraction decomposition. Fortunately, a large number of integrals involving complicated techniques appear in tables of integrals such as Table A.3 in the appendix to the text. Let us begin by demonstrating how to use the table of integrals effectively.

Example 1.   Evaluate

$$\int \frac{dx}{\sqrt{9x^2 + 1}} \,.$$

We see from Table A.3 that this is of the form of entry 14, provided that we factor the coefficient of x^2, namely 9, out of the radical sign. Thus, we first write

$$\int \frac{dx}{\sqrt{9x^2 + 1}} = \frac{1}{3} \int \frac{dx}{\sqrt{x^2 + \frac{1}{9}}} \,.$$

We then use entry 14 with $a = \frac{1}{3}$. Hence,

$$\int \frac{dx}{\sqrt{9x^2 + 1}} = \frac{1}{3} \log \frac{x + \sqrt{x^2 + \frac{1}{9}}}{\frac{1}{3}} + C$$

$$= \tfrac{1}{3} \log[3x + \sqrt{9x^2 + 1}] + C,$$

where C is an arbitrary constant.

Example 2.   Evaluate

$$\int \sqrt{\frac{x+1}{x-1}}\, dx.$$

We first note that this is not in a form suitable for the tables. However, if we rational-ize the expression $\sqrt{(x+1)/(x-1)}$ by multiplying numerator and denominator by $\sqrt{x+1}$ to obtain

$$\sqrt{\frac{x+1}{x-1}} = \frac{x+1}{\sqrt{x^2-1}},$$

we see that

$$\int \sqrt{\frac{x+1}{x-1}}\, dx = \int \frac{x\, dx}{\sqrt{x^2-1}} + \int \frac{dx}{\sqrt{x^2-1}}.$$

The first integral can be evaluated by setting $x^2 - 1 = u^2$ (or $u = \sqrt{x^2-1}$) so that

$$\int \frac{x\, dx}{\sqrt{x^2-1}} = \int \frac{u\, du}{u} = u + C_1 = \sqrt{x^2-1} + C_1.$$

From entry 15 in Table A.3 with $a = 1$, we obtain

$$\int \frac{dx}{\sqrt{x^2-1}} = \log\left[\frac{x+\sqrt{x^2-1}}{1}\right] + C_2 = \log[x + \sqrt{x^2-1}] + C_2$$

and thus

$$\int \sqrt{\frac{x+1}{x-1}} = \sqrt{x^2-1} + \log[x + \sqrt{x^2-1}] + C,$$

where C is the sum of the two arbitrary constants C_1 and C_2.

Example 3.   Evaluate $\int x^4 e^{2x}\, dx$.

Using entry 34 in Table A.3 with $n = 4$ and $a = 2$, we obtain

$$\int x^4 e^{2x}\, dx = \frac{x^4 e^{2x}}{2} - 2\int x^3 e^{2x} + C.$$

We again use entry 34 repeatedly to obtain

$$\int x^3 e^{2x}\, dx = \frac{x^3 e^{2x}}{2} - \frac{3}{2}\int x^2 e^{2x}\, dx = \frac{x^3 e^{2x}}{2} - \frac{3}{2}\left[\frac{x^2 e^{2x}}{2} - \int x e^{2x}\, dx\right]$$

$$= \frac{x^3 e^{2x}}{2} - \frac{3}{2}\left[\frac{x^2 e^{2x}}{2} - \frac{x e^{2x}}{2} + \frac{1}{2}\int e^{2x}\, dx\right]$$

$$= \frac{x^3 e^{2x}}{2} - \frac{3}{2}\left[\frac{x^2 e^{2x}}{2} - \frac{x e^{2x}}{2} + \frac{1}{4} e^{2x}\right] + C_1.$$

Hence,

$$\int x^4 e^{2x} \, dx = \frac{x^4 e^{2x}}{2} - x^3 e^{2x} + \tfrac{3}{2} x^2 e^{2x} - \tfrac{3}{2} x e^{2x} + \tfrac{3}{4} e^{2x} + C,$$

where C is an arbitrary constant.

In the above example we could have repeatedly used the integration by parts formula, but that is certainly more cumbersome than using the integral tables.

Example 4.   Evaluate

$$\int_0^{1/2} \frac{dx}{1 - x^2}.$$

We first observe that an antiderivative of $1/(1 - x^2)$ can be found by means of entry 16 in Table A.3 with $a = 1$. Thus,

$$\int_0^{1/2} \frac{dx}{1 - x^2} = \frac{1}{2} \log\left(\frac{1 + x}{1 - x}\right)\Big|_0^{1/2} = \frac{1}{2} \log\left(\frac{1 + \frac{1}{2}}{1 - \frac{1}{2}}\right) - \frac{1}{2} \log 1 = \frac{1}{2} \log 3.$$

In evaluating integrals by use of Table A.3, first see if the integral is of the form of one of the entries in the table. If not, try to get the integral into one of these forms by either the method of substitution or integration by parts.

As a final example, we show how useful the technique of integration by parts can be in using Table A.3.

Example 5.   Evaluate

$$\int \frac{x^4 \, dx}{\sqrt{1 - x^2}}.$$

As we scan Table A.3, we do not find an entry of the form of this integral. Suppose we rewrite the above integral as

$$\int \frac{x^4 \, dx}{\sqrt{1 - x^2}} = \int x^3 \cdot \frac{x \, dx}{\sqrt{1 - x^2}}.$$

Now, if we perform an integration by parts with $u = x^3$ and

$$dv = \frac{x}{\sqrt{1 - x^2}} \, dx$$

—yielding $v = -\sqrt{1 - x^2}$—we note that

$$\int x^3 \cdot \frac{x \, dx}{\sqrt{1 - x^2}} = -x^3 \sqrt{1 - x^2} + 3 \int x^2 \sqrt{1 - x^2} \, dx.$$

Now from entry 26 with $a = 1$, we find that

$$\int x^2 \sqrt{1 - x^2} \, dx = -\frac{1}{4} x(1 - x^2)^{3/2} + \frac{x}{8} \sqrt{1 - x^2} + \frac{1}{8} \arcsin x + C_1,$$

where C_1 is an arbitrary constant. Thus,

$$\int \frac{x^4\,dx}{\sqrt{1-x^2}} = -x^3\sqrt{1-x^2} - \frac{3}{4}x(1-x^2)^{3/2} + \frac{3x}{8}\sqrt{1-x^2} + \frac{3}{8}\text{ arc sin } x + C,$$

where C is an arbitrary constant.

Exercises

1. By the use of tables, evaluate the following integrals.

   (a) $\int \sqrt{4x^2 + 3}\,dx$ $\qquad$ (f) $\int x^4 e^{2x}\,dx$

   (b) $\int \dfrac{dx}{x\sqrt{9 - 2x^2}}$ $\qquad$ (g) $\int \sqrt{\dfrac{x-3}{x+3}}\,dx$

   (c) $\int \sqrt{\dfrac{x-1}{x+2}}\,dx$ $\qquad$ (h) $\int \dfrac{dx}{x\sqrt{9 - x^2}}$

   (d) $\int x^3 e^{8x}\,dx$ $\qquad$ (i) $\int \dfrac{x\,dx}{5x+4}$

   (e) $\int \dfrac{e^x\,dx}{(3e^{2x} + 5)^2}$ $\qquad$ (j) $\int x^2\sqrt{25 - x^2}\,dx$

2. Evaluate the following.

   (a) $\displaystyle\int_3^4 \dfrac{dx}{x^2 - 2}$ $\qquad$ (d) $\displaystyle\int_1^2 x^3 \log x\,dx$

   (b) $\displaystyle\int_0^1 x^4 e^{2x}\,dx$ $\qquad$ (e)* $\displaystyle\int_0^1 e^x(\sin e^x)^2\,dx$

   (c) $\displaystyle\int_0^2 \dfrac{x\,dx}{2x+3}$

3. By first reducing to a form suitable for using Table A.3 of the appendix, evaluate the following.

   (a) $\int \dfrac{x^2\,dx}{\sqrt{2x^2 + 3}}$ $\qquad$ (c) $\int \dfrac{x^9\,dx}{\sqrt{1 - x^4}}$

   (b) $\displaystyle\int_{\log 3/2}^{\log 2} \dfrac{e^x\,dx}{e^{2x} - 1}$

4. Find the average value of the function $f(x) = x^3 e^x$ on the interval $[0, 1]$.
5. Find the volume of the solid generated by revolving about the x axis the region bounded by $y = x^2 e^x$, $x = 0$, and $x = 1$.

7.3* Additional Comments on Integration. Improper Integrals

We recall that in Chapter 5, we always assumed that in order to evaluate definite integrals on an interval $[a, b]$, the function in the integrand had to have an antiderivative on $[a, b]$—or had to be continuous on $[a, b]$. (See Section 5.1.) Actually, this is really more than one needs. It turns out that if f is bounded on $[a, b]$—that is, if there exists a number M such that $|f(x)| \leq M$ for $x \in [a, b]$—and if f is continuous except at a finite number of points, say $c_1, c_2, \ldots, c_n$ (with $a \leq c_1 \leq c_2 \leq \cdots \leq b$), we can still define $\int_a^b f(x)\, dx$ in a reasonable manner.

Let us begin by considering the following example. (See Figure 7.3.1.) The function f

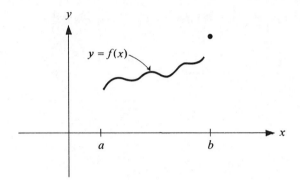

FIGURE 7.3.1

is defined on $[a, b]$ and has an antiderivative $F(x)$ on $[a, b)$. We also assume that f is bounded on $[a, b]$. Thus f is Newton integrable on $[a, b - h]$ for $h > 0$ and $\int_a^{b-h} f(x)\, dx = F(b - h) - F(a)$. It is an obvious and attractive strategy to define

$$\int_a^b f(x)\, dx = \lim_{h \to 0^+} \int_a^{b-h} f(x)\, dx = \lim_{h \to 0^+} (F(b - h) - F(a)) = \lim_{h \to 0^+} F(b - h) - F(a).$$

Clearly this definition works only if $\lim_{h \to 0^+} F(b - h)$ is defined. It is our good fortune that it always is. To repeat, if f is bounded on $[a, b]$ and has an antiderivative F on $[a, b)$, then the $\lim_{h \to 0^+} F(b - h)$ *always* exists.

We now expand this method slightly to cover a number of situations that arise frequently in practice.

We first consider a function f, bounded on $[a, b]$ and having a discontinuity at one point, say $c_1 \in [a, b]$. In this case one can prove that $\int_a^b f(x)\, dx$ exists and has the value

$$\int_a^b f(x)\, dx = \lim_{h \to 0^+} \left[\int_a^{c_1 - h} f(x)\, dx + \int_{c_1 + h}^b f(x)\, dx \right].$$

We are really saying that if f is continuous except at c_1 and bounded on $[a, b]$, then the limit on the right exists. Note that f is certainly Newton integrable on $[a, c_1 - h]$ and $[c_1 + h, b]$, but not on $[a, b]$. Thus,

$$\int_a^{c_1 - h} f(x)\, dx = F(c_1 - h) - F(a) \qquad \text{and} \qquad \int_{c_1 + h}^b f(x)\, dx = G(b) - G(c_1 + h),$$

where F is any antiderivative of f on $[a, c_1 - h]$ and G any antiderivative of f on $[c_1 + h, b]$. In the event that the discontinuities occur at a finite number of points $c_1, c_2, \ldots, c_m$, the value of the integral is given by

$$\int_a^b f(x)\, dx = \lim_{h \to 0^+} \left[\int_a^{c_1 - h} f(x)\, dx + \int_{c_1 + h}^{c_2 - h} f(x)\, dx + \cdots \right.$$

$$\left. + \int_{c_{m-1}+h}^{c_m - h} f(x)\, dx + \int_{c_m + h}^b f(x)\, dx \right].$$

(See Example 3 below.) One can always show that if $|f(x)| \le M$, a constant, for $x \in [a, b]$, then the limit on the right exists. Let us consider some examples.

Example 1. Suppose f is defined on $[0, 2]$ as follows:

$$f(x) = \begin{cases} x, & 0 \le x < 1 \\ 2, & 1 \le x \le 2. \end{cases}$$

The graph of f is shown in Figure 7.3.2. Find $\int_0^2 f(x)\, dx$.

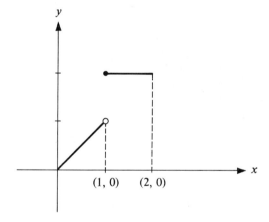

FIGURE 7.3.2

Solution. It is clear that on $[0, 2]$, $|f(x)| \le 2$ and the only discontinuity occurs at $x = 1$. Hence, $\int_0^2 f(x)\, dx$ exists and is given by

$$\int_0^2 f(x)\, dx = \lim_{h \to 0^+} \left[\int_0^{1-h} f(x)\, dx + \int_{1+h}^2 f(x)\, dx \right]$$

$$= \lim_{h \to 0^+} \left[\int_0^{1-h} x\, dx + \int_{1+h}^2 2\, dx \right]$$

$$= \lim_{h \to 0^+} \left[\frac{x^2}{2} \Big|_0^{1-h} + 2x \Big|_{1+h}^2 \right]$$

$$= \lim_{h \to 0^+} \left[\frac{(1-h)^2}{2} + 4 - 2(1+h) \right]$$

$$= \frac{1}{2} + 4 - 2 = \frac{5}{2}.$$

Example 2.   Consider the function defined on $[0, 2]$ by

$$f(x) = \begin{cases} 1, & 0 \leq x < 1 \\ \frac{1}{2}, & x = 1 \\ x - 1, & 1 < x \leq 2. \end{cases}$$

The graph of f is shown in Figure 7.3.3.   Find $\int_0^2 f(x)\, dx$.

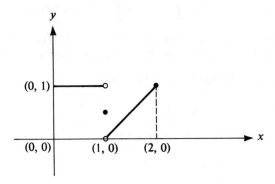

FIGURE 7.3.3

Solution.   We immediately observe that $|f(x)| \leq 1$ for $x \in [0, 2]$ and the only point of discontinuity occurs at $x = 1$.   Hence,

$$\int_0^2 f(x)\, dx = \lim_{h \to 0^+} \left[\int_0^{1-h} f(x)\, dx + \int_{1+h}^2 f(x)\, dx \right]$$

$$= \lim_{h \to 0^+} \left[\int_0^{1-h} dx + \int_{1+h}^2 (x - 1)\, dx \right]$$

$$= \lim_{h \to 0^+} \left[x \Big|_0^{1-h} + \left(\frac{x^2}{2} - x \right) \Big|_{1+h}^2 \right]$$

$$= \lim_{h \to 0^+} \left[1 - h - \frac{(1 + h)^2}{2} + (1 + h) \right] = \frac{3}{2}.$$

Note the important fact that the value of the integral is independent of the value of f at the point of discontinuity, $x = 1$.

We consider one more example.

Example 3.   Let f be defined as follows:

$$f(x) = \begin{cases} 2, & 0 \leq x < 1 \\ 1, & 1 < x \leq 2 \\ \frac{1}{2}, & 2 < x < 3 \\ \frac{1}{4}, & x = 3 \\ x - 3, & 3 < x < 4 \\ 5 - x, & 4 < x \leq 5. \end{cases}$$

(See Figure 7.3.4.)   Find $\int_0^5 f(x)\, dx$.

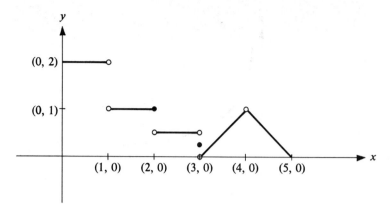

FIGURE 7.3.4

Solution. It is clear in this case that the points of discontinuity occur at 1, 2, 3, and 4. In fact, f is not even defined at 1 and 4. Still, on $[0, 5]$, excluding 1 and 4, $|f(x)| \leq 2$ and we can evaluate $\int_0^5 f(x)\, dx$. (The value of the integral is completely independent of the value of the function at the points of discontinuity.) We easily extend the method used for one point of discontinuity in the preceding examples.

$$\int_0^5 f(x)\, dx = \lim_{h \to 0^+} \left[\int_0^{1-h} 2\, dx + \int_{1+h}^{2-h} dx + \int_{2+h}^{3-h} \frac{1}{2}\, dx \right.$$

$$\left. + \int_{3+h}^{4-h} (x-3)\, dx + \int_{4+h}^{5} (5-x)\, dx \right]$$

$$= \lim_{h \to 0^+} \left[2(1-h) + (2-h) - (1+h) + \frac{1}{2}(3-h) - \frac{1}{2}(2+h) \right.$$

$$\left. + \left(\frac{x^2}{2} - 3x \right) \Big|_{3+h}^{4-h} + \left(5x - \frac{x^2}{2} \right) \Big|_{4+h}^{5} \right]$$

$$= \frac{9}{2}.$$

A far more serious problem arises when we consider the integral of a function $f(x)$ that has a finite number of discontinuities on $[a, b]$ and for which $|f(x)|$ assumes arbitrarily large values near these points of discontinuity. That is, f is no longer bounded on $[a, b]$. Suppose we consider the function $f(x) = 1/x$ for $0 < x \leq 1$. We note that as $x \to 0$, $f(x)$ becomes very large. We now ask whether we can assign a meaning to $\int_0^1 x^{-1}\, dx$? Let us first observe that $\int_b^1 x^{-1}\, dx = \log 1 - \log b = \log 1/b$, where $0 < b < 1$. Now, as $b \to 0$, $\log 1/b \to \infty$. In this case, we say that $\int_0^1 dx/x$ has *no* meaning—that is, the function $1/x$ is not integrable on $[0, 1]$. Note, that for any $\alpha, 0 < \alpha < 1$, the integral $\int_\alpha^1 x^{-1}\, dx$ exists and has the value $\log 1/\alpha$.

It is remarkable, however, that if we consider the function $f(x) = x^{-1/2}$, $x > 0$, we can assign a meaning to $\int_0^1 x^{-1/2}\, dx$. Both $1/x$ and $x^{-1/2}$ become infinite as $x \to 0$. Thus both have one discontinuity in the interval $[0, 1]$. (See Figure 7.3.5.) Now, we investigate $\int_h^1 x^{-1/2}\, dx = 2x^{1/2} \big|_h^1 = 2 - 2h^{1/2}$. As $h \to 0$, $\int_h^1 x^{-1/2}\, dx \to 2$, since $h^{1/2} \to 0$.

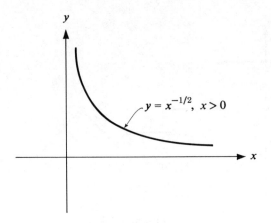

FIGURE 7.3.5

We say that $\int_0^1 x^{-1/2}\, dx$ exists or converges as an improper integral even though the function is not defined at 0 and is not continuous in the closed interval $[0, 1]$.

In general, consider a function f continuous on $a < x \leq b$. Thus, for every positive number h such that $a + h < b$, f is continuous on $[a + h, b]$. It is also assumed that as $x \to a$, $|f(x)| \to \infty$. We then form the integral, $\int_{a+h}^b f(x)\, dx$. Since f is Newton integrable on $[a + h, b]$, $\int_{a+h}^b f(x)\, dx = F(b) - F(a + h)$, where F is any antiderivative of f on $[a + h, b]$.

DEFINITION.   *We say that $\int_a^b f(x)\, dx$ exists or converges as an improper integral if and only if $\lim_{h \to 0^+} [F(b) - F(a + h)]$ exists, where F is an antiderivative of f.*

In this case, we set $\int_a^b f(x)\, dx = F(b) - \lim_{h \to 0^+} F(a + h)$ by definition. If $\lim_{h \to 0^+} [F(b) - F(a + h)]$ does not exist, we say that $\int_a^b f(x)\, dx$ is *divergent*.

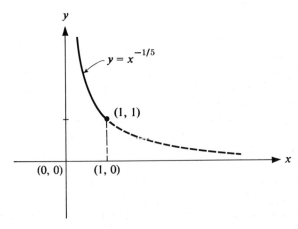

FIGURE 7.3.6

Example 1. Determine whether or not the improper integral $\int_0^1 x^{-1/5}\,dx$ converges. If it does converge, evaluate the integral. (See Figure 7.3.6.)

Solution. First observe that the function $f(x)$ is continuous on $(0, 1]$ but not on $[0, 1]$. Hence, the integral is an improper integral. We then look at

$$\int_h^1 x^{-1/5}\,dx = \frac{5}{4}x^{4/5}\Big|_h^1 = \frac{5}{4} - \frac{5}{4}h^{4/5}.$$

Now, $\lim_{h\to 0^+}\left[\frac{5}{4} - \frac{5}{4}h^{4/5}\right] = \frac{5}{4}$. Hence, the improper integral converges and $\int_0^1 x^{-1/5}\,dx = \frac{5}{4}$.

Example 2. Determine whether or not the improper integral $\int_0^1 x^{-2}\,dx$ converges. If so, evaluate it. (See Figure 7.3.7.)

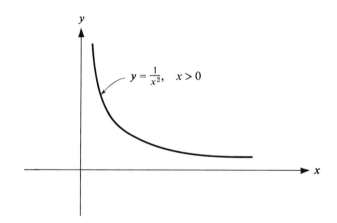

FIGURE 7.3.7

Solution. Again, since $x^{-2} \to \infty$ as $x \to 0$, $\int_0^1 x^{-2}\,dx$ is *improper*. Now, for $h > 0$, x^{-2} is continuous on $[h, 1]$. Hence,

$$\int_h^1 \frac{dx}{x^2} = -x^{-1}\Big|_h^1 = \frac{1}{h} - 1.$$

However, $\lim_{h\to 0^+}\left[1/h - 1\right]$ does not exist. Thus, $\int_0^1 x^{-2}\,dx$ is divergent.

Suppose we consider a function f continuous on $[a, b)$ but not continuous on $[a, b]$ and assume that as $x \to b^-$, $|f(x)| \to \infty$. Thus, for $h > 0$ and $a < (b - h)$, f is continuous on $[a, b - h]$. We say that the *improper* integral $\int_a^b f(x)\,dx$ is convergent if and only if $\lim_{h\to 0^+}[F(b - h) - F(a)]$ exists, where F is any antiderivative of f on $[a, b - h]$.

Example 3. Determine whether $\int_{1/2}^{1} (x-1)^{-1/3}\, dx$ converges. If so, evaluate it. (See Figure 7.3.8.)

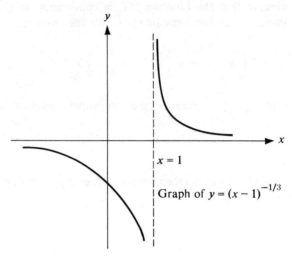

$x = 1$

Graph of $y = (x-1)^{-1/3}$

FIGURE 7.3.8

Solution. Clearly, the function $f(x) = (x-1)^{-1/3}$ is continuous on $[\tfrac{1}{2}, 1)$ but not on $[\tfrac{1}{2}, 1]$. Hence, the integral is improper. Now,

$$\int_{1/2}^{1-h} \frac{dx}{(x-1)^{1/3}} = \int_{-1/2}^{-h} u^{-1/3}\, du = \frac{3}{2} u^{2/3}\Big|_{-1/2}^{-h}$$

$$= \frac{3}{2}(-h)^{2/3} - \frac{3}{2^{5/3}}.$$

However,

$$\lim_{h \to 0^+} \left[\frac{3}{2}(-h)^{2/3} - \frac{3}{2^{5/3}} \right] = -\frac{3}{2^{5/3}}.$$

Thus, $\int_{1/2}^{1} (x-1)^{1/3}\, dx$ converges and has the value $-3/2^{5/3}$.

It is also possible that f is continuous on (a, b) but not on $[a, b]$. Thus, if $h > 0$ with $a + h < b - h$, f is continuous on $[a + h, b - h]$. We say that $\int_a^b f(x)\, dx$ is convergent if and only if $\lim_{h \to 0^+} F(b - h)$ and $\lim_{h \to 0^+} F(a + h)$ exist, where F is any antiderivative of f on $[a + h, b - h]$. In this case,

$$\int_a^b f(x)\, dx = \lim_{h \to 0^+} F(b - h) - \lim_{h \to 0^+} F(a + h).$$

Here we are assuming that as $x \to a^+$ and $x \to b^-$, $|f(x)| \to \infty$.

Example 4.   Does the integral $\int_0^1 [x(x-1)]^{-1}\, dx$ converge? If so, evaluate it. (See Figure 7.3.9.)

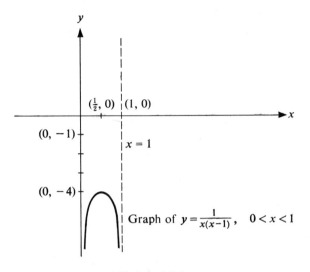

$$\text{Graph of } y = \frac{1}{x(x-1)}, \quad 0 < x < 1$$

FIGURE 7.3.9

Solution.   It is clear that the function $1/x(x-1)$ is continuous on $(0, 1)$ but not on $[0, 1]$. Thus, the integral is improper. First note that

$$\frac{1}{x(x-1)} = \frac{1}{x-1} - \frac{1}{x}.$$

Thus,

$$\int_0^1 \frac{dx}{x(x-1)} = \int_0^1 \frac{dx}{x-1} - \int_0^1 \frac{dx}{x}.$$

Now,

$$\int_0^{1-h} \frac{dx}{x-1} = \log|x-1|\Big|_0^{1-h} = \log|-h|$$

and

$$\int_h^1 \frac{dx}{x} = \log|x|\Big|_h^1 = -\log|h|.$$

Observe that $\lim_{h\to 0^+} \log|-h|$ and $\lim_{h\to 0^+} \log|h|$ do not exist. Hence, since $\lim_{h\to 0^+} \log|-h|$ and $\lim_{h\to 0^+} \log|h|$ do not exist, $\int_0^1 [x(x-1)]^{-1}\, dx$ is *divergent*.

Example 5.   Let

$$f(x) = \begin{cases} x^{-1/2}, & 0 < x < \frac{1}{2} \\ (1-x)^{-1/2}, & \frac{1}{2} \le x < 1. \end{cases}$$

(See Figure 7.3.10.) Does $\int_0^1 f(x)\,dx$ exist? If so, evaluate it.

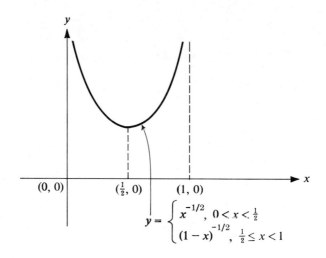

FIGURE 7.3.10

Solution.   It is clear from Figure 7.3.10 that as $x \to 0$ and $x \to 1$, $|f(x)| \to \infty$. Hence, the integral is certainly improper. However, f is certainly continuous on $[h, 1-h]$, where $0 < h < 1$. It is clear that

$$\int_h^{1-h} f(x)\,dx = \int_h^{1/2} x^{-1/2}\,dx + \int_{1/2}^{1-h} (1-x)^{-1/2}\,dx.$$

We must check to see if

$$\lim_{h \to 0^+} \left[\int_h^{1/2} x^{-1/2}\,dx \right] + \lim_{h \to 0^+} \left[\int_{1/2}^{1-h} (1-x)^{-1/2}\,dx \right]$$

exists. Now,

$$\int_h^{1/2} x^{-1/2}\,dx = 2x^{1/2} \Big|_h^{1/2} = \frac{2}{\sqrt{2}} - 2\sqrt{h}$$

and

$$\int_{1/2}^{1-h} (1-x)^{-1/2}\,dx = -2(1-x)^{1/2} \Big|_{1/2}^{1-h} = -2h^{1/2} + \frac{2}{\sqrt{2}}.$$

Since

$$\lim_{h\to 0^+}\left[\frac{2}{\sqrt{2}} - 2\sqrt{h}\right] = \frac{2}{\sqrt{2}} \quad \text{and} \quad \lim_{h\to 0^+}\left[\frac{2}{\sqrt{2}} - 2\sqrt{h}\right] = \frac{2}{\sqrt{2}},$$

$\int_0^1 f(x)\, dx$ exists and has the value $2/\sqrt{2} + 2/\sqrt{2} = 4/\sqrt{2}$.

There are other ways in which $\int_a^b f(x)\, dx$ can be improper. Suppose that f has unbounded discontinuities at a finite set of numbers $c_1, c_2, c_3, \ldots, c_n$, where $a < c_1 < c_2 < c_3 < \cdots < c_n < b$. That is, $|f(x)| \to \infty$ as $x \to c_i$, $i = 1, 2, \ldots, n$. The problem now is to determine whether or not $\int_a^b f(x)\, dx$ exists. We assert that $\int_a^b f(x)\, dx$ is convergent if and only if

$$\int_a^{c_1} f(x)\, dx + \int_{c_1}^{c_2} f(x)\, dx + \cdots + \int_{c_{n-1}}^{c_n} f(x)\, dx + \int_{c_n}^b f(x)\, dx$$

exists. In the event that the sum does *not* exist, we say that $\int_a^b f(x)\, dx$ *diverges*. We consider some examples.

Example 6.   Does $\int_{-1}^1 x^{-2}\, dx$ converge? If so, evaluate it.

Solution.   Clearly, $1/x^2$ has an unbounded discontinuity at $x = 0$. $\int_{-1}^1 x^{-2}\, dx$ converges if and only if

$$\lim_{h\to 0^+}\left[\int_{-1}^{0-h}\frac{dx}{x^2}\right] + \lim_{h\to 0^+}\left[\int_{0+h}^1 \frac{dx}{x^2}\right]$$

exists.   Now,

$$\int_{-1}^{-h}\frac{dx}{x^2} = -x^{-1}\Big|_{-1}^{-h} = \frac{1}{h} - 1$$

and

$$\int_h^1 \frac{dx}{x^2} = -x^{-1}\Big|_h^1 = -1 + \frac{1}{h}.$$

Since $\lim_{h\to 0^+}[1/h - 1]$ and $\lim_{h\to 0^+}[-1 + 1/h]$ do not exist, the integral $\int_{-1}^1 x^{-2}\, dx$ is divergent.

Suppose we were given $\int_{-1}^1 x^{-2}\, dx$ and neglected to see that f has an unbounded discontinuity at $x = 0$. We would then be tempted to blindly use the formula

$$\int_{-1}^1 \frac{dx}{x^2} = -x^{-1}\Big|_{-1}^1 = -1 - 1 = -2$$

and say that $\int_{-1}^1 x^{-2}\, dx$ certainly exists. However, this is false (as you have seen above)! You should now realize that it is essential to check to see if the integrand has any unbounded discontinuities within the range of integration. Otherwise, you may be led to false conclusions and thus easily misinterpret any data involving such integrals. This is one reason why it is not sufficient simply to memorize formulas without understanding what is happening!

Let us consider one more example.

Example 7.   Determine whether the integral $\int_{-1}^{3} |x - 1|^{-1} \, dx$ converges. If so, evaluate it. (See Figure 7.3.11.)

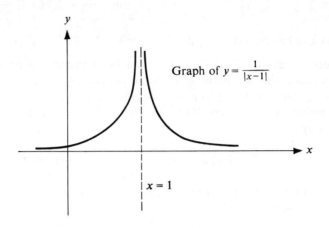

Graph of $y = \dfrac{1}{|x-1|}$

$x = 1$

FIGURE 7.3.11

Solution.   First, note that $1/|x - 1|$ has an unbounded discontinuity at $x = 1$. Also recall that if $x > 1$, $1/|x - 1| = 1/(x - 1)$, and if $x < 1$, $1/|x - 1| = -1/(x - 1) = 1/(1 - x)$. Now, $\int_{-1}^{3} |x - 1|^{-1} \, dx$ exists if and only if

$$\lim_{h \to 0^+} \left[\int_{-1}^{1-h} \frac{dx}{1 - x} \right] + \lim_{h \to 0^+} \left[\int_{1+h}^{3} \frac{dx}{x - 1} \right]$$

exists. But,

$$\int_{-1}^{1-h} \frac{dx}{1 - x} = -\log|1 - x| \Big|_{-1}^{1-h} = \log 2 - \log|h|$$

while

$$\int_{1+h}^{3} \frac{dx}{x - 1} = \log|x - 1| \Big|_{1+h}^{3} = \log 2 - \log|h|.$$

Neither the limit, $\lim_{h \to 0^+} [\log 2 - \log|h|]$, nor $\lim_{h \to 0^+} [\log 2 - \log|h|]$ exists and thus $\int_{-1}^{3} |x - 1|^{-1} \, dx$ is divergent.

There is one other type of improper integral, dealing with large values. Suppose f is a continuous function on $[a, \infty)$. We say that $\int_{a}^{\infty} f(x) \, dx$ exists or is convergent if and only if $\lim_{N \to \infty} [F(N) \quad F(a)]$ exists, where F is any antiderivative of f on $[a, N]$. We may write

$$\int_{a}^{\infty} f(x) \, dx = \lim_{N \to \infty} \int_{a}^{N} f(x) \, dx.$$

Note that f is certainly Newton integrable on $[a, N]$.

In like manner if f is continuous on $(-\infty, b]$, we say that $\int_{-\infty}^{b} f(x) \, dx$ exists or is convergent if and only if $\lim_{M \to -\infty}[\int_M^b f(x)\,dx]$ exists. If it exists we write

$$\int_{-\infty}^b f(x)\,dx = \lim_{M \to -\infty} \int_M^b f(x)\,dx.$$

Example 8. Determine whether the improper integral $\int_0^\infty e^{-x}\,dx$ is convergent. If so, evaluate it.

Solution. We first evaluate

$$\int_0^N e^{-x}\,dx = -e^{-x}\Big|_0^N = 1 - e^{-N}.$$

Now, $\lim_{N \to \infty}[1 - e^{-N}] = 1$. Hence, $\int_0^\infty e^{-x}\,dx$ is convergent and $\int_0^\infty e^{-x}\,dx = 1$.

Example 9. Determine whether the improper integral $\int_1^\infty x^{-1}\,dx$ is convergent.

Solution. First, evaluate $\int_1^N x^{-1}\,dx = \log N$. Now, $\lim_{N \to \infty} \log N$ does not exist and hence the integral is divergent.

Example 10. Determine whether the improper integral $\int_{-\infty}^0 e^x\,dx$ converges. If so, evaluate it.

Solution. First evaluate

$$\int_M^0 e^x\,dx = e^x\Big|_M^0 = 1 - e^M.$$

Since $\lim_{M \to -\infty}(1 - e^M) = 1$, we see that $\int_{-\infty}^0 e^x\,dx$ is convergent and has the value 1.

Example 11. Determine whether the integral $\int_{-\infty}^\infty xe^{-x^2}\,dx$ converges. If so, evaluate it.

Solution. In the case that both limits are infinite, we say that $\int_{-\infty}^\infty xe^{-x^2}\,dx$ converges if and only if $\lim_{B \to -\infty} \int_B^0 xe^{-x^2}\,dx$ and $\lim_{A \to \infty} \int_0^A xe^{-x^2}\,dx$ exist. Letting $u = x^2$,

$$\int_0^A xe^{-x^2}\,dx = \frac{1}{2}\int_0^{A^2} e^{-u}\,du = -\frac{1}{2}e^{-A^2} + \frac{1}{2}$$

and

$$\int_B^0 xe^{-x^2}\,dx = \frac{1}{2}\int_{B^2}^0 e^{-u}\,du = -\frac{1}{2} + \frac{1}{2}e^{-B^2}$$

Now,

$$\lim_{A \to \infty}\left[-\tfrac{1}{2}e^{-A^2} + \tfrac{1}{2}\right] = \tfrac{1}{2} \qquad \text{and} \qquad \lim_{B \to -\infty}\left[\tfrac{1}{2}e^{-B^2} - \tfrac{1}{2}\right] = -\tfrac{1}{2}.$$

Hence, $\int_{-\infty}^\infty xe^{-x^2}\,dx$ exists and has the value, $-\tfrac{1}{2} + \tfrac{1}{2} = 0$.

It is frequently possible to determine the convergence of an improper integral without computing it, by comparing it with another integral that is known to converge. We omit a discussion of this topic, but refer the reader to any of the more complete calculus texts for additional details. In the social and biological sciences, it is often necessary to compute the value of convergent improper integrals.

Exercises

1. Let $f(x)$ be defined by

$$f(x) = \begin{cases} x, & 0 \le x < 1 \\ 10, & x = 1 \\ -x + 2, & 1 < x < 2. \end{cases}$$

   (a) Sketch a graph of f.

   (b) Evaluate $\int_0^2 f(x)\, dx$. Why does this integral exist?

2. Define a function $h(x)$ as follows:

$$h(x) = \begin{cases} 1, & x = 0 \\ x, & 0 < x < 1 \\ 0, & x = 1 \\ -x + 2, & 1 < x < 2 \\ 1, & x = 2. \end{cases}$$

   (a) Graph the function.

   (b) Evaluate $\int_0^2 h(x)\, dx$. Compare this value with the integral in Exercise 1. Does the result make sense? Why?

3. Define a function $f(x)$ as follows:

$$f(x) = \begin{cases} x^{-1/2}, & 0 < x < 1 \\ x, & 1 \le x < 2 \\ 3, & x = 2. \end{cases}$$

   (a) Sketch the graph of f.

   (b) Does $\int_0^2 f(x)\, dx$ exist? If so, evaluate it.

4. Determine whether or not the following integrals converge. If so, evaluate them.

   (a) $\displaystyle \int_0^1 \frac{1}{x^{1/6}}\, dx$

   (d) $\displaystyle \int_0^2 \frac{dx}{x^2 - 2x}$

   (b) $\displaystyle \int_{-1}^1 \frac{dx}{x^3}$

   (e) $\displaystyle \int_0^5 \frac{dx}{5 - x}$

   (c) $\displaystyle \int_0^1 x \log x \, dx$

   (f) $\displaystyle \int_0^2 \frac{dx}{|x - 2|}$

5. Determine whether the following integrals are convergent. If so, evaluate them.

   (a) $\displaystyle \int_0^1 \frac{x\, dx}{\sqrt{1 - x^2}}$

   (c) $\displaystyle \int_1^2 \frac{x\, dx}{\sqrt{1 - x^2}}$

   (b) $\displaystyle \int_{-1}^1 \frac{dx}{\sqrt{|x|}}$

   (d) $\displaystyle \int_0^1 \frac{x\, dx}{1 - x^2}$

6. Determine whether the following integrals are convergent. If so, evaluate them.

(a) $\int_1^\infty \dfrac{dx}{x^{2/3}}$

(d) $\int_{-\infty}^\infty xe^{-|x|}\,dx$

(b) $\int_0^\infty x^2 e^{-x}\,dx$

(e) $\int_1^\infty \dfrac{dx}{x^{.98}}$

(c) $\int_{-\infty}^0 xe^x\,dx$

(f) $\int_{-1}^\infty \dfrac{(3x^2+2)\,dx}{(x^3+2x+1)^8}$

7. (a) For what numbers n does $\int_0^1 x^{-n}\,dx$ exist? Evaluate the integral for those n.

   (b) For what numbers n does $\int_1^\infty x^{-n}\,dx$ exist? Evaluate the integral for those n.

8. (a) Show that $\int_2^\infty (x \log x)^{-1}\,dx$ is divergent, but $\int_2^\infty x^{-1}(\log x)^{-2}\,dx$ is convergent, and find its value.

   (b) For what values of n does $\int_2^\infty x^{-1}(\log x)^{-n}\,dx$ exist?
   [HINT: Use Exercise 7(b).]

9. Show that the area bounded by the x axis, the curve $y = 1/x$, and the ordinate $x = 1$ is infinite, whereas the volume generated by revolving this area about the x axis is finite. Find this volume.

7.4 Numerical Integration

We have now seen that if a function is continuous except at a finite number of points over $[a, b]$ then its Newton integral can be defined. However, not all definite integrals can be evaluated by any of the techniques we have introduced previously. Examples of such integrals are $\int_0^1 e^{-x^2}\,dx$ and $\int_0^3 \sqrt{x^3 + 1}\,dx$. In such instances methods of approximating integrals have considerable practical value.

Let us investigate a numerical scheme to find an approximate value for $\int_0^1 e^{-x^2}\,dx$. First, we graph $y = e^{-x^2}$ and observe that the integral represents the area under the curve $y = e^{-x^2}$ from $x = 0$ to $x = 1$. (See Figure 7.4.1.) We try to find an approximation to

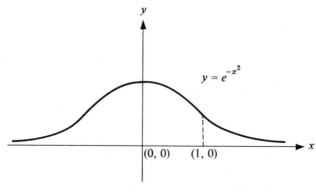

FIGURE 7.4.1

this area. First, divide the interval $[0, 1]$ into four parts, say $[0, \frac{1}{4}]$; $[\frac{1}{4}, \frac{1}{2}]$; $[\frac{1}{2}, \frac{3}{4}]$; $[\frac{3}{4}, 1]$ and construct ordinates to the curve at the end points of each subdivision. (See Figure 7.4.2.) Next, replace the curve by a broken line by drawing chords as indicated in Figure 7.4.2. Then, we calculate the area of each one of these trapezoids

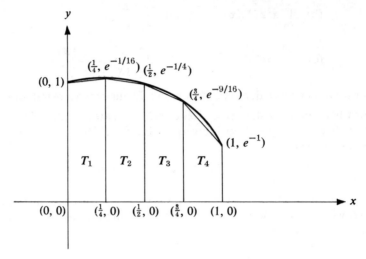

FIGURE 7.4.2

and add the results. Recall that the area of a trapezoid is equal to one-half the sum of the parallel sides multiplied by the distance between them. If we label the four resulting trapezoids, T_1, T_2, T_3, and T_4, then the area of each is given by

$$\text{area of } T_1 = A(T_1) = \tfrac{1}{2}(1 + e^{-1/16}) \cdot \tfrac{1}{4}$$
$$A(T_2) = \tfrac{1}{2}(e^{-1/16} + e^{-1/4}) \cdot \tfrac{1}{4}$$
$$A(T_3) = \tfrac{1}{2}(e^{-1/4} + e^{-9/16}) \cdot \tfrac{1}{4}$$
$$A(T_4) = \tfrac{1}{2}(e^{-9/16} + e^{-1}) \cdot \tfrac{1}{4}.$$

From Table A.1 in the appendix, $e^{-1/16} \cong e^{-0.07} = 0.93$, $e^{-1/4} = e^{-0.25} = 0.78$, $e^{-9/16} \cong 0.58$, $e^{-1} = 0.37$. Thus,

$$A(T_1) + A(T_2) + A(T_3) + A(T_4) = \frac{1}{8}[1.93 + 1.71 + 1.36 + 0.95] \cong \frac{5.95}{8} \cong 0.74.$$

We then might say that to two decimal places $\int_0^1 e^{-x^2}\, dx = 0.74$. (We use the symbol $\cong$ to indicate "is approximately equal to.") Obviously, we are making some errors in our method of computation. It turns out that the more subdivisions we take for $[0, 1]$, the better our approximate value will be. The technique outlined above is called the trapezoid rule for approximating integrals and can obviously be extended to any continuous function defined on $[a, b]$. The number of subdivisions of $[a, b]$ we choose depends on how accurate we wish our computation. If we are interested in only seeing a very approximate value of integral, then often four subdivisions is enough. We could estimate carefully the error introduced by using the trapezoidal rule, but in the present discussion we choose to omit it. We summarize by stating that to compute the value of $\int_a^b f(x)\,dx$ approximately by the trapezoidal rule we first subdivide $[a, b]$ into n subintervals, say

$$[a, b] = [a, x_1] \cup [x_1, x_2] \cup \cdots \cup [x_{n-2}, x_{n-1}] \cup [x_{n-1}, b]$$

(where the symbol "$\cup$" is read as "joined with") and replace the curve by the chords joining $(x_{i-1}, f(x_{i-1}))$ to $(x_i, f(x_i))$ and then sum up the areas of the resulting trapezoids. Our choice of n is dictated by the accuracy we desire.

The trapezoidal rule is based essentially on the principle of approximating the curve by a series of linear segments. If the curve is concave upward on $[a, b]$, then the broken-line segments will always lie above the curve. This rule is equivalent to approximating the function to be integrated by a series of linear functions—or polynomials of the first degree. We now consider a method based on approximation by polynomials of the second degree. It turns out in practice that this gives us a better approximation to $\int_a^b f(x)\,dx$ and is very easy to apply. Nowadays, it is convenient to use digital computers to compute definite integrals. The next technique, known as Simpson's rule, is readily adapted for use on computers.

This method is based on the simple formula

$$A_p = \frac{h}{3}(y_0 + 4y_1 + y_2) \tag{7.4.1}$$

for the area under the arc of the parabola $y = Ax^2 + Bx + C$ between $x = -h$ and $x = h$ (see Figure 7.4.3), where $y_0 = Ah^2 - Bh + C$, $y_1 = C$, and $y_2 = Ah^2 + Bh + C$. Formula

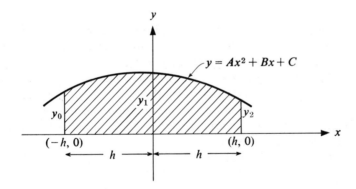

FIGURE 7.4.3

(7.4.1) is easily obtained by computing the area under the parabola and observing that the curve goes through the points $(-h, y_0)$, $(0, y_1)$, and (h, y_2).

Simpson's rule follows by applying (7.4.1) to successive pieces of the curve $y = f(x)$ between $x = a$ and $x = b$. Each separate piece of the curve, covering an x subinterval of length $2h$, is approximated by an arc of a parabola through its ends and its midpoint. (See Figure 7.4.4.) The area under each parabolic arc is given by an expression like

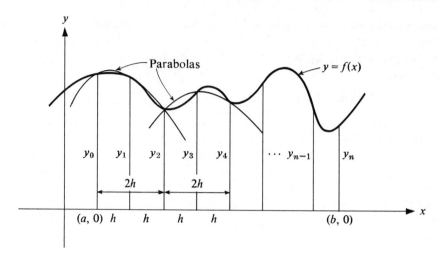

FIGURE 7.4.4

(7.4.1). If we sum all areas of these type we obtain,

$$A_s = \frac{h}{3} [(y_0 + 4y_1 + y_2) + (y_2 + 4y_3 + y_4)$$

$$+ (y_4 + 4y_5 + y_6) + \cdots + (y_{n-2} + 4y_{n-1} + y_n)]$$

$$= \frac{h}{3} [y_0 + 4y_1 + 2y_2 + 4y_3 + 2y_4 + 4y_5 + 2y_6 + \cdots + 2y_{n-2} + 4y_{n-1} + y_n] \quad (7.4.2)$$

(Note that the coefficient of the first and last y_i is one and the coefficients of the other y_i's are alternately 4's and 2's which is taken as an approximate value to $\int_a^b f(x)\, dx$.) In (7.4.2), $y_0, y_1, \ldots, y_n$ are given by $y_0 = f(a)$, $y_1 = f(a + h)$, $y_2 = f(a + 2h)$, $\ldots$, $y_n = f(a + nh) = f(b)$ corresponding to a subdivision of the interval $a \le x \le b$ into n, when n is an *even* integer, equal subintervals each of width $h = (b - a)/n$. To apply this rule, we do not have to graph the function f nor determine the approximating parabolic arcs. The formula (7.4.2) does this for us. It turns out that if f is continuous the approximations get better as n increases and h gets smaller.

We now give an example of the above.

Example.   Approximate $\int_0^1 e^{-x^2}\, dx$ by Simpson's rule using

(a) four subdivisions, that is, $n = 4$

(b) six subdivisions, that is, $n = 6$.

Solution.   (a) We wish to use formula (7.4.2) with $n = 4$, $a = 0$, $b = 1$, and $h = \frac{1}{4}$. Thus, $y_0 = 1$, $y_1 = e^{-(0.25)^2}$, $y_2 = e^{-(0.50)^2}$, $y_3 = e^{-(0.75)^2}$, $y_4 = e^{-1}$.

$$\int_0^1 e^{-x^2} \cong \frac{1}{12} [e^{-0} + 4e^{-(0.25)^2} + 2e^{-(0.50)^2} + 4e^{-(0.75)^2} + e^{-1}].$$

Using Table A.1 we obtain

$$\int_0^1 e^{-x^2}\, dx \cong \frac{1}{12}[1 + 4(0.940) + 2(0.779) + 4(0.570) + (0.368)]$$

$$\cong 0.747.$$

Recall that using the trapezoidal rule, we obtained 0.72. Thus, by taking $n = 4$ in Simpson's rule, we did gain a little. (Note that we should expect our result to be greater than 0.72.)

(b) In this case we use formula (7.4.2) with $n = 6$, $a = 0$, $b = 1$ and $h = \frac{1}{6}$. Thus, $y_0 = 1$, $y_1 = e^{-(1/6)^2}$, $y_2 = e^{-(1/3)^2}$, $y_3 = e^{-(1/2)^2}$, $y_4 = e^{-(2/3)^2}$, $y_5 = e^{-(5/6)^2}$, $y_6 = e^{-1}$. Hence,

$$\int_0^1 e^{-x^2} \cong \frac{1}{18}[1 + 4e^{-(0.0256)} + 2e^{-(0.0999)} + 4e^{-(0.2500)}$$

$$+ 2e^{-(0.4296)} + 4e^{-(0.6739)} + e^{-1}]$$

$$\cong \frac{1}{18}[1 + 4(0.974) + 2(0.906) + 4(0.779) + 2(0.652)$$

$$+ 4(0.510) + (0.368)]$$

$$\cong 0.758.$$

If we choose more subdivisions, we should expect to get a better result. It is possible to get an estimate for our error by making use of inequalities involving the fourth derivative of f. However, we omit a discussion of this here.

Exercises

1. Approximate the following integral by means of the trapezoidal rule using 4 subdivisions of $[0, 2]$: $\int_0^2 e^{-0.5x^2}\, dx$. (Use Table A.1 in the appendix.)
2. Use Simpson's rule instead of the trapezoidal rule in Exercise 1 and compare the results.
3. Approximate $\int_0^1 (1 + x^2)^{-1}\, dx$ by

    (a) The trapezoidal rule with five subdivisions of $[0, 1]$.

    (b) Simpson's rule with four subdivisions.

    Check your results by evaluating the integral with the aid of Table A.3 in the appendix.
4. A certain curve is given by the following pairs of rectangular coordinates

x	1	2	3	4	5	6	7	8	9
y	0	0.6	0.9	1.2	1.4	1.5	1.7	1.8	2

    (a) Approximate the area between the curve, the x axis, and the ordinates $x = 1$ and $x = 9$, using Simpson's rule.

(b) Approximate the volume of the solid generated by revolving the area in (a) about the x axis, using Simpson's rule.

5. About how much crushed rock will be required to make a roadbed 1 mile long having the cross section shown in Figure 7.4.5 below. [HINT: Use Simpson's rule; first find the area, $A(x)$, and then the volume $V = \int A(x)\, dx$.]

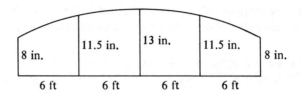

8 in.   11.5 in.   13 in.   11.5 in.   8 in.

6 ft   6 ft   6 ft   6 ft

FIGURE 7.4.5

7.5   Summary of Chapter 7

(1) Familiarize yourself, once again, with the table of integrals, Table A.3 in the back of the book.

(2)* *Improper Integrals*

(a) Let f be continuous on (a, b) but *not* on $[a, b]$. Then, $\int_a^b f(x)\, dx$ is *convergent* if and only if $\lim_{h \to 0^+} F(b - h)$ and $\lim_{h \to 0^+} F(a + h)$ exist, where F is any antiderivative of f on $[a + h, b - h]$. In this case,

$$\int_a^b f(x)\, dx = \lim_{h \to 0^+} F(b - h) - \lim_{h \to 0^+} F(a + h).$$

(b) If f has unbounded discontinuities at a finite set of numbers $c_1, c_2, \ldots, c_m$, where $a < c_1 < c_2 < \cdots < c_m < b$ (that is, $\lim_{x \to c^i} |f(x)| = \infty$, $i = 1, 2, \ldots, m$), then $\int_a^b f(x)\, dx$ is *convergent* if and only if the sum

$$\int_a^{c_1} f(x)\, dx + \int_{c_1}^{c_2} f(x)\, dx + \cdots + \int_{c_{n-1}}^{c_n} f(x)\, dx + \int_{c_n}^b f(x)\, dx$$

exists [according to the definition given in (a)].

(c) $\int_a^\infty f(x)\, dx$ is *convergent* if and only if $\lim_{M \to \infty} \int_a^M f(x)\, dx$ exists. In that case, $\int_a^\infty f(x)\, dx = \lim_{M \to \infty} \int_a^M f(x)\, dx$.

(3) *Numerical integration*

(a) *Trapezoidal rule.* To compute the value of $\int_a^b f(x)\, dx$, first divide $[a, b]$ into n subintervals, say $[a, x_1]$, $[x_1, x_2]$, $\ldots$, $[x_{n-2}, x_{n-1}]$, $[x_{n-1}, b]$; replace the curve by the chords joining $(x_{i-1}, f(x_{i-1}))$ to $(x_i, f(x_i))$, $i = 1, 2, \ldots, n$, and then sum up the areas of the resulting trapezoids. The choice of n is dictated by the accuracy we desire.

(b) *Simpson's rule.* $\int_a^b f(x)\, dx = h/3\,[y_0 + 4y_1 + 2y_2 + 4y_3 + 2y_4 + \cdots + 2y_{n-2} + 4y_{n-1} + y_n]$, where $y_0 = f(a)$, $y_1 = f(a + h)$, $y_2 = f(a + 2h)$, $\ldots$, $y_n = f(a + nh) = f(b)$, n is an **even** integer, and $h = (b - a)/n$. Again the choice of n is determined by the accuracy we desire.

REVIEW EXERCISES

1. Evaluate the following integrals (use Table A.3).

(a) $\int \dfrac{dx}{9 - x^2}$

(f) $\int \dfrac{dx}{e^x + 1}$

(b) $\int \dfrac{dx}{\sqrt{4x^2 + 9}}$

(g) $\int (x + 1)\log(2x^2 + 4x + 1)\, dx$

(c) $\int \dfrac{x + 2}{\sqrt{x^2 + 2x - 3}}\, dx$

(h) $\int \dfrac{dx}{x(3x + 6)^2}$

(d) $\int \dfrac{x + 2}{\sqrt{x^2 + 9}}\, dx$

(i) $\int x^2 \sqrt{25 - x^2}\, dx$

(e) $\int \sqrt{3x^2 + 5}\, dx$

(j) $\int \dfrac{e^x\, dx}{\sqrt{e^{2x} + 9}}$

2. Evaluate the following (if they exist):

(a) $\int_0^4 \dfrac{dx}{(x - 1)^2}$

(d) $\int_{-\infty}^{\infty} \dfrac{dx}{e^x + e^{-x}}$

(b) $\int_0^4 \dfrac{dx}{\sqrt[3]{x - 1}}$

(e) $\int_{-1}^{1} x^{-2/3}\, dx$

(c) $\int_{-\infty}^{0} e^{3x}\, dx$

(f) $\int_2^{\infty} \dfrac{x^2\, dx}{(x^3 - 1)^{3/2}}$

3. Find the area bounded by the curve $f(x) = xe^{-x^2}$ and the x axis. (Assume $x \geq 0$.)

4. Show that $\int_0^1 \log x\, dx$ is a convergent integral and find its value. Sketch the integrand for $0 < x \leq 1$.

5. Evaluate:

$$\lim_{h \to 0} \frac{1}{h} \int_2^{2+h} e^{-x^2}\, dx.$$

[HINT: Define $F(x) = \int_0^x e^{-u^2}\, du$ and observe that $F(2 + h) - F(2) = \int_2^{2+h} e^{-u^2}\, du$.]

6. A plot of land lies between a straight fence and a stream. At distances x yards from one end of the fence, the width of the plot was measured as follows:

x	0	20	40	60	80	100	120
y	0	22	41	53	38	17	0

Use Simpson's rule to approximate the area of the plot.

7.   Approximate $\int_0^1 \sqrt{1 + x^3}\, dx$ using

    (a) the trapezoidal rule with $n = 5$.

    (b) Simpson's rule with $n = 4$.

    Compare the results.

8.   From our geometric interpretation of the integral, we know that $\int_0^1 \sqrt{1 - x^2}\, dx = \pi/4$. Find the approximate values of π by using this equation, $n = 4$ and (a) the trapezoidal rule; (b) Simpson's rule.

8

Elementary Differential Equations

8.1 Introduction

Suppose we denote by $P(t)$ the function that gives the size of a population at time t. The simplest type of population growth is obtained by assuming that the rate of change of the size of a population is proportional to the population size. In terms of the function $P(t)$ this assumption implies that there exists a constant k, called the *coefficient of growth* such that

$$\frac{dP(t)}{dt} = kP(t) \qquad\qquad (8.1.1)$$

Thus, in order to find the population at any time t, we must find a solution to Equation (8.1.1). This is an example of a differential equation—that is, an equation involving a derivative. Equation (8.1.1) is a type of equation that arises very often in the discussion of physical and social science phenomena. We show shortly that the function

$$P(t) = Ce^{kt}, \qquad (C \text{ is an arbitrary constant})$$

is a solution of Equation (8.1.1) since

$$\frac{dP}{dt} = Cke^{kt} = k(Ce^{kt}) = kP(t).$$

Thus, a function P is called a solution of Equation (8.1.1) provided that $P(t)$ satisfies Equation (8.1.1) for every $t \in$ domain P. In general, an equation involving x, y, and the derivatives dy/dx, d^2y/dx^2, $\ldots$, d^ny/dx^n is called an *ordinary differential equation* of order n. Equation (8.1.1) is an ordinary differential equation of order 1. A function $y = f(x)$ is called a solution of the differential equation if for every x in the *domain of f*, it satisfies the equation. For example, if our equation is of the form

$$x + y + \frac{dy}{dx} + \frac{d^2y}{dx^2} = 0,$$

then $y = f(x)$ is a solution of this second order equation provided that $x + f(x) + df/dx + d^2f/dx^2 = 0$.

In computing antiderivatives of functions in Chapter 5, we were actually finding the solution to certain differential equations. For example, we found previously that $\int (x^2 + 1)\, dx = x^3/3 + x + C$. Thus, the function $f(x) = x^3/3 + x + C$ must be a solution of the differential equation $dy/dx = x^2 + 1$, since $f'(x) = x^2 + 1$. This is an example of a first order differential equation. Consider the following differential equation

$$\frac{dy}{dx} = x^{1/2} + 2.$$

To find a solution to this equation, we must simply find an antiderivative of $x^{1/2} + 2$. Thus a solution is given by

$$y(x) = \int (x^{1/2} + 2)\, dx = \frac{2}{3} x^{3/2} + 2x + C,$$

where C is an arbitrary constant. More generally, we can say that a solution to the differential equation

$$\frac{dy}{dx} = f(x), \tag{8.1.2}$$

where f is continuous on some interval is given by

$$y(x) = \int f(x)\, dx.$$

Notice that in all of our examples above, one arbitrary constant appeared in the solution. In solving problems from physical, biological, and social sciences, we usually have a condition on y that allows us to evaluate that constant. For example, in the first problem considered, suppose we know that at time $t = 0$, the population size is $P(0) = 200$. We see that if $P(t) = Ce^{kt}$, then $P(0) = 200 = C$ and the particular solution we want is $P(t) = 200e^{kt}$.

We have observed that the solution to Equation (8.1.1) involved one arbitrary constant. This is a particular case of the fact that the solution of every first order differential equation contains one arbitrary constant. It turns out that the solution of every second order equation contains two arbitrary constants. In fact, we can state that every nth order equation has a solution containing n arbitrary constants. Such a solution will be called a *general solution* of the differential equation. A solution obtained by giving particular values to the arbitrary constants is called a *particular solution*. Let us now consider some examples to illustrate these statements.

Example 1.   Find the general solution of the differential equation

$$\frac{d^2y}{dx^2} = x.$$

Solution. We observe that one integration of the equation yields

$$\frac{dy}{dx} = \frac{x^2}{2} + C_1$$

and thus integrating once more, we obtain

$$y(x) = \frac{x^3}{6} + C_1 x + C_2.$$

We expected two arbitrary constants since our original equation is a second order differential equation. This is a general solution to the equation. To check, we simply observe that $dy/dx = 3x^2/6 + C_1$ and $d^2y/dx^2 = x$.

Example 2. Find the particular solution of the equation $d^2y/dx^2 = 5x^3$ containing the points $(1, 0)$ and $(0, 1)$.

Solution. First, we find the general solution (containing two arbitrary constants) and then evaluate the constants using the facts that $y(1) = 0$ and $y(0) = 1$. Integrating once, we obtain

$$\frac{dy}{dx} = \frac{5x^4}{4} + C_1.$$

Hence, another integration yields

$$y(x) = \frac{5x^5}{20} + C_1 x + C_2 = \frac{x^5}{4} + C_1 x + C_2.$$

Since $y(0) = 1$, we see that $y(0) = 0 + 0 + C_2$ and thus $C_2 = 1$. To determine C_1, we use the fact that $y(1) = 0 = \frac{1}{4} + C_1 + 1$ and thus, $C_1 = -\frac{5}{4}$. Hence, the particular solution we seek is given by

$$y(x) = \frac{x^5}{4} - \frac{5x}{4} + 1.$$

Example 3. Find the particular solution of the second order equation

$$\frac{d^2y}{dx^2} = 4x^3 + 3x^2$$

passing through the point $(0, 1)$ and satisfying the condition $y'(0) = 2$.

Solution. First, we find the general solution of the equation. We know that this solution must contain two arbitrary constants. One integration of the equation yields

$$\frac{dy}{dx} = x^4 + x^3 + C_1, \qquad (C_1 \text{ an arbitrary constant}).$$

In order to find $y(x)$, we must integrate once more. Thus,

$$y(x) = \frac{x^5}{5} + \frac{x^4}{4} + C_1 x + C_2, \qquad (C_2 \text{ an arbitrary constant}).$$

We now proceed to evaluate the C_1 and C_2. The condition that the solution must pass through $(0, 1)$ means that $y(0) = 1$. Hence, $y(0) = C_2 = 1$. Now, $y'(0) = C_1$ and thus $C_1 = 2$. Our particular solution is given by

$$y(x) = \frac{x^5}{5} + \frac{x^4}{4} + 2x + 1.$$

Exercises

1. Verify that the function $y(x) = x^4/4 + C_1 x + C_2$ is a solution of the differential equation $y''(x) = 3x^2$.
2. Show that $y(x) = C/x$ is a solution to the differential equation $dy/dx = -y/x$.
3. Verify that $y(x) = C_1 e^x + C_2 x e^x$ is a solution of the differential equation $y''(x) - 2y'(x) + y = 0$.
4. Find the general solution to the following differential equations.

$$\text{(a)} \quad y'(x) = x^3 - 3x$$
$$\text{(b)} \quad y'(x) = x e^{x^2}$$
$$\text{(c)} \quad y''(x) = x^2 + 1$$
$$\text{(d)} \quad y''(x) = 5x^2 + e^x$$
$$\text{(e)} \quad y'(x) = x/(x^2 + 1)$$

5. Find the particular solution of the following differential equations satisfying the given conditions.

$$\text{(a)} \quad y'(x) = x^4 - 3x^2 + 1, \quad y(0) = 1$$
$$\text{(b)} \quad y''(x) = 5x^4 + e^x, \quad y'(0) = 1, y(0) = 3$$
$$\text{(c)} \quad y'(x) = x e^{-x^2}, \quad y(0) = \tfrac{1}{2}$$
$$\text{(d)} \quad y''(x) = \log x, \quad y(1) = y'(1) = 0.$$

6. Consider the equation $y' + 5y = 2$.

   (a) Show that the function f given by $f(x) = \tfrac{2}{5} + Ce^{-5x}$ is a solution, where C is any constant.
   (b) Assuming every solution has this form find that solution satisfying $f(1) = 2$.
   (c) Find the solution f that satisfies the condition $f(1) = 3f(0)$.

7. Consider the differential equation $y'' = 3x + 1$.

   (a) Find all solutions on the interval $0 \le x \le 1$.
   (b) Find the solution f that satisfies $f(0) = 1, f'(0) = 2$.
   (c) Find that solution f that satisfies $f(0) = 0, f'(1) = 3$.

8.2   Separation of Variables

In this section, we are concerned with a certain type of first order differential equations, namely, one of the form

$$\frac{dy}{dx} = \frac{g(x)}{h(y)}, \tag{8.2.1}$$

where g and h are functions only of x and y, respectively. There is no known technique that leads to the solution of all first order equations. Fortunately, a wide class of equations [such as Equation (8.2.1)], many of which appear in applications, can be tackled by a method known as separation of variables. Any first order equation that can be written in the form of Equation (8.2.1) is said to have variables separated. In this case our equation may be written in the form

$$h(y)\frac{dy}{dx} = g(x). \tag{8.2.2}$$

For example, consider the differential equation

$$\frac{dy}{dx} = y. \tag{8.2.3}$$

We see that

$$\frac{1}{y}\frac{dy}{dx} = 1$$

and thus the equation is one with variables separated, with $h(y) = 1/y$ and $g(x) = 1$. Observe that

$$\int \frac{1}{y}\frac{dy}{dx}\,dx = \int dx.$$

Thus,

$$\int \frac{1}{y}\frac{dy}{dx}\,dx = x + C_1.$$

Now, in the first integral, let $u = y(x)$; formally, we see that $du = y'(x)\,dx$ and by our change of variable theorem in Chapter 5, the right-hand side becomes $\int 1/u\,du = x + C_1$. Thus, $\log u = x + C_1$ or $\log y(x) = x + C_1$.

We can now find $y(x)$ by noting that

$$y(x) = Ke^x, \tag{8.2.4}$$

where $K = e^{C_1}$, a constant. It is a simple matter to show that this function is indeed a solution to (8.2.3), since $dy/dx = Ke^x = y(x)$. This example suggests that to find a solution to Equation (8.2.2) we simply integrate both sides of Equation (8.2.2), use a change of variable, and then obtain a solution. Let us now do this.

If y satisfies Equation (8.2.2), we see that

$$\int h(y)\frac{dy}{dx}\,dx = \int g(x)\,dx. \tag{8.2.5}$$

Letting $u = y(x)$, we obtain via the change of variable theorem

$$\int h(u)\,du = \int g(x)\,dx, \tag{8.2.6}$$

and thus a solution to the differential equation is given by any function $u = y(x)$ satisfying Equation (8.2.6).

Formally, the Leibnitz notation for derivatives leads in a suggestive way to Equation (8.2.6). We begin with Equation (8.2.1), $dy/dx = g(x)/h(x)$ and rewrite this as $h(y)\,dy = g(x)\,dx$ where we emphasize, as in Chapter 5, that this manipulation is strictly a mnemonic device. Integrating formally, we obtain Equation (8.2.6), namely, $\int h(y)\,dy = \int g(x)\,dx$. Thus, we may now state the following theorem that yields the general solution to Equation (8.2.1).

THEOREM.   *Let h and g be given continuous functions. Then the general solution of the differential equation*

$$y' = \frac{g(x)}{h(y)} \tag{8.2.7}$$

is given by

$$\int h(y)\, dy = \int g(x)\, dx \tag{8.2.8}$$

in the sense that any solution $y = f(x)$ must satisfy Equation (8.2.8).

In all cases, we can check that we really have a solution to the given differential equation by seeing if any solution $y = f(x)$ really satisfies the given differential equation. Let us now proceed to some examples.

Example 1.   Solve the differential equation $y' = 2x/3y^2$.

Solution.   Using the Leibnitz notation, we write the differential equation as

$$3y^2 \frac{dy}{dx} = 2x,$$

and hence

$$3y^2\, dy = 2x\, dx.$$

Thus our theorem tells us that the general solution is given by

$$\int 3y^2\, dy = \int 2x\, dx$$

or $y^3 = x^2 + C$, where C is any arbitrary constant. We might check our solution by observing that

$$3y^2 \frac{dy}{dx} = 2x \qquad \text{or} \qquad \frac{dy}{dx} = \frac{2x}{3y^2}.$$

Hence, $y^3 = x^2 + C$ really yields the solution.

We might remark at this point why in Example 1 we obtain only one arbitrary constant, C. We note that $\int 3y^2\, dy = y^3 + C_1$ and $\int 2x\, dx = x^2 + C_2$, where C_1 and C_2 are arbitrary constants. Hence, we see that $y^3 + C_1 = x^2 + C_2$ or equivalently $y^3 = x^2 + C_2 - C_1$. Now if C_1 and C_2 are arbitrary constants, then so is $C_2 - C_1$. We simply set $C = C_2 - C_1$. In the future, we will always insert only one arbitrary constant into our solution of a separable first order equation.

Example 2.   Find all solutions to the differential equation

$$y' = e^{x-y}.$$

Solution.   Using the Leibnitz notation, we wish to solve $dy/dx = e^{x-y}$ or equivalently, $e^y dy = e^x dx$ (since $e^{x-y} = e^x e^{-y}$). Thus, all solutions are given by $\int e^y dy = \int e^x dx$ or $e^y = e^x + C$. In order to find y, we take the logarithm of both sides and recall that $\log e^y = y$. Thus, $y = \log(e^x + C)$. It is not difficult to check this solution. Differentiating, we obtain $dy/dx = e^x/(e^x + C)$. But, $e^x + C = e^y$, and thus, $dy/dx = e^x/e^y = e^{x-y}$.

Example 3.   We now return to the problem first considered in Section 8.1. In that example, it was necessary for us to solve the differential equation

$$\frac{dP(t)}{dt} = kP(t).$$

Hence, we see that a solution must be given by

$$\int \frac{dP}{P} = \int k\, dt \qquad \text{or} \qquad \log P = kt + C_1,$$

where C_1 is any constant. Thus,

$$P = e^{kt + C_1} = Ce^{kt},$$

where $C = e^{C_1}$ is again an arbitrary constant. If the population at time $t = 0$ is denoted by $P(0)$, we see that $C = P(0)$ and the particular solution we seek is given by

$$P(t) = P(0)e^{kt}.$$

Exercises

1.   Find all solutions to the following differential equations.

(a) $y' = x^2 y$

(d) $y' = \dfrac{e^{x-y}}{1 + e^x}$

(b) $yy' = x$

(e) $y' = x^3 y - 4x^3$

(c) $y' = \dfrac{x + x^2}{y - y^2}$

2.   (a)   Find the solution of the differential equation $y' = 2y^{1/2}$ passing through the point (x_0, y_0), where $y_0 \geq 0$.
     (b)   Find all solutions of this equation passing through $(x_0, 0)$.

3.   A firm observes that the rate of decrease of sales with respect to price is directly proportional to the sales volume and inversely proportional to the sales price plus a constant; that is,

$$\frac{ds}{dp} = -\frac{Bs}{A + p},$$

where s is the sales volume, p is the unit sales price, and A and B are constants. Find the relationship between sales volume and price.

4. Find the particular solutions to the following differential equations.

(a) $\dfrac{dy}{dx} = \dfrac{x^2}{y^2}$,   where $y(0) = 1$.

(b) $\sqrt{1 + x^2}\,\dfrac{dy}{dx} + 3xy^2 = 0$,   where $y(1) = 1$.

(c) $\dfrac{dy}{dx} = \dfrac{xy + y}{x + xy}$,   where $y(1) = 1$.

8.3 First Order Equations Solved with Integrating Factors

Another type of differential equation that often arises in applications is

$$\frac{dy}{dx} + p(x)y = q(x), \tag{8.3.1}$$

where $p(x)$ and $q(x)$ are known functions. An equation of this form is a first order linear differential equation. You do not need to know the name, but you should recognize the form.

It is usually not possible to separate variables in (8.3.1) as you can readily check. Let us try a different approach. If we could find a function $f(x)$ with the property that

$$\frac{d}{dx}[f(x)y(x)] = f(x)\frac{dy}{dx} + f(x)p(x)y,$$

then it would be an easy matter to solve (8.3.1). Indeed, we would just multiply both sides of (8.3.1) by $f(x)$ to obtain

$$f(x)\frac{dy}{dx} + f(x)p(x)y = f(x)q(x)$$

and then rewrite this as

$$\frac{d}{dx}[f(x)y(x)] = f(x)q(x). \tag{8.3.2}$$

Now we can solve Equation (8.3.2) by integrating both sides to obtain

$$f(x)y(x) = \int f(x)q(x)\,dx$$

or equivalently

$$y(x) = \frac{1}{f(x)}\int f(x)q(x)\,dx. \tag{8.3.3}$$

The big question is: Can we find a function $f(x)$ with these magical properties? The answer is yes. Indeed the function

$$f(x) = e^{\int p(x)\,dx}$$

does all that we require. For those who do not regard this "rabbit-out-of-a-hat" technique kindly, we now track the elusive function to its lair. Let us begin with a specific example. Consider

$$\frac{dy}{dx} + xy = x. \qquad (8.3.4)$$

In this case $p(x) = q(x) = x$. First, we multiply both sides of Equation (8.3.4) by a function $f(x)$ to obtain

$$f(x)\frac{dy}{dx} + xyf(x) = xf(x). \qquad (8.3.5)$$

Now we wish to write the left-hand side of Equation (8.3.5) as follows:

$$f(x)\frac{dy}{dx} + xyf(x) = \frac{d}{dx}[f(x)y].$$

Thus, we want

$$f(x)\frac{dy}{dx} + xyf(x) = \frac{df}{dx}(x)y + f(x)\frac{dy}{dx}.$$

In order to have this equality it is clear that we must set

$$xf(x) = \frac{df}{dx}(x). \qquad (8.3.6)$$

Therefore, if we can find any function f satisfying the first order equation (8.3.6), we can immediately solve our given differential equation. Observe that Equation (8.3.6) is really an equation whose variables are separated, since we may write Equation (8.3.6) as

$$\frac{1}{f(x)}\frac{df}{dx} = x.$$

Hence, we see that the solution to the above is obtained by integration. Namely,

$$\int \frac{df}{f} = \int x\,dx, \qquad \text{and thus} \qquad \log f = \frac{x^2}{2} + C.$$

However, since *any* solution of Equation (8.3.6) will do, we can take $C = 0$. The function f we seek is just given by

$$\log f = \frac{x^2}{2} \qquad \text{or equivalently,} \qquad f(x) = e^{x^2/2}.$$

Returning to Equation (8.3.4), we can obtain y by observing that

$$\left(\frac{dy}{dx} + xy\right)e^{x^2/2} = \frac{d}{dx}[e^{x^2/2}y] = xe^{x^2/2}.$$

Integrating once again, we see that

$$e^{x^2/2}y = \int xe^{x^2/2}\,dx = e^{x^2/2} + C.$$

Thus,

$$y(x) = 1 + Ce^{-x^2/2}$$

is a general solution of Equation (8.3.4). To check that our results are correct, we just show that the above form of the solution satisfies Equation (8.3.4). Note that $dy/dx = -Cxe^{-x^2/2}$ and thus

$$\frac{dy}{dx} + xy = -Cxe^{-x^2/2} + x(1 + Ce^{-x^2/2}) = x.$$

We can easily generalize the scheme we have used in this example. We must first find a function f, such that $f(x)\,y'(x) + p(x)f(x)y = f(x)\,y'(x) + yf'(x)$. In order that this equality may hold, it follows that

$$p(x)f(x) = \frac{df}{dx}. \tag{8.3.7}$$

As before, a solution to Equation (8.3.7) is obtained by integration

$$\int \frac{df}{f} = \int p(x)\,dx.$$

Thus,

$$\log f(x) = \int p(x)\,dx \qquad \text{and} \qquad f(x) = e^{\int p(x)\,dx}. \tag{8.3.8}$$

The function $f(x) = e^{\int p(x)\,dx}$ is called an *integrating factor* of the differential equation

$$\frac{dy}{dx} + p(x)y = q(x).$$

The general solution is found by first multiplying both sides of the equation by $e^{\int p(x)\,dx}$ and then integrating the result. After we multiply by $e^{\int p(x)\,dx}$ our equation becomes

$$e^{\int p(x)\,dx}\frac{dy}{dx} + e^{\int p(x)\,dx}p(x)y = e^{\int p(x)\,dx}q(x).$$

We now rewrite the left-hand side to obtain

$$\frac{d}{dx}[e^{\int p(x)\,dx}y] = e^{\int p(x)\,dx}q(x).$$

Thus,

$$e^{\int p(x)\,dx}y = \int [e^{\int p(x)\,dx}q(x)]\,dx + C.$$

Finally we multiply by $e^{-\int p(x)\,dx}$ to obtain

$$y = e^{-\int p(x)\,dx} \cdot \int [e^{\int p(x)\,dx}q(x)]\,dx + Ce^{-\int p(x)\,dx},$$

where C is an arbitrary constant.

Let us now consider several examples.

Example 1.   Find the general solution to the differential equation $y' = ky$, where k is a fixed number.

Solution.   This differential equation has separable variables and could be solved by methods of Section 8.2. However, we choose to use the techniques in this section and show that the result is, of course, equivalent. Rewrite the equation in the form $y' - ky = 0$. In this case, $p(x) = -k$ and $q(x) = 0$. Hence, an integrating factor is given by

$$f(x) = e^{-\int k\, dx} = e^{-kx}.$$

Multiplying both sides of the equation by $f(x)$, we have

$$\left(\frac{dy}{dx} - ky\right)e^{-kx} = \frac{d}{dx}[ye^{-kx}] = 0.$$

Hence,

$$ye^{-kx} = C \qquad \text{and} \qquad y = Ce^{kx},$$

where C is an arbitrary constant. This is the result obtained in Section 8.2.

As you can see, a given differential equation may be solved by more than one method. When this occurs, equivalent solutions are always obtained.

Example 2.   Find the general solution of the differential equation

$$\frac{dy}{dx} - 2y = 1.$$

Solution.   In this example, $p(x) = -2$ and $q(x) = 1$. Hence, an integrating factor is given by

$$f(x) = e^{-\int 2\, dx} = e^{-2x}.$$

Multiplying both sides of the equation by $f(x)$, we have

$$\left(\frac{dy}{dx} - 2y\right)e^{-2x} = \frac{d}{dx}(ye^{-2x}) = e^{-2x}$$

and hence,

$$ye^{-2x} = \int e^{-2x}\, dx + C.$$

However,

$$\int e^{-2x}\, dx = -\frac{1}{2}e^{-2x}.$$

Therefore,

$$ye^{-2x} = -\tfrac{1}{2}e^{-2x} + C \qquad \text{and} \qquad y = -\tfrac{1}{2} + Ce^{2x},$$

where C is an arbitrary constant. It is a simple matter to check that this is a solution simply by substituting this form for y into the left-hand side of the differential equation and observing that the result is 1.

Example 3.   A differential equation relating net profits P and advertising effort x is given by

$$\frac{dP}{dx} + aP = b - ax,$$

where a and b are constants. Suppose that at $x = 0$, $P = 2 \cdot (b + 1)/a$. Find the net profit in terms of the advertising effort x.

  Solution.   In this example, $p(x) = a$ and $q(x) = b - ax$. Hence an integrating factor is

$$f(x) = e^{\int a\,dx} = e^{ax}.$$

Multiplying both sides of the equation by $f(x)$, we have

$$\left(\frac{dP}{dx} + aP\right)e^{ax} = \frac{d}{dx}(Pe^{ax}) = (b - ax)e^{ax},$$

and hence

$$Pe^{ax} = \int (b - ax)e^{ax}\,dx + C.$$

Now,

$$\int (b - ax)e^{ax}\,dx = \frac{b}{a} \cdot e^{ax} - a\left[\frac{x}{a} \cdot e^{ax} - \frac{1}{a^2} \cdot e^{ax}\right]$$

$$= \frac{1 + b}{a} \cdot e^{ax} - xe^{ax} + C = \left(\frac{1 + b}{a} - x\right)e^{ax} + C.$$

Therefore,

$$P = \left(\frac{1 + b}{a} - x\right) + Ce^{-ax},$$

where C is an arbitrary constant. Using the condition $P(0) = 2 \cdot (b + 1)/a$ we can evaluate the constant C.

$$P(0) = \frac{2(1 + b)}{a} = \frac{1 + b}{a} + C.$$

Thus, $C = (1 + b)/a$ and the particular solution is given by

$$P(x) = \frac{1 + b}{u} - x + \frac{1 + b}{a}e^{-ax}.$$

Example 4.   Find the particular solution of the differential equation

$$\frac{dy}{dx} + xy = (2 + x)e^{2x},$$

satisfying the condition $y(0) = 2$.

Solution. For this example, $p(x) = x$ and $q(x) = (2 + x)e^{2x}$. Thus, an integrating factor is given by

$$f(x) = e^{\int x\,dx} = e^{x^2/2}.$$

Therefore,

$$\left(\frac{dy}{dx} + xy\right)e^{x^2/2} = \frac{d}{dx}(ye^{x^2/2}) = (2 + x)e^{2x} \cdot e^{x^2/2},$$

and hence

$$ye^{x^2/2} = \int (2 + x)e^{(2x+x^2/2)}\,dx = e^{(2x+x^2/2)} + C$$

and

$$y(x) = e^{2x} + Ce^{-x^2/2},$$

where C is an arbitrary constant. The condition $y(0) = 2$ yields $2 = 1 + C$, implying that $C = 1$. The particular solution is then given by

$$y = e^{2x} + e^{-x^2/2}.$$

The methods presented in Sections 8.2 and 8.3 apply only to differential equations of the appropriate form [the form of Equation (8.3.1) or Equation (8.2.1)]. Differential equations of these two forms occur frequently in applications to the physical, social, and biological sciences. This is well illustrated in Section 8.4. However, differential equations in many other forms arise in various applications and have been thoroughly studied by mathematicians. The reader interested in learning more about these differential equations is referred to the many elementary textbooks on the subject.

Exercises

1. Find the general solution to the following differential equations.

    (a) $\dfrac{dy}{dx} - y = -2e^{-x}$

    (b) $\dfrac{dy}{dx} + 2xy = 4x$

    (c) $\dfrac{dy}{dx} - \dfrac{y}{x} = x^2 + 3x - 2$

    (d) $(x - 2)\dfrac{dy}{dx} = y + 2(x - 2)^3$

    [HINT: Assuming $x \neq 2$, first divide through by $(x - 2)$ and then write the equation in the form $dy/dx + p(x)y = q(x)$.]

    (e) $x^3\dfrac{dy}{dx} + (2 - 3x^2)y = x^3$

(f) $\dfrac{dy}{dx} + \dfrac{1}{x \log x} y = \dfrac{1}{x}$

(g) $\dfrac{dy}{dx} = x^3 - 2xy$

2. Find the particular solution to the following differential equations.

(a) $\dfrac{dy}{dx} = x^3 - 2xy,$ where $y(1) = 2$.

(b) $\dfrac{dy}{dx} = ay + b,$ $(a, b$ constants$),$ $y(0) = -\dfrac{b}{a}.$

(c) $2e^{-x}\dfrac{dy}{dx} + e^x y = 2e^x,$ $y(0) = e + 2$.

(d) $x\dfrac{dy}{dx} + y = 3x^3 - 1,$ (for $x > 0$), $y(1) = 1$.

(e) $\dfrac{dy}{dx} + 2xy = xe^{-x^2},$ $y(0) = 1$.

(f) $\dfrac{dy}{dx} - 2y = x^2 + x,$ $y(0) = \dfrac{1}{2}.$

3. Manufacturing costs K are related to the number of items produced x by the equation

$$\frac{dK}{dx} + aK = b + cx,$$

where a, b, and c are constants. If $K = 0$ when $x = 0$, find K as a function of x.

4. Consider the differential equation

$$x^2 \frac{dy}{dx} + 2xy = 1,   \text{where } 0 < x < \infty.$$

(a) Show that every solution tends to zero as $x \to \infty$.

(b) Find the solution f that satisfies the condition $f(2) = 2f(1)$.

5.* Various equations can be made linear by suitable changes of variables. For example, the equation

$$2xy\frac{dy}{dx} + y^2 = x \tag{5a}$$

becomes, if we let $Y = y^2$, the linear equation

$$x\frac{dY}{dx} + Y = x. \tag{5b}$$

(a) Check that the change of variable $Y = y^2$ transforms (5a) into (5b).

(b) Find the general solution to (5a) by first solving (5b).

6.* Consider the differential equation

$$\frac{dy}{dx} + p(x) = q(x)e^{my}. \tag{6a}$$

(a)   Show that the change of variable $Y = e^{-my}$ transforms the equation (6a) into

$$-\frac{1}{m}\frac{dY}{dx} + p(x)Y = q(x). \tag{6b}$$

(b)   Now find a form for the general solution of (6a) by first solving (6b).

(c)   Using the technique indicated in (a), find the general solution to the differential equation:

$$x\frac{dy}{dx} + 1 = x^2 e^{2y}.$$

8.4   Applications of Differential Equations

In this section we are concerned with various applications of first order differential equations to problems in business and the social and biological sciences.

APPLICATION 1. (GROWTH).   Any important bit of news spreads quickly. The number of people who know it grows with time. In fact, the number $N = N(t)$ of people who know the news at time t should be proportional to the original number N_0 who were told the news at time t_0 and who could not help but spread the news. Thus, the older the news, the more people know it. If $N(t)$ people know the news at time t, a natural question to ask is how many know it at a slightly later time, say $t = t_1$. It is not unreasonable to expect that the number $N(t_1) - N(t)$ of people who learn the news in the time interval $[t, t_1]$ is approximately proportional to both $N(t)$ and the lapse of time $t_1 - t$. If we accept this idea as our initial assumption, its mathematical reformulation becomes

$$N(t_1) - N(t) = kN(t)[t_1 - t], \qquad N(t_0) = N_0, \tag{8.4.1}$$

where k is a positive constant called the *growth coefficient*. We now wish to discover how N is related to the initial number N_0. With (8.4.1) as our model, we let the time lapse $t_1 - t$ tend to zero and obtain a differential equation for N.

Rewriting (8.4.1) as

$$\frac{N(t_1) - N(t)}{t_1 - t} = kN(t),$$

we see that

$$\lim_{t_1 \to t} \frac{N(t_1) - N(t)}{t_1 - t} = kN(t).$$

Hence, we obtain the differential equation

$$\frac{dN}{dt} = kN(t), \qquad \text{with } N(t_0) = N_0. \tag{8.4.2}$$

Equation (8.4.2) tells us simply that the rate of change of N is proportional to N. This is the basic mathematical model for growth. By solving Equation (8.4.2), we can determine N in terms of the original number of people N_0 who were first told the news. If k should be negative, Equation (8.4.2) becomes the basic equation for decay. If $k = 0$, then N must be constant and there is no spread of news. For convenience, we shall always take $t_0 = 0$ so that the condition $N(t_0) = N_0$ becomes $N(0) = N_0$.

From our results in Section 8.2, we see that

$$N(t) = N_0 e^{kt}, \tag{8.4.3}$$

where t is time elapsed since the beginning of the process.

The differential equation (8.4.2) and its solution (8.4.3) obviously have many other applications than the spreading of news. They represent a model for the growth of timber and vegetation, population growth (both people and bacteria), the growth of money in bonds (those that credit the interest to the capital continuously), the growth of a substance in the course of a chemical reaction, radioactive decay, and other processes too numerous to mention.

Example 1. Assume that the rate of change of the world population is proportional to the population size (which unfortunately is true). Suppose that in 1950 the world population was 2 billion (2×10^6) and in 1960 the world population was 3 billion (3×10^6). What will be the world population in the year 2000?

Solution. Our basic differential equation is again Equation (8.4.2). Now, let $t = 0$ correspond to the year 1950 and $t = 1$ correspond to the year 1960. We are taking our unit as 10 years. $N(0) = 2 \times 10^6$. Thus, from Equation (8.4.3),

$$N(t) = [2 \times 10^6]e^{kt}.$$

However, the fact that the population was 3 billion in 1960 implies that $N(1) = 3 \times 10^5$. Hence,

$$N(1) = [3 \times 10^6] = [2 \times 10^6]e^k \quad \text{and} \quad e^k = \tfrac{3}{2} \quad \text{or} \quad k = \log \tfrac{3}{2}.$$

It is essential for us to correctly determine k in order to solve the problem. The population at any time t (where t is measured in units of 10 years) is given by

$$N(t) = (2 \times 10^6)e^{t \log 3/2}.$$

We wish to know what the population will be in the year 2000. This corresponds to $t = 5$

$$N(5) = (2 \times 10^6)e^{5 \log 3/2}$$
$$= (2 \times 10^6)e^{\log(3/2)^5}$$
$$= (2 \times 10^6)(\tfrac{3}{2})^5 = {}^{243}_{16} \times 10^6.$$

Thus, the population in the year 2000 should be approximately 15,200,000,000.

Example 2. If we deposit $10 in a bank that offers interest and this interest is compounded instantaneously at 5%, how many years will it take for the total to reach $20? (Various banks compute interest yearly, semiannually, monthly, or daily. Compounding interest instantaneously is a natural extension of the above.)

Solution.   We have not defined what we mean by compounding interest instantaneously. One reasonable definition states that $dN(t)/dt = kN(t)$, where $N(t)$ is the amount of money at time t, and k is the (instantaneous) interest rate. This equation is just our old friend, Equation (8.4.2). It is also possible to approach the notion of instantaneous compound interest in a different way. Suppose we compute interest at k per cent compounded n times per year. If we start with the amount P_0, then our principal P at the end of a year is given by the expression $P = P_0(1 + k/n)^n$. If we let n tend to infinity, which corresponds to our intuitive notion of compounding interest instantaneously; we discover that both definitions lead to the same answer! Since they agree, we will not pursue the discussion of the definitions further but instead return to our original problem. Using the first definition, we see from Equation (8.4.3) that $N(t) = N_0 e^{kt}$. Since we are given $k = 0.05$ and $N_0 = 10$, our equation becomes $N(t) = 10e^{0.05t}$. We now want to find when $N(t)$ will equal 20. We thus set $20 = 10e^{0.05t}$ or $e^{0.05t} = 2$. Taking the logarithm of both sides, we find that $t = 20 \log 2$. Since by Table A.2, $\log 2 = 0.69$, this means $t = 13.9$, the number of years it takes for the total to reach \$20.

In Chapter 5, we represented the solution to this problem in terms of an integral. That integral is just another form of the solution to the differential equation studied here.

Example 3.   In the chemical reaction that produces phlogiston, the amount produced at a given moment is proportional to the amount present. If there is one gram of phlogiston at time zero and 10 grams 2 minutes later, find the equation for the amount at time t.

Solution.   The basic differential equation is again Equation (8.4.2) where N_0 represents the initial amount of phlogiston, and $N(t)$ represents the amount of phlogiston at time t. First we must find the growth coefficient. Since $N_0 = 1$, our equation is $N(t) = e^{kt}$. Since $N(2) = 10$, we find that $10 = e^{2k}$. Now we take the logarithm of both sides of this equation and obtain $2k = \log 10$ or $k = \frac{1}{2} \log 10$. The answer to the problem can thus be written as $N(t) = e^{(1/2 \log 10)t}$.

APPLICATION 2. (DECAY).   Various radioactive elements show marked differences in their rate of decay, but they all share the property that the rate at which a given substance decomposes at any instant is proportional to the amount present at that instant. If we denote by $N(t)$ the amount present at time t, then the "law of decay" states that

$$\frac{dN}{dt} = -kN(t), \tag{8.4.4}$$

where k is a positive constant (called the decay constant) whose actual value depends upon the particular element that is decomposing. Note that the minus sign appears since N decreases as t increases and hence dN/dt must always be negative.

If $N(0) = N_0$, then every solution of Equation (8.4.4) must be of the form

$$N(t) = N_0 e^{-kt}. \tag{8.4.5}$$

Observe that there can be *no* finite time at which $N(t) = 0$ since e^{-kt} is never zero for finite t. Thus, there is no use in studying the "total lifetime" of a radioactive substance. It is of course possible to determine the time required for any particular *fraction* of a sample to decay. For convenience, the fraction $\frac{1}{2}$ is chosen and the time T at which $N(T)/N_0 = \frac{1}{2}$ is called the *half-life* of the substance. This is determined by solving the equation

$$e^{-kT} = \tfrac{1}{2}$$

for T. We see immediately by using logarithms that

$$-kT = \log \tfrac{1}{2} = -\log 2,$$

and therefore the equation

$$T = \frac{1}{k} \log 2$$

relates the half-life of the decay constant k.

Example 4.   The half-life for radium is approximately 1600 years. Find what percentage of a given quantity of radium disintegrates in 100 years.

Solution.   From our previous discussion we see that we can determine k from the equation

$$T = 1600 = \frac{1}{k} \log 2 = \frac{1}{k} (0.6931).$$

Thus,

$$k = \frac{0.6931}{1600} = 0.000433.$$

Now, at the end of 100 years,

$$N(100) = N_0 \, e^{-0.000433(100)} \qquad \text{or} \qquad N(100) \cong N\,(0.958)$$

where $\cong$ is read "is approximately." We have resorted to Table A.1 to determine these numbers. Thus, at the end or 100 years we have approximately $0.958N_0$ amount of radium left. This tells us that only 4.2% disappears in 100 years.

This simple decay model that we have considered above describes the essential features of many other phenomena as well. We need only change the names of the characters in our script in order that the results may apply to other situations.

APPLICATION 3. (BOUNDED GROWTH).   If for some reason the population cannot exceed a certain maximum number, say M (for example, after M people the food supply is exhausted), we may reasonably assume that the rate of growth is jointly proportional to both $N(t)$ and $M - N(t)$. Thus, we have a second type of growth law

$$\frac{dN}{dt}(t) = kN(t)[M - N(t)], \qquad N(0) = N_0, \qquad (8.4.6)$$

where k is the "growth" constant and, say, M is a constant.

We solve Equation (8.4.6) using the separation of variables technique discussed in Section 8.2. After the first step we obtain

$$\int \frac{dN}{N(M - N)} = k \int dt.$$

We note that

$$\frac{1}{N(M - N)} = \frac{1}{M} \left[\frac{1}{N} + \frac{1}{M - N} \right]$$

and hence,

$$\int \frac{dN}{N(M - N)} = \frac{1}{M} \left[\log N - \log(M - N) \right].$$

Thus,

$$\frac{1}{M} \left[\log N - \log(M - N) \right] = kt + C, \tag{8.4.7}$$

where C is an arbitrary constant. From Equation (8.4.7) we see that

$$\log \frac{N}{M - N} = Mkt + C_1,$$

where C_1 is the constant CM. The above result yields

$$\frac{N(t)}{M - N(t)} = C_2 e^{Mkt},$$

where C_2 is the constant e^{C_1}. Using the fact that $N(0) = N_0$ and $C_2 = N_0/(M - N_0)$, we see that

$$N(t) = [M - N(t)] \frac{N_0}{M - N_0} e^{Mkt},$$

or

$$N(t) \left[1 + \frac{N_0}{M - N_0} \overline{e^{Mkt}} \right] = \frac{MN_0}{M - N_0} e^{Mkt}.$$

A final simplification yields

$$N(t) = \frac{MN_0 e^{Mkt}}{M + N_0(e^{Mkt} - 1)}. \tag{8.4.8}$$

We notice the following facts about Equation (8.4.8). First, it may be rewritten in the form

$$N(t) = \frac{MN_0}{N_0 + (M - N_0)e^{-Mkt}}.$$

This is particularly useful if we wish to determine the behavior of the population as t becomes very large. In fact, the solution tells us that as $t \to \infty$, $N(t) \to M$, This is certainly in agreement with our original statement of the problem that said the population can never be larger than M.

APPLICATION 4. (PRODUCTION CAPACITY).* The Red, White, and Blue Company manufactures a variety of products, one of which is an American flag. The price is competitive with respect to flags of the same size and quality, and owing to increased emphasis on patriotism there has been a large demand for these flags. Since the company has reason to expect this demand to continue, they wish to expand their production capacity. At present, 500 machines are capable of supplying current demand. However, since the demand is expected to double over the next 10 years, the management wants to institute plans for gradual plant expansion. Toward that end, Red, White, and Blue has decided to purchase 20 machines in the first year of the expansion program, 40 in the second year, 60 in the third year and so on. At first glance, this rate of expansion seems too great since more than 500 new machines will be purchased by the end of the eighth year. However, the management is well aware that machines tend to wear out, and that productivity decreases with time. In fact, according to the manufacturer's own data, it can be expected that productivity will decrease by 5% per year. In order to compensate for this, more machines must be purchased each year. Moreover, by the tenth year, Red, White, and Blue must double its plant capacity to keep up with increased demand. A main source of concern is the validity of the machine manufacturer's estimate of the annual productivity decrease factor. Obviously, if this figure were 4% or 6% Red, White, and Blue might find itself overexpanded or underexpanded at the end of the 10 year period. Thus, management wishes to obtain a relationship for plant capacity in any given year of the expansion program in terms of the decay factor in order to analyze the effect of possible errors in the estimate of the annual decrease in machine productivity.

The simplest method of determining the number of years required to double plant capacity assumes that all *new* equipment is purchased at the end of the year. Thus, if year 1 is begun with 500 machines, and 5% of them will become unproductive during the year, then only 475 machines will be producing at full capacity at the end of year 1. With the purchase of 20 new machines the year end total will be 495. (See table below.)

Plant Capacity

Year	Begin	Loss	Gain	End
1	500	25	20	495
2	495	25	40	510
3	510	26	60	544
4	544	27	80	597
5	597	30	100	667
6	667	33	120	754
7	754	38	140	856
8	856	43	160	973
9	973	49	180	1104
10	1104	55	200	1249

---

* Example taken from Stern, *Mathematics for Management*, (Prentice-Hall, Englewood Cliffs, New Jersey, 1963), Chapter 6.

Of the 495 machines producing at the beginning of year 2, 95% of them, or 470, will be producing at the end of the year. After 40 new machines are purchased, the year end total will be 510. In this manner, the year-end production capacity may be computed for as many years as desired. From the table, it appears as if the objective of doubling machine capacity is very nearly achieved at the end of year 8. There are several difficulties that arise in this tabular method of determining plant capacity. The most important deficiencies are: (i) productivity decrease is not considered as a continuous phenomenon and (ii) it does not lend itself to the derivation of a formula by which the plant capacity can be represented as a function of the decay factor and the year of the expansion program. We now show how to derive such a formula.

Let C represent the plant capacity that depends on time t (that is, the plant capacity $C = C(t)$, a function of time t). The time rate of change of capacity resulting from the productivity decay factor λ is $-\lambda C(t)$. This is equivalent to the statement that the capacity is continuously decaying at a rate $\lambda C(t)$. In order to find the continuous rate at which additions to capacity are being made, note that, in the discrete case, the added capacity after the year t is

$$20 + 40 + 60 + \cdots + 20t = 20(1 + 2 + 3 + \cdots + t).$$

Now, it can be shown that if t is a positive integer, then

$$1 + 2 + 3 + \cdots + t = \frac{t(t + 1)}{2}.$$

(You can check this for various values of t.) Hence,

$$20(1 + 2 + 3 + \cdots + t) = 20\,\frac{t(t + 1)}{2} = 10t^2 + 10t.$$

The instantaneous rate of plant expansion is then given by the derivative with respect to t of the above—namely, $20t + 10$. Since the rate of change of plant capacity is equal to the loss through decrease of productivity plus the instantaneous rate of expansion, we have

$$\frac{dC}{dt} = -\lambda C + 20t + 10,$$

or

$$\frac{dC}{dt} + \lambda C = 20t + 10, \tag{8.4.9}$$

where C = plant capacity and λ = the productivity decay factor.

Equation (8.4.9) is a first order differential equation that can be solved via the methods given in Section 8.3. An integrating factor $f(t)$ is

$$f(t) = e^{\int \lambda \, dt} = e^{\lambda t}.$$

Hence, from Section 8.3, we have

$$C(t)e^{\lambda t} = \int (20t + 10)e^{\lambda t} \, dt = 20\left[\frac{t}{\lambda} - \frac{1}{\lambda^2}\right]e^{\lambda t} + \frac{10}{\lambda}e^{\lambda t} + k,$$

where k is an arbitrary constant. Therefore,

$$C(t) = ke^{-\lambda t} + \frac{10}{\lambda}[2t + 1] - \frac{20}{\lambda^2} = ke^{-\lambda t} + \frac{20t}{\lambda} + \frac{10\lambda - 20}{\lambda^2}. \qquad (8.4.10)$$

Using the condition that $C(0) = 500$ (our assumption was that we began with 500 machines), we can evaluate k.

$$500 = k + \frac{10\lambda - 20}{\lambda^2}, \qquad \text{and hence} \qquad k = 500 - \frac{10\lambda - 20}{\lambda^2}.$$

Thus, the complete solution is given by

$$C(t) = \left(500 - \frac{10\lambda - 20}{\lambda^2}\right)e^{-\lambda t} + \frac{20t}{\lambda} + \frac{10\lambda - 20}{\lambda^2} \qquad (8.4.11)$$

This allows us to compute the plant capacity at any time $t \geq 0$ for a given value of λ.

If the manufacturer's statement on decay is correct—that is, $\lambda = 0.05$—then (8.4.11) becomes

$$C(t) = 8300e^{-0.05t} + 400t - 7800.$$

For $t = 8$, we find that

$$C(8) = 8300e^{-0.4} + 3200 - 7800 = 8300(0.67032) - 4600 = 963.$$

A closer approximation to the point in time when $C = 1000$ (double the initial capacity) can be found by trying $t = 8.3$. $C(8.3) = 1000$. Thus it is reasonable to expect that the objective of doubling plant capacity will be achieved at the beginning of the second quarter of year 9 of the expansion program. Note that $C(8) = 963$ is 12 units less than the estimate made in the discrete table. The main difference in the results can be attributed to the fact that we consider a continuous rather than discrete analysis of the table.

After reading through the previous four applications of first order differential equations, you may suspect that as we apply mathematics to the various sciences we reach a point where there appear to be only a few different kinds of processes occurring at the present mathematical level. The equations remain the same, only the names of the functions and variables change from science to science. You would be right.

Exercises

1. The rate of growth of a colony of bacteria is proportional to the population. Initially, there are 2000 bacteria and one day later, 5000. What should the population of the colony be after five days if it grows without restrictions?
2. Write the differential equation for the growth of a bacterial population which increases 2.5% every hour. If there are N_0 bacteria at the start, how many are there at the end of ten hours?
3. A debt of $300 will be paid in 10 years. Find its present value if money can earn 6% per year compounded instantaneously.

4. A firm conducts pricing experiments and finds that an increase of $1 in the unit sales price results in a constant decrease of b units in sales volume. What is the relationship between sales and price? Find the relationship if we are told that the sales volume is V when the price is $5.
    [HINT: Suppose there exists a relationship between the sales volume s and price p. Then, the rate of change of sales with respect to price is ds/dp, and we are told that $ds/dp = -b$].

5. An airline hires 100 stewardesses and a constant number quit each month. If 82 stewardesses are still employed at the end of one year, how many will be employed after two years? After x months? What is the maximum length of service of the stewardesses?
    [HINT: We approximate the discrete case with a continuous variable. Let $y =$ the number of stewardesses and $x =$ the number of months since first employed; then $dy/dx = -a$, where a is the number who quit each month.]

6. If airline stewardesses quit at rates proportional to the number of months of service and if 82 out of 100 stewardesses are still employed at the end of one year, what is the expected number remaining after two years? What is the maximum length of service of the stewardesses?
    [HINT: Again approximate the discrete case with a continuous variable. In this case $dy/dx = -ax$ (using the same notation as in 5).]

7. If $2000 is deposited at 7% interest per year and the interest is compounded instantaneously, what is the amount in 25 years? How many years will it take for the total to reach $20,000?

8. In a certain first order chemical reaction, half of the original substance has decomposed in 10 seconds. How long from the beginning is required for 90% of the substance to decompose?

9. Carbon-14, C^{14}, is a radioactive isotope of carbon that has a half-life of approximately 5600 years. Find what percentage of a given quantity of radium disintegrates in 500 years. What percent of the original substance is left at the end of 50,000 years?

10. In a certain stimulus-sampling theory, it is assumed that there is a continuous function N such that if T is the number of trials, then $N(T)$ is the expected number of elements conditioned. Let S be the number of stimulus elements effective at any given time (which is assumed to be constant). Moreover, it is assumed that

$$\frac{dN}{dx} = \frac{s}{S}[S - N(x)]$$

for x in the domain of N. Find N for any number of trials x.

11. Let N be the number of persons each with an income of M or more dollars. Clearly N decreases as M increases. More specifically, it has been estimated that

$$\frac{dN}{dM} = \frac{-1.5}{M}.$$

Find the relationship between N and M if $N = 100$ when $M = 10,000$ (Pareto's law).

12. The rate of increase in profits per dollar increase in sales price dP/dp is given by

$$\frac{dP}{dp} = (a - 1) + \frac{c}{p} - \log p,$$

where P is the profit, p the sales price, and a and c are constants. Show that profits are given by

$$P = (a - 1)p + c \log p + p \log p - p + \text{constant}.$$

13. Show that the solution to the differential equation for bounded growth Equation (8.4.6) can be expresed in the form

$$N(t) = \frac{M}{1 + e^{-\alpha(t - t_0)}}$$

where α is a constant and t_0 is the time at which $N(t) = M/2$.

14. Suppose in the fourth application to production capacity, the Red, White, and Blue Company decides to increase its capacity by purchasing n machines at the end of each year, where n is a number to be determined. (The management has decided it is not feasible to increase its purchase order by 20 machines each year.) Write a differential equation relating C to n, λ, and t. Solve this differential equation assuming that initially the Red, White, and Blue Company has 500 machines. If in 10 years the Red, White, and Blue Company wants to double its capacity, what number n of machines should it add at the end of each year? (Always assume that manufacturer's statement on decay is correct—that is, $\lambda = 0.05$.)

8.5   Summary of Chapter 8

In this section we summarize the most important results of this chapter.

(1) The solution to the differential equation

$$\frac{dy}{dx} = f(x),$$

where f is continuous on some interval is given by

$$y = \int f(x) \, dx.$$

A differential equation of the form

$$\frac{d^n y}{dx^n} = f(x)$$

(n a positive integer) requires n integrations to determine $f(x)$. In this case the solution will contain n arbitrary constants.

(2) The general solution to the differential equation

$$h(y) \frac{dy}{dx} = g(x), \tag{2a}$$

where h and g are continuous functions, is given by

$$\int h(y)\,dy = \int g(x)\,dx \tag{2b}$$

in the sense that any solution $y = f(x)$ must satisfy Equation (2b). One arbitrary constant is contained in the solution.

(3) Given the differential equation

$$\frac{dy}{dx} + p(x)y = q(x),$$

we call the function $f(x) = e^{\int p(x)\,dx}$ an *integrating factor* of the equation. The general solution is given by

$$y(x) = e^{-\int p(x)\,dx}\left[\int q(x)e^{\int p(x)\,dx}\,dx\right] + Ce^{-\int p(x)\,dx},$$

where C is an arbitrary constant. (We incorporate all of the constants involved in the integrations into C.)

(4) The basic differential equations for various applications are given as follows:

(a) *Growth*:

$$\frac{dN}{dt} = kN,$$

where k is called the "growth" coefficient.

(b) *Decay*:

$$\frac{dN}{dt} = -aN,$$

where a is called the "decay" factor.

In both (a) and (b), the constants k and a are determined by the initial condition $N(0) = N_0$.

(c) *Bounded growth*:
   If the function $N(t)$ cannot exceed a certain number M, then the equation for bounded growth is given by

$$\frac{dN}{dt} = kN(t)[M - N(t)].$$

k is again a growth constant determined from condition $N(0) = N_0$.

(5) Remember that in applying differential equations to various fields of social and natural sciences, the basic mathematical equations are the same; only the names of the functions and variables change from one field to the other.

REVIEW EXERCISES

1. Find the general solution to the following differential equations.

    (a) $y''(x) = x^2 + x^3 + 3e^x$

    (b) $y' + xy = 1$

    (c) $\dfrac{dy}{dx} - \dfrac{y}{x} = 1 + \sqrt{x}$

    (d) $y' - \dfrac{2y}{x} = 3x^3$

    (e) $\log y \dfrac{dy}{dx} = \dfrac{y}{x}$

    (f) $(x-1)^3 \dfrac{dy}{dx} + 4(x-1)^2 y = x + 1$

    (g) $\dfrac{dy}{dx} = \dfrac{y^2 - y - 2}{x^2 + x}$

    (h) $2xy' + y = \dfrac{2ax^2 + 1}{x}$,

    where a is a constant.

2. Find the particular solution of the following.

    (a) $xy' - 2y = x^5$,   with $y(1) = 1$

    (b) $y' + xy = x^3$,   with $y(0) = 0$

    (c) $\dfrac{dp}{dt} + p = e^{2t}$,   with $p(0) = 1$

    (d) $xy' + (1-x)y = e^{2x}$,   with $y(1) = b$

    (e) $yy' = 5x$,   with $y(0) = 1$

    (f) $y' + 2xe^y = 0$,   with $y(0) = 0$.

3. Find all solutions of $x(x+1)y' + y = x(x+1)^2 e^{-x^2}$ on the interval $(-1, 0)$. Prove that all solutions approach 0 as $x \to -1$ but only one of them has a finite limit as $x \to 0$.

4. Solve the differential equation $(1 + y^2 e^{2x})y' + y = 0$ by introducing a change of variable of the form $y = u(x)e^{mx}$, where m is a constant and u is a new unknown function.

5. Scientists at an atomic works station isolated one gram of a new radioactive element called Deteriorum. It was found to decay at a rate proportional to the square of the amount present. After one year, one half gram remained. Set up and solve the differential equation for the amount of Deteriorum remaining at time t.

6. (a) Discuss the growth of a population in which the birth rate and death rate per year remain constant.

    (b) Discuss the growth if the birth rate per thousand, which is now higher than the death rate, decreases uniformly with time.

7. (a) What is the present value of $1000 when the interest is compounded instantaneously at 5% for 10 years?

   (b) At the end of what year will the principal have doubled if it is invested at 5% compounded instantaneously?

8. Equipment maintenance and operating costs C are related to the overhaul interval x by the equation

$$x^2 \frac{dC}{dx} - (n-1)xC = -na,$$

where a and n are constants and $C = C_0$ when $x = x_0$. Find C as a function of x.

9. Consider the following generalization of the bounded growth equation. Suppose that the population increases at a rate proportional to the product of two factors,

$$\frac{dN}{dt} = aN\left(1 - \frac{N}{b}\right),$$

where a, b are constants. This equation is often referred to as the "logistic" equation. Solve the differential equation and sketch a graph of its solution.

10. Find all differentiable functions f that satisfy the equation

$$f(x) = \int_0^x f(t)\, dt.$$

[HINT: Differentiate both sides to obtain a differential equation.]

9

Functions of
Several
Variables

9.1   Introduction

Although many situations can be described with the use of functions of one variable (such as those applications considered in Chapters 4, 5, and 8), there are also problems in which we must consider more than one variable. For instance, the cost of a product may depend on both the price of raw materials and the cost of labor; the profits of a resort hotel may depend on the state of the economy in general, prices charged by competing hotels, the weather, and numerous other factors; supply may depend on the size of its potential market, its retail price, and also the price of competing markets. To be more specific, we consider the equation $z = x + 2y - 1$, which relates the cost of a product to x, the cost of raw materials, and y, the cost of labor. A typical problem is to determine the minimum price the manufacturer can charge and still make a reasonable profit. In order to solve such problems, it is necessary for us first to discuss carefully the concept of a function of two variables (x, y in the above examples) and the rate of change of the function with respect to both of these variables. In order to graph or represent pictorially functions of two variables, we need a spatial, three-dimensional coordinate system

We now describe a method for attaching to every point in three-dimensional space an ordered triple (x, y, z). This has already been discussed briefly in an appendix to Chapter 5. First, we introduce the x, y, and z axes as in Figure 9.1.1. The three axes are mutually perpendicular and form a right-handed system. (This means that if we turn the x axis toward the y axis, the positive z axis points in the direction the z axis would go if it were a right-hand screw.) The point of intersection of the x, y, and z axes is called the origin 0. The plane containing the x axis and the y axis is called the

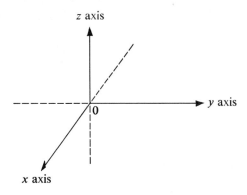

FIGURE 9.1.1

x–y plane. (Two distinct lines determine a plane.) The plane containing the x axis and the z axis is called the x–z plane. The plane containing the y axis and the z axis is called the y–z plane. (See Figure 9.1.2.)

To every ordered triple (a, b, c) we will associate a point P in three-space. Consider the triple $(3, 1, 2)$. To find the associated point we first count off three units on the positive x axis; then move one unit in the direction of the positive y axis (one unit in the direction parallel to the y axis and perpendicular to the x–z plane); and finally we count up two units in the direction of the positive z axis (two units up in the direction perpendicular to the x–y plane).

See Figure 9.1.3 for a pictorial version of this counting system. Also examine Figure 9.1.1 again.

Thus, if the triple (a, b, c) represents the point P, then P is a units from the y–z plane, b units from the x–z plane, and c units from the x–y plane.

Not only does every triple (a, b, c) represent a point, but every point can be represented as a triple. Given P, to find its coordinates (a, b, c), one simply measures the perpendicular distances to the y–z, x–z, and x–y planes, respectively. This correspondence

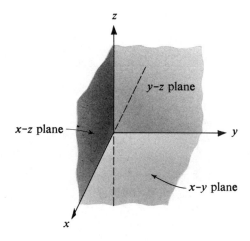

FIGURE 9.1.2

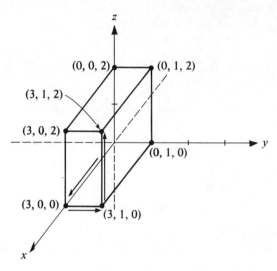

FIGURE 9.1.3

between points and ordered triples is unique. Thus, there is exactly one ordered triple for each point and exactly one point for each ordered triple. From now on we identify points and ordered triples by writing $P = (a, b, c)$.

The planes in three-space, which are perpendicular to one of the coordinate axes, are of particular interest. By the plane $x = 1$, one means the set of points $\{(x, y, z) \mid x = 1\}$, that is, all points of the form $(1, y, z)$. This set of points is a plane, perpendicular to the x axis and passing through $(1, 0, 0)$. (See Figure 9.1.4.)

Let us now consider some examples illustrating the remarks made in this section.

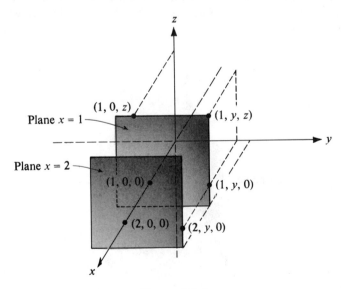

FIGURE 9.1.4

Example 1. Draw a three-dimensional coordinate system and plot the points (3, 2, 4) and (−4, −3, 5).

Solution.   To plot the point (3, 2, 4), first measure three units along the positive *x* axis, two units parallel to the positive *y* axis, and then measure four units up in the direction of the positive *z* axis. (See Figure 9.1.5.) The point (−4, −3, 5) is plotted in Figure 9.1.6. In this case, first measure four units along the negative *x* axis, then three units in the direction of the negative *y* axis, and then five units up in the direction of the positive *z* axis.

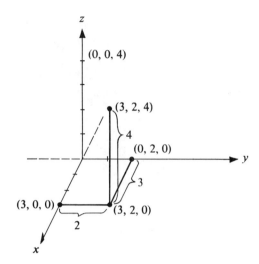

FIGURE 9.1.5

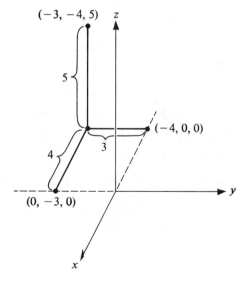

FIGURE 9.1.6

Example 2.   Sketch the plane $y = 2$.

Solution.   We are really asked to plot the set of all triples of the form $(x, 2, z)$. First draw the line $y = 2$ in the *x–y* plane and then construct a plane containing the line $y = 2$ parallel to the *x–z* plane. (See Figure 9.1.7.)

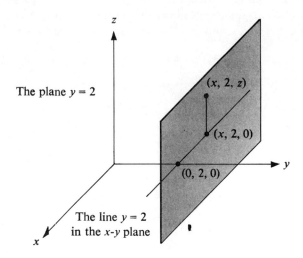

FIGURE 9.1.7

Exercises

1.   Graph the following points.

(a) $(1, -1, 3)$ (e) $(5, 4, 6)$

(b) $(0, 1, -4)$ (f) $(-2, 1, 2)$

(c) $(-1, -1, 1)$ (g) $(4, -1, 1)$

(d) $(3, -1, 0)$ (h) $(1, 1, 1)$.

2.   Sketch the following.

(a) the plane $x = -1$ (e) the plane $y = -2$

(b) the plane $x = 0$ (f) the plane $z = 2$

(c) the plane $y = 1$ (g) the plane $z = 0$

(d) the plane $y = 0$ (h) the plane $z = -3$

3.   Using a three-dimensional system of coordinate axes, plot the points $(3, 0, 0)$, $(0, 4, 0)$, and $(0, 0, 5)$ and connect them with straight lines to indicate the plane on which they all lie. According to this diagram, does the plane also contain the point $(0, 0, 1)$? Does it seem to contain the point $(1\frac{1}{2}, 0, 2\frac{1}{2})$?

4.* Use the Pythagorean theorem to show that the distance (geometric) between the points $P = (x_1, y_1, z_1)$ and $Q = (x_2, y_2, z_2)$ is given by

$$\text{distance } [P, Q] = [(x_1 - x_2)^2 + (y_1 - y_2)^2 + (z_1 - z_2)^2]^{1/2}.$$

5. Use Exercise 4 to show that the point

$$\left(\frac{x_1 + x_2}{2}, \frac{y_1 + y_2}{2}, \frac{z_1 + z_2}{2}\right)$$

is the midpoint of the line segment joining the points (x_1, y_1, z_1) and (x_2, y_2, z_2).

9.2   Definition of a Function of Two Variables and Some Graphing

By a function of two variables we mean (1) a region D in the x–y plane called the domain and (2) a rule that associates with every point in D exactly one real number. Customarily, such functions are written as $z = f(x, y)$. For example, if we write $z = 1 + x + y^2$, then the domain is understood to be the x–y plane. We associate the real number $1 + (2) + (1)^2 = 4$ with the point $(2, 1)$ in the x–y plane. With the point $(-5, 9)$ in the x–y plane we associate the real number $1 + (-5) + (9)^2 = 77$.

Geometrically we can think of the graph of a function $z = f(x, y)$ as a surface. If the domain of the function is D, then to obtain the graph of f think of D lifted up as in Figure 9.2.1 and stretched into an irregular shape. In Figure 9.2.1 we have represented

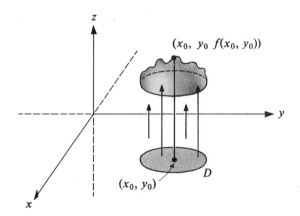

FIGURE 9.2.1

$z = f(x, y)$ by its graph. The graph of $f(x, y)$ is simply $\{(x, y, z) \mid z = f(x, y)\}$. One can think of the function f as sending the point (x_0, y_0) in the x–y plane into the point $(x_0, y_0, f(x_0, y_0))$ in three-space. If $z = f(x, y)$ is a function, then f can lift, push down, stretch, wrinkle, or tear D. It can *not* fold D. (If it did, then f would associate more than one real number with some point in the domain.)

It is natural to ask how one goes about graphing a function $z = f(x, y)$. There is no easy answer to this question. We first give several examples and then briefly discuss one technique. If no domain is specified, then it is understood to be the x–y plane.

Example 1. $z = \sqrt{1 - (x^2 + y^2)}$. The graph of this function is the top half of a sphere. Domain $= \{(x, y) \mid x^2 + y^2 \le 1\}$. (See Figure 9.2.2.)

Example 2. $z = 1 - x - y$. The graph of this function z is a plane. (See Figure 9.2.3.)

We now discuss a technique of graphing, known as the profile method, in some detail.

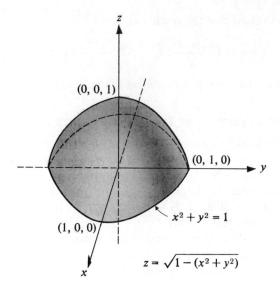

$x^2 + y^2 = 1$

$z = \sqrt{1 - (x^2 + y^2)}$

FIGURE 9.2.2

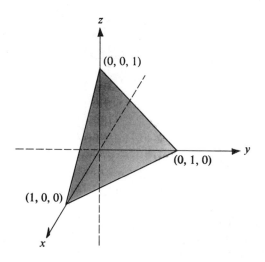

FIGURE 9.2.3

Example 3. Consider the function $z = x^2 + y^2$, where the domain is the x–y plane. Since x^2 and y^2 are both positive, clearly $z = x^2 + y^2 \geq 0$, so the graph of the surface will lie above the x–y plane. We now fix one of the variables and see what we get under these circumstances. For example, let $z = 5$. Then $5 = x^2 + y^2$ is the equation of a circle. Setting $z = 5$ is the same as intersecting the surface with the plane $z = 5$. (See Figure 9.2.4.) We now know that the circle in Figure 9.2.4 is part of the surface $z = x^2 + y^2$. This may not seem like much information, but you can consider it to be one clue in tracking down the actual shape of the surface. Note that for any $z_0 > 0$, the intersection of the plane $z = z_0$ with the surface $z = x^2 + y^2$ is a circle; in fact, it is the circle $z_0 = x^2 + y^2$ ($z = z_0$). Now we know that the surface is formed by a bunch of circles piled on top of one another. However, there are many different ways they could be arranged. Several such arrangements appear in Figure 9.2.5. To determine the correct arrangement, we examine the intersection of the plane $x = 0$ (the y–z plane) with the surface. To do this we simply set $x = 0$ in the equation $z = x^2 + y^2$ and obtain the parabola $z = y^2$ in the y–z plane. The profile of the surface in the plane $x = 0$ is given in Figure 9.2.6. If we combine our two sets of profiles, we see that the surface has the shape given in Figure 9.2.7. The surface $z = x^2 + y^2$ thus bears some resemblance to the end of a football. This surface is fairly simple, and we were able to determine it with only a few profiles. For a complicated surface, many profiles may be needed.

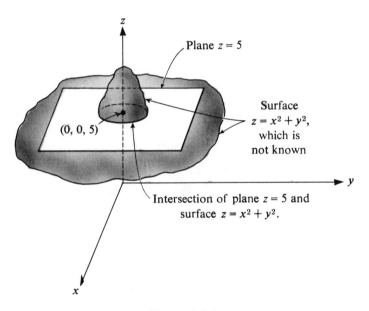

FIGURE 9.2.4

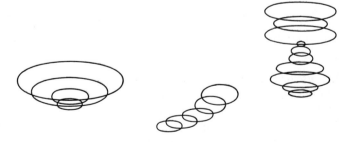

FIGURE 9.2.5

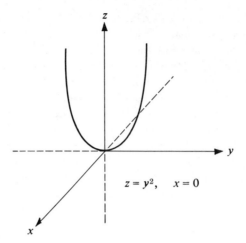

FIGURE 9.2.6

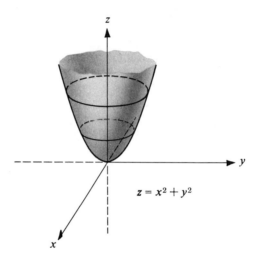

FIGURE 9.2.7

Example 4. Consider the surface $z = y^2 - x^2$ where the domain is the x–y plane. Sketch a graph of this surface.

Solution. Let us first consider the profile obtained by intersecting the surface $z = y^2 - x^2$ with the planes $y = y_0$. To do this, we simply substitute $y = y_0$ to obtain $z = -x^2 + y_0^2$. The curve $z = -x^2 + y_0^2$ is a parabola in the $y = y_0$ plane with its vertex pointing up. The vertices of the parabola lie in the plane $x = 0$. For $x = 0$ the profile we obtain is $z = y^2$, another parabola. Putting this information into a picture, we obtain Figure 9.2.8. Figure 9.2.8 already gives us a pretty good idea of what the graph will look

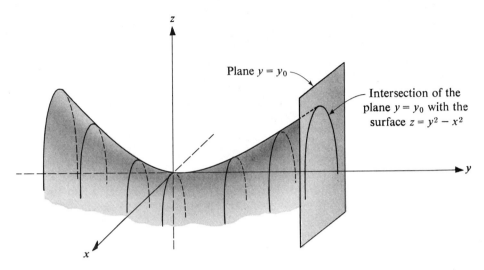

FIGURE 9.2.8

like. Let us now consider the profile obtained by intersecting the surface $z = y^2 - x^2$ with the planes $x = x_0$. We thus obtain $z = y^2 - x_0^2$. This is just the equation of a parabola in the $x = x_0$ plane with vertex pointing down. (See Figure 9.2.9.) The vertices of these parabolas lie along the curve $z = -x^2$ in the plane $x = x_0$. We now can draw a reasonable picture of this surface. (See Figure 9.2.10.) It is saddle-shaped and for this compelling reason is sometimes called a saddle surface. A person standing at the origin on this surface would see himself as crossing a mountain pass between two higher peaks. This concludes our discussion of graphing.

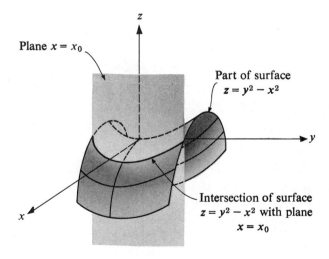

FIGURE 9.2.9

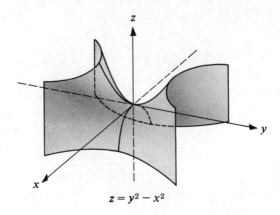

$$z = y^2 - x^2$$

FIGURE 9.2.10

Exercises

1. Can a sphere be represented as the graph of a function $z = f(x, y)$? Why?
2. Can a torus (donut or inner tube) be represented as the graph of a function $z = f(x, y)$? Why?
3. Graph the function $z = -x + 2y + 5$, given that it is a plane.
4. If a company spends x dollars on research and development and y dollars on advertising, its profit (in dollars) is given by

$$P(x, y) = 40{,}000 + 50x + 30y + \frac{xy}{100}$$

    for all positive integers x and y less than 25,000.

    (a)   What is the company's profit if it spends \$2000 on research and development and \$5000 on advertising?

    (b)   What will be the company's profit if it spends \$8000 on research and development and \$6000 on advertising?

    (c)   If the company is planning to spend \$4000 on research and development and hopes to make a profit of \$590,000, how much should it spend on advertising?

5. If $g(x, y) = x^2 + y^2 + 2$, for all real values of x and y, find:

    (a) $g(1, 0)$ (d) $g(1, 1)$

    (b) $g(-1, -1)$ (e) $g(-3, -4)$

    (c) $g(3, 4)$ (f) $g(-2, 0)$

6. Graph the following functions by the profile method.

    (a) $z = \sqrt{x^2 + y^2}$,   domain x–y plane

    (b) $z = -\sqrt{1 - x^2 - y^2}$,   domain $\{(x, y) \mid x^2 + y^2 \leq 1\}$

    (c) $z = (x^2 + y^2)^{1/4}$,   domain x–y plane (Bugle surface)

    (d) $z = \sqrt{1 - y^2}$,   domain $-\infty < x < \infty$, $|y| \leq 1$

(e) $z = x + y^2$,   domain x–y plane

(f) $z = |x| + |y|$,   domain x–y plane (inverted pyramid)

(g) $z = 2x - y + 1$ [HINT: This surface is a plane. Since 3 points determine a plane, you should have no trouble.]

(h) $z = 1 + 2x^2 + y^2$ [HINT: $a^2x^2 + b^2y^2 = c^2$ is the equation of an ellipse.]

(i) $z = -x - 4y + 2$,   domain x–y plane

(j) $z = x^2 - y^2$,   domain x–y plane

(k) $z = -2 + x^2 + y^2$,   domain x–y plane.

9.3 Limits and Continuity. A Brief Discussion

We now discuss the notion of limit for a function of two variables. When we write $\lim_{(x,y) \to (a,b)} f(x, y) = L$, in a very crude sense we mean that for points (x, y) very close to but not equal to (a, b), the value $f(x, y)$ of the function is very close to L [or as (x, y) approaches (a, b), $f(x, y)$ approaches L.]

An equivalent interpretation of the expression $\lim_{(x,y) \to (a,b)} f(x, y) = L$, which sounds more precise but is still not completely satisfactory, goes as follows: For $d > 0$, let C_d be the disc $\{(x, y) \mid (x - a)^2 + (y - b)^2 < d^2\}$. (See Figure 9.3.1.) By moving C_d up and down, we sweep out a cylindrical region as in the figure. Think of the walls of the cylinder as if they were the walls of a tin can of infinite height. We will now "can" the part of the graph of f over C_d. For convenience we assume that (a, b) is not an element of domain f. Choose a bottom B and top T for the can so that the graph of f for points in C_d lies above B and below T, that is, the graph of f for points in C_d is contained in the tin can we have just constructed. (See Figure 9.3.1.) If this process can be undertaken for every disc C_d in such a way that as d tends to zero then the height of the corresponding can tends to zero, we say $\lim_{(x,y) \to (a,b)} f(x, y)$ exists. The value L of the limit will be the z value where the top and bottom of the cans meet.

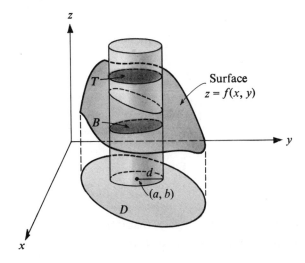

FIGURE 9.3.1

We assume that the reader now has a working definition of limits for a function of two variables. To obtain a precise definition of $\lim_{(x, y)\to(a, b)} f(x, y) = L$, one must reverse the procedure above. Thus, for any $D > 0$, the top of the can is specified to be at a height $L + D$ and the bottom of the can at a height $L - D$. Then the disc C_d with center at (a, b) must be chosen so that the graph of the function over C_d is " inside " the tin can. This definition is given in Exercise 3*. It is now possible to talk about continuity for a function of two variables. Thus, $f(x, y)$ is continuous at (a, b) if: (i) $(a, b) \in$ domain of f and (ii) $\lim_{(x, y)\to(a, b)} f(x, y) = f(a, b)$. The function f is continuous in the domain D if it is continuous at every point of D. Geometrically speaking, f is continuous on D if the graph of f has no tears. This is analogous to the situation where a function of one variable is continuous if there are no jumps in its graph. All the functions graphed in Section 9.2 are continuous. Let us give an example of a function that is not continuous.

Example:

$$z = \begin{cases} 1, & 0 \le x \le 1; 0 \le y < \tfrac{1}{2} \\ 2, & 0 \le x \le 1; \tfrac{1}{2} \le y \le 1. \end{cases}$$

Domain $= \{(x, y) \mid 0 \le x \le 1; 0 \le y \le 1\}$.

The graph of z is shown in Figure 9.3.2. This function is not continuous at any point of the form $(x, \tfrac{1}{2})$ for $0 \le x \le 1$. (See Figure 9.3.2.)

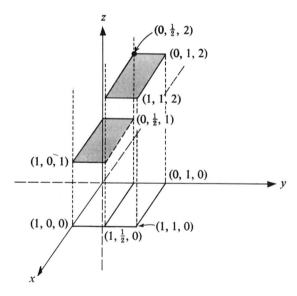

FIGURE 9.3.2

Exercises

1.   Draw a picture of a function $z = f(x, y)$ that is discontinuous at $(1, 2)$.

2.   Is the following function continuous at $(0, 0)$? (See Figure 9.3.3.)

$$z = \begin{cases} \sqrt{1 - (x^2 + y^2)}, & 0 < x^2 + y^2 \le 1 \\ 2, & x = y = 0 \end{cases}$$

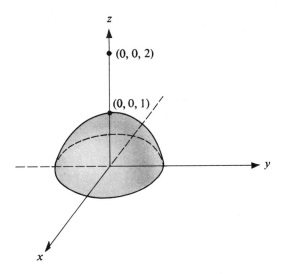

FIGURE 9.3.3

3.* Let $f(x, y)$ be a function with domain R. Then $\lim_{(x, y) \to (a, b)} f(x, y) = L$ if for every $D > 0$ there exists $d > 0$ such that

$$|f(x, y) - L| < D \qquad \text{for} \qquad 0 < (x - a)^2 + (y - b)^2 < d^2.$$

Use this definition to show that the function in Exercise 2 has a limit at the point $(0, 0)$ but is not continuous there.

4.* Let

$$f(x, y) = \begin{cases} 0, & x = y = 0 \\ \dfrac{xy}{x^2 + y^2}, & \text{for } x^2 + y^2 > 0. \end{cases}$$

Is $f(x, y)$ continuous at $(0, 0)$?

[HINT: Consider $\lim f(x, y)$ when $y = 0$ and $x \to 0$. Then consider $\lim f(x, y)$ when $x = y$ and $x \to 0$.]

5.* Graph the function $f(x) = \sqrt{x^2 + y^2 - 1}$ for $x^2 + y^2 \geq 1$.

[HINT: Write $z = \sqrt{x^2 + y^2 - 1}$ as $z^2 = x^2 + y^2 - 1$. Now look at what f does to the points $x^2 + y^2 = R^2$, R fixed. Finally consider what happens when $x = 0$. This surface is sometimes called the bugle surface.]

9.4 Partial Derivatives

As we have already seen, it is very informative when studying curves of the form $y = f(x)$ to examine the slope of the curve at every point. This was done by looking at the derivative. Let us try to generalize this notion to surfaces.

A mountain climber pauses at point P as in Figure 9.4.1. There are three paths to the summit of varying steepness. As before, we can measure the steepness. The important thing is to notice that the slope of the mountain at point P depends on the direction taken

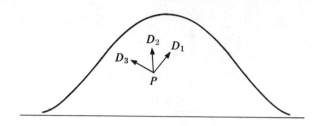

FIGURE 9.4.1

(that is, whether we go in the direction D_1, D_2, or D_3 in Figure 9.4.1). Let us give another example that illustrates this behavior more clearly.

In Figure 9.4.2, the path in direction D_2 is quite steep whereas the path in direction D_1 is level, that is, has zero slope. (Anyone who skis is acutely aware of this phenomenon.) By means of a mathematical version of this simple notion (called the partial derivative), we shall investigate surfaces.

Let $z = f(x, y)$ be a function with domain D, and let (x_0, y_0) be a point in D. Consider the intersection of the surface $z = f(x, y)$ with the plane $y = y_0$. (See Figure 9.4.3.) The

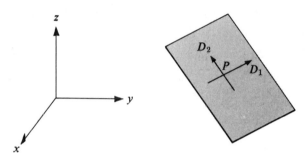

FIGURE 9.4.2

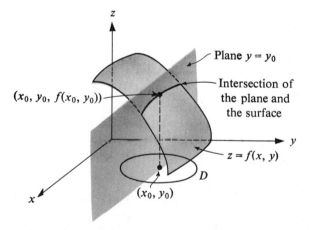

FIGURE 9.4.3

intersection of the plane and the surface is a curve. Indeed, it is a curve in the plane $y = y_0$, and its equation is $z = f(x, y_0)$. Let us pull the plane $y = y_0$ out of Figure 9.4.4.

Now we can easily find the slope of the curve $z = f(x, y_0) = g(x)$ at the point x_0. It is just $g'(x_0)$. By definition,

$$g'(x_0) = \lim_{x \to x_0} \frac{g(x) - g(x_0)}{x - x_0} = \lim_{x \to x_0} \frac{f(x, y_0) - f(x_0, y_0)}{x - x_0}.$$

We have arrived at the definition of the partial derivative!

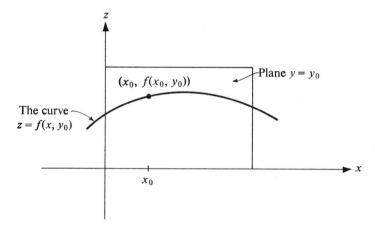

FIGURE 9.4.4

DEFINITION.   *Let* $z = f(x, y)$. *Then*

$$\left.\frac{\partial f}{\partial x}\right|_{(x_0, y_0)} = \left.\frac{\partial z}{\partial x}\right|_{(x_0, y_0)} = \frac{\partial f}{\partial x}(x_0, y_0) = \lim_{x \to x_0} \frac{f(x, y_0) - f(x_0, y_0)}{x - x_0}$$

(read the partial of f *with respect to* x *at the point* (x_0, y_0).) *Similarly*,

$$\left.\frac{\partial f}{\partial y}\right|_{(x_0, y_0)} = \left.\frac{\partial z}{\partial y}\right|_{(x_0, y_0)} = \frac{\partial f}{\partial y}(x_0, y_0) = \lim_{y \to y_0} \frac{f(x_0, y) - f(x_0, y_0)}{y - y_0}.$$

To summarize, $\left.\partial f/\partial x\right|_{(x_0, y_0)}$ is the slope of the surface $z = f(x, y)$ in the x direction at (x_0, y_0) (or the slope of the intersection of the surface and the plane $y = y_0$), and $\left.\partial f/\partial y\right|_{(x_0, y_0)}$ is the slope of the surface in the y direction at (x_0, y_0). (See Figure 9.4.5.)

At this point, you might think that it is going to be difficult to calculate $\partial f/\partial x$ or $\partial f/\partial y$ given $z = f(x, y)$. The next pleasant surprise is that it is very easy.

RULE.   *Let* $z = f(x, y)$.
  To find $\left.\partial f/\partial x\right|_{(x_0, y_0)}$
Step 1.   Differentiate f *with respect to* x *treating all the* y's *as though they were constants.*
Step 2.   Substitute x_0 *for* x, *and* y_0 *for* y.

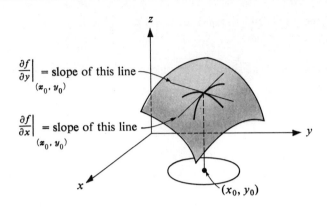

FIGURE 9.4.5

To find $\partial f/\partial y|_{(x_0,\ y_0)}$:
Step 1.   Differentiate f with respect to y treating all the x's as though they were constants.
Step 2.   Substitute x_0 for x and y_0 for y.

Example 1.   If $z = f(x, y) = x^2y + e^y$, find $\partial f/\partial x|_{(1,\ 0)}$ and $\partial f/\partial y|_{(2,\ 0)}$.

Solution.   Following the rule by steps, we obtain

Step 1.
$$\frac{\partial f}{\partial x} = \frac{\partial}{\partial x}(x^2y) + \frac{\partial}{\partial x}(e^y) \quad \text{so} \quad \frac{\partial f}{\partial x} = 2xy + 0.$$

Step 2.
$$\left.\frac{\partial f}{\partial x}\right|_{(1,\ 0)} = 2(1)(0) = 0.$$

Similarly for the second half of the problem.

Step 1.
$$\frac{\partial f}{\partial y} = x^2 + e^y.$$

Step 2.
$$\left.\frac{\partial f}{\partial y}\right|_{(2,\ 0)} = (2)^2 + e^0 = 5.$$

[handwritten: y is a constant = to 1, derivative of a constant is 0]

Example 2.   If $z = f(x, y) = x^3y^2 + \log(xy)$, find $\partial f/\partial x|_{(1,\ 2)}$ and $df/dy|_{(1,\ 2)}$.

Solution.   Applying our rule for partial differentiation we obtain

Step 1.
$$\frac{\partial f}{\partial x} = \frac{\partial}{\partial x}(x^3y^2) + \frac{\partial}{\partial x}\log(xy) = 3x^2y^2 + \frac{1}{xy}\frac{\partial}{\partial x}(xy)$$

$$= 3x^2y^2 + \frac{1}{xy}\,y = 3x^2y^2 + \frac{1}{x}.$$

Step 2.
$$\left.\frac{\partial f}{\partial x}\right|_{(1,\ 2)} = 3(1)^2(2)^2 + \frac{1}{1} = 13.$$

Repeating this process for the second half of the problem, we find

Step 1.
$$\frac{\partial f}{\partial y} = \frac{\partial}{\partial y}(x^3 y^2) + \frac{\partial}{\partial y} \log (xy),$$

so

$$\frac{\partial f}{\partial y} = 2x^3 y + \frac{1}{xy}\frac{\partial}{\partial y}(xy) = 2x^3 y + \frac{1}{xy} \cdot x = 2x^3 y + \frac{1}{y}.$$

Step 2.
$$\frac{\partial f}{\partial y}\bigg|_{(1, 2)} = 2(1)^3(2) + \frac{1}{2} = 4\frac{1}{2}.$$

Of course, we can find $\partial f/\partial x$ or $\partial f/\partial y$ at a general point (x, y).

Example 3. If $z = f(x, y) = \sqrt{x^2 + y^2} + xe^y$, find $\partial f/\partial x$ and $\partial f/\partial y$.

Solution. We simply evaluate $\partial f/\partial x$ by our rule above

$$\frac{\partial f}{\partial x} = \frac{\partial}{\partial x}\sqrt{x^2 + y^2} + \frac{\partial}{\partial x}(xe^y) = \frac{1}{2}(x^2 + y^2)^{-1/2}\frac{\partial}{\partial x}(x^2 + y^2) + e^y$$

$$= \frac{1}{2}\frac{1}{\sqrt{x^2 + y^2}} \cdot 2x + e^y = \frac{x}{\sqrt{x^2 + y^2}} + e^y.$$

Similarly, we see that

$$\frac{\partial f}{\partial y} = \frac{\partial}{\partial y}\left(\sqrt{x^2 + y^2}\right) + \frac{\partial}{\partial y}(xe^y) = \frac{1}{2}(x^2 + y^2)^{-1/2}\frac{\partial}{\partial y}(x^2 + y^2) + xe^y$$

$$= \frac{y}{\sqrt{x^2 + y^2}} + xe^y.$$

Having defined partial derivatives, we can extend our definitions to higher order derivatives. Thus, if $z = f(x, y)$,

$$\frac{\partial^2 f}{\partial x^2} = \frac{\partial^2 z}{\partial x^2} = \frac{\partial}{\partial x}\left(\frac{\partial f}{\partial x}\right) \tag{9.4.1}$$

$$\frac{\partial^2 f}{\partial x\,\partial y} = \frac{\partial}{\partial x}\left(\frac{\partial f}{\partial y}\right) \tag{9.4.2}$$

$$\frac{\partial^2 f}{\partial y\,\partial x} = \frac{\partial}{\partial y}\left(\frac{\partial f}{\partial x}\right) \tag{9.4.3}$$

$$\frac{\partial^3 f}{\partial y^3} = \frac{\partial}{\partial y}\left[\frac{\partial}{\partial y}\left(\frac{\partial f}{\partial y}\right)\right] \tag{9.4.4}$$

We remark that for all functions we are likely to encounter, Equations (9.4.2) and (9.4.3) are equal. Let us consider the following example.

Example 4. If $f(x, y) = x^2 y + ye^x$, find $\partial^2 f/\partial x^2$, $\partial^2 f/\partial y^2$, $\partial^2 f/\partial x\, \partial y$, and $\partial^2 f/\partial y\, \partial x$.

Solution. First we compute $\partial f/\partial x$ and $\partial f/\partial y$. We have $\partial f/\partial x = 2xy + ye^x$ and $\partial f/\partial y = x^2 + e^x$. Now,

$$\frac{\partial^2 f}{\partial x^2} = \frac{\partial}{\partial x}\left[\frac{\partial f}{\partial x}\right] = 2y + ye^x \qquad \text{and} \qquad \frac{\partial^2 f}{\partial y^2} = \frac{\partial}{\partial y}\left[\frac{\partial f}{\partial x}\right] = 0,$$

(since no y term appears in $\partial f/\partial y$). Thus,

$$\frac{\partial^2 f}{\partial y\, \partial x} = \frac{\partial}{\partial y}\left[\frac{\partial f}{\partial x}\right] = 2x + e^x$$

and

$$\frac{\partial^2 f}{\partial x\, \partial y} = \frac{\partial}{\partial x}\left[\frac{\partial f}{\partial y}\right] = 2x + e^x.$$

Observe that $\partial^2 f/\partial x\, \partial y = \partial^2 f/\partial y\, \partial x$. This is not just a coincidence. It turns out that if we consider a function $f(x, y)$ that is continuous and if $\partial^2 f/\partial x\, \partial y$ and $\partial^2 f/\partial y\, \partial x$ are also continuous, then $\partial^2 f/\partial x\, \partial y = \partial^2 f/\partial y\, \partial x$.

We conclude this section by giving several examples that show how one interprets partial derivatives in economic applictions.

Example 5. Consider the equation

$$z = 100 + 6x + 10y,$$

which relates z, the cost of a certain product in dollars, to x, the cost of raw materials in dollars per pound, and y, the cost of labor in dollars per hour. Note that

$$\frac{\partial z}{\partial x} = 6 \qquad \text{and} \qquad \frac{\partial z}{\partial y} = 10$$

for all x and y. This means that when the cost of labor is held fixed, an increase of $1.00 per pound in the cost of raw materials causes an increase of $6.00 in the cost of the product, and that when the cost of raw materials is held fixed, an increase of $1.00 in the hourly cost of labor brings about an increase of $10.00 in the cost of the product.

Example 6. (*Marginal Demands*). Let the demands for two different commodities be d_1 and d_2 and let the respective prices be x and y. If the demand functions for two related commodities are given in the form

$$d_1 = f(x, y) \qquad \text{and} \qquad d_2 = g(x, y) \tag{9.4.5}$$

then

$\dfrac{\partial f}{dx}$ is the (partial) marginal demand of d_1 with respect to x,

$\dfrac{\partial f}{\partial y}$ is the (partial) marginal demand of d_1 with respect to y,

Although this theorem has many hypotheses, it is easy to apply in practice, as the following example shows.

Example 1.   Find the absolute maximum of the function

$$f(x, y) = 2x + 4y - 2x^2 - 3y^2$$

when the domain of f is the disc $x^2 + y^2 \leq 100$. Note that f is certainly continuous for $(x, y) \in D$, where $D = \{(x, y) \mid x^2 + y^2 \leq 100\}$. The boundary of the domain is the circle $x^2 + y^2 = 100$.

Solution.   First we find the common zeros of the partial derivatives

$$\frac{\partial f}{\partial x} = 2 - 4x \qquad \text{and} \qquad \frac{\partial f}{\partial y} = 4 - 6y.$$

Setting these equal to zero, we obtain

$$2 - 4x = 0 \qquad \text{or} \qquad x = \tfrac{1}{2}$$

and

$$4 - 6y = 0 \qquad \text{or} \qquad y = \tfrac{2}{3}.$$

Thus the only point where $\partial f / \partial x = \partial f / \partial y = 0$ is $(\tfrac{1}{2}, \tfrac{2}{3})$. Moreover, $f(\tfrac{1}{2}, \tfrac{2}{3}) = 2(\tfrac{1}{2}) + 4(\tfrac{2}{3})$ $- 2(\tfrac{1}{2})^2 - 3(\tfrac{2}{3})^2 = \tfrac{11}{6}$. (See Figure 9.6.1.) We now have to check f on the boundary. Now

$$f(x, y) = 2x + 4y - (2x^2 + 3y^2), \qquad 2x^2 + 3y^2 \geq x^2 + y^2 = 100$$

on the boundary; thus $-(2x^2 + 3y^2) \leq -100$ on the boundary. (Why?) For (x, y) on the boundary of D, $|x| \leq 10$ and $|y| \leq 10$. (Why?) Hence,

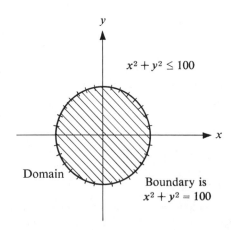

FIGURE 9.6.1

Exercises

1. Find the regression line for the following points.

x	1	3	5	7	9
y	1	4	4	6	10

Sketch your results.

2. Find the regression line for the following points.

x	-3	-2	-1	0	1	2
y	4	2	1	-1	-2	-3

3. Find the regression line for the following points.

x	2	-1	2	-2	3
y	-1	3	0	5	-1

4. A manufacturer keeps a record of sales per month and advertising costs. For the first five months of 1975, he finds they are running as follows:

Advertising expenditures	1	2	3	4	5
Sales	2	4	7	9	11

Find the regression line for this data. On the basis of the regression line, predict his sales if advertising expenditures are set at 6 and 10 units per month.

5.* Assume we are given the following four points: $(1, 3)$, $(1, 1)$, $(5, 3)$, and $(5, 1)$. Let $y = mx + b$, and set $y_1 = m \cdot 1 + b$ and $y_2 = m \cdot 5 + b$. Draw a picture of the four points. Convince yourself that

$$|y_1 - 1| + |y_1 - 3| + |y_2 - 1| + |y_2 - 3|$$

is constant, as long as the line $y = mx + b$ lies below $(1, 3)$ and $(5, 3)$ and above $(1, 1)$ and $(5, 1)$. Convince yourself that the minimum value of

$$|y_1 - 1| + |y_1 - 3| + |y_2 - 1| + |y_2 - 3|$$

is 4 over all possible choices of m and b. (This is most easily done by considering a picture.) What does all this imply about fitting a line to the four points given above when the test of a good fit is the smallness of

$$[|y_1 - \hat{y}_1| + |y_2 - \hat{y}_2| + |y_3 - \hat{y}_3| + |y_4 - \hat{y}_4|]?$$

9.9   Summary of Chapter 9

In this chapter we introduced functions of the form $z = f(x, y)$, that is, functions of two variables, and developed techniques for finding the absolute maxima and minima of such functions. As a first step we defined the partial derivatives

$$\left.\frac{\partial f}{\partial x}\right|_{(x_0,\, y_0)} = \lim_{x \to x_0} \frac{f(x, y_0) - f(x_0, y_0)}{x - x_0}$$

and

$$\left.\frac{\partial f}{\partial y}\right|_{(x_0,\, y_0)} = \lim_{y \to y_0} \frac{f(x_0, y) - f(x_0, y_0)}{y - y_0}.$$

point of D. If f has a *relative* maximum or *relative* minimum at (x_0, y_0), $\partial f/\partial x|_{(x_0, y_0)}$ and $\partial f/\partial y|_{(x_0, y_0)}$ exist, then $\partial f/\partial x|_{(x_0, y_0)} = \partial f/\partial y|_{(x_0, y_0)} = 0$. This means that as long as the partial derivatives exist, the relative maximum and minimum points are found among those points such that $\partial f/\partial x = \partial f/\partial y = 0$.

If $z = f(x, y)$ is defined on a bounded domain D with boundary and either f is continuous on D or $|\partial f/\partial x| \leq M$ and $|\partial f/\partial y| \leq M$ at all points of D for some constant M, then f has both an absolute maximum and an absolute minimum on D. If $\partial f/\partial x$ and $\partial f/\partial y$ exist at every point of D, then the absolute maximum [minimum] occurs either on the boundary of D or at a point where both $\partial f/\partial x$ and $\partial f/\partial y$ equal zero. It should be clear that every absolute maximum or minimum point is also a relative maximum or minimum point.

If $z = f(x, y)$ is defined on a domain without its boundary, f may not have an absolute maximum or minimum. If $\partial f/\partial x$ and $\partial f/\partial y$ exist at all points of D, then an absolute maximum [minimum] must occur at a point where both $\partial f/\partial x$ and $\partial f/\partial y$ equal zero.

The method of Lagrange multipliers is particularly helpful if we wish to maximize or minimize a quantity w when $w = F(x, y, z)$ and the variables x, y, and z must satisfy a side condition $G(x, y, z) = 0$. Thus we wish to find the point (x_0, y_0, z_0) in the domain of F, where $w = F(x, y, z)$ is the largest [smallest] and $G(x_0, y_0, z_0) = 0$ as well. To do this, we set up the auxiliary equation

$$F(x, y, z) - \lambda G(x, y, z) = 0$$

and find $\partial/\partial x$, $\partial/\partial y$, $\partial/\partial z$, and $\partial/\partial \lambda$. This yields four equations,

$$\text{(1)} \quad \frac{\partial F}{\partial x} - \lambda \frac{\partial G}{\partial x} = 0$$

$$\text{(2)} \quad \frac{\partial F}{\partial y} - \lambda \frac{\partial G}{\partial y} = 0$$

$$\text{(3)} \quad \frac{\partial F}{\partial z} - \lambda \frac{\partial G}{\partial y} = 0$$

$$\text{(4)} \quad G(x, y, z) = 0.$$

in the four unknowns x, y, z, and λ. We now solve Equations (1), (2), (3), and (4) simultaneously. The point (x_0, y_0, z_0) we are looking for will usually be found among the set of solutions to Equations (1), (2), (3), and (4). The decision as to which solution is the right one must be made on the basis of information particular to each problem. Usually this is not difficult.

REVIEW EXERCISES

1. Find $\partial f/\partial x$ and $\partial f/\partial y$ for the following functions. The domain of the function is the x-y plane unless otherwise stated.

(a) $z = 2x^3y + e^{xy}$

(f) $z = e^{[1/(x+y)-(x/y)]}$ for $y \neq 0$

(b) $z = \log \dfrac{1}{\sqrt{1+xy}}$

(g) $z = xy^2 + y^2x$

(c) $z = \log(x^2y + e^y)$

(h) $z = [1 - (x^2 + y^2)]^{1/2}$, $x^2 + y^2 \leq 1$

(d) $z = \dfrac{xy}{x+y}$ for $x \neq -y$ (i) $z = e^x + \log y$

(e) $z = \dfrac{(1+x^2)^2}{\sqrt{x^2+y^2}}$, $|x| + |y| > 0$

2.   Let $z = \sqrt{1 - (x^2 + y^2)}$ when the domain is the disc $x^2 + y^2 \leq 1$. Find the absolute maximum and minimum of the function.

3.   Let $z = x(1 - x) + y^2(1 - y)$. Find the absolute maximum if the domain is the square $\{(x, y) \mid 0 \leq |x| \leq 1 \text{ and } 0 \leq |y| \leq 1\}$.

4.   Let $z = 1 + x + 2y$. Find the absolute maximum and absolute minimum if the domain is the triangle $\{(x, y) \mid 0 \leq x, 0 \leq y, \text{ and } 2x + y \leq 2\}$.

5.   Let $z = x^2 + y^2$, when the domain is the square $\{(x, y) \mid |x| \leq 1 \text{ and } |y| \leq 1\}$. Find the absolute maximum and minimum.

6.   Find the distance from the point $(1, 1, 1)$ to the plane $x - 2y + 3z = 10$; that is, minimize the quantity

$$(1 - x)^2 + (1 - y)^2 + (1 - z)^2 \quad \text{when} \quad x - 2y + 3z = 10.$$

7.   The Schlock China Company manufactures two kinds of china, economy and prestige. Let x be the price of the economy line per set and y be the price of the prestige line per set. Then $E(x, y) = 100 - 4x + y$ is the number of economy sets sold as a function of pricing and $P(x, y) = 80 - y + x$ is the number of prestige sets sold as a function of price. Let $G(x, y) = $ total gross sales in dollars as a function of the price. Then,

$$G(x, y) = xE(x, y) + yP(x, y).$$

If the production capacity of the company is for all intents unlimited, and if the sole interest of the company is to maximize gross sales, how should it price each line.

(*Note:* $G(x, y)$ can be rewritten as $100x + 80y - (y - 2x)^2 - 2xy$ by means of a little algebra. Thus one can conclude that for either x or y (or both) very large, the gross sales will be small, in fact negative. Hence on heuristic grounds one may conclude that an absolute maximum does exist.)

8.* A trough is to be constructed out of wood in the shape indicated in Figure 9.9.1. It is to have ends and a bottom but no top. There are 60 sq ft of lumber available. What is the maximum volume of the trough?

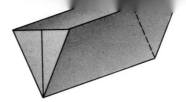

FIGURE 9.9.1

[HINT: Introduce variables as indicated. Then the volume $V(x, y, z) = xyz$ and the surface area

$$S(x, y, z) = 2xy + 2\sqrt{x^2 + y^2} \cdot z.$$

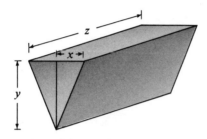

FIGURE 9.9.2

Justify these equations. Thus we wish to maximize the quantity $V(x, y, z) = xyz$ subject to the side condition $2xy + 2\sqrt{x^2 + y^2} \cdot z - 60 = 0$. Complete the problem.]

9. The Super Six Motor Company produces three models of their automobile: the Atlas, the Apollo, and the Hercules. Total production is limited in that $2x + y + 5z = 103$ (when $x =$ number of Atlas's per day, $y =$ number of Apollos per day, and $z =$ number of Hercules's per day). The total profit $P(x, y, z)$ is given by the equation

$$P(x, y, z) = x^2 - xy + y^2 + z^2.$$

How many cars per day in each of the model lines should Super Six produce to make its profits largest?

10. Find the least squares regression line for the data:

x	1	3	4	6	9	10
y	−3	2	3	7	12	14

Graph your answer.

Appendix

Table A.1. Exponential Functions

x	e^x	e^{-x}	x	e^x	e^{-x}
0.00	1.0000	1.0000	1.0	2.7183	0.3679
0.05	1.0513	0.9512	1.1	3.0042	0.3329
0.10	1.1052	0.9048	1.2	3.3201	0.3012
0.15	1.1618	0.8607	1.3	3.6693	0.2725
0.20	1.2214	0.8187	1.4	4.0552	0.2466
0.25	1.2840	0.7788	1.5	4.4817	0.2231
0.30	1.3499	0.7408	1.6	4.9530	0.2019
0.35	1.4191	0.7047	1.7	5.4739	0.1827
0.40	1.4918	0.6703	1.8	6.0496	0.1653
0.45	1.5683	0.6376	1.9	6.6859	0.1496
0.50	1.6487	0.6065	2.0	7.3891	0.1353
0.55	1.7333	0.5769	2.1	8.1662	0.1225
0.60	1.8221	0.5488	2.2	9.0250	0.1108
0.65	1.9155	0.5220	2.3	9.9742	0.1003
0.70	2.0138	0.4966	2.4	11.023	0.0907
0.75	2.1170	0.4724	2.5	12.182	0.0821
0.80	2.2255	0.4493	2.6	13.464	0.0743
0.85	2.3396	0.4274	2.7	14.880	0.0672
0.90	2.4596	0.4066	2.8	16.445	0.0608
0.95	2.5857	0.3867	2.9	18.174	0.0550

x	e^x	e^{-x}	x	e^x	e^{-x}
3.0	20.086	0.0498	4.5	90.017	0.0111
3.1	22.198	0.0450	4.6	99.484	0.0101
3.2	24.533	0.0408	4.7	109.95	0.0091
3.3	27.113	0.0369	4.8	121.51	0.0082
3.4	29.964	0.0334	4.9	134.29	0.0074
3.5	33.115	0.0302	5	148.41	0.0067
3.6	36.598	0.0273	6	403.43	0.0025
3.7	40.447	0.0247	7	1096.6	0.0009
3.8	44.701	0.0224	8	2981.0	0.0003
3.9	49.402	0.0202	9	8103.1	0.0001
4.0	54.598	0.0183	10	22026	0.00005
4.1	60.340	0.0166			
4.2	66.686	0.0150			
4.3	73.700	0.0136			
4.4	81.451	0.0123			

Table A.2. Natural Logarithm Function

x	$\log x$	x	$\log x$	x	$\log x$
0.0	......	2.0	0.6913	4.0	1.3863
0.1	−2.303	2.1	0.7419	4.1	1.4110
0.2	−1.609	2.2	0.7885	4.2	1.4351
0.3	−1.204	2.3	0.8329	4.3	1.4586
0.4	−0.916	2.4	0.8755	4.4	1.4816
0.5	−0.693	2.5	0.9163	4.5	1.5041
0.6	−0.511	2.6	0.9555	4.6	1.5261
0.7	−0.357	2.7	0.9933	4.7	1.5476
0.8	−0.223	2.8	1.0296	4.8	1.5686
0.9	−0.105	2.9	1.0647	4.9	1.5892
1.0	0.0000	3.0	1.0986	5.0	1.6094
1.1	0.0953	3.1	1.1314	5.1	1.6292
1.2	0.1823	3.2	1.1632	5.2	1.6487
1.3	0.2624	3.3	1.1939	5.3	1.6677
1.4	0.3365	3.4	1.2238	5.4	1.6864
1.5	0.4055	3.5	1.2528	5.5	1.7047
1.6	0.4700	3.6	1.2809	5.6	1.7228
1.7	0.5306	3.7	1.3083	5.7	1.7405
1.8	0.5878	3.8	1.3350	5.8	1.7579
1.9	0.6419	3.9	1.3610	5.9	1.7750

Table A.2.—*Continued*

x	log x	x	log x	x	log x
6.0	1.7918	7.5	2.0149	9.0	2.1972
6.1	1.8083	7.6	2.0281	9.1	2.2083
6.2	1.8245	7.7	2.0412	9.2	2.2192
6.3	1.8406	7.8	2.0541	9.3	2.2300
6.4	1.8563	7.9	2.0669	9.4	2.2407
6.5	1.8718	8.0	2.0794	9.5	2.2513
6.6	1.8871	8.1	2.0919	9.6	2.2618
6.7	1.9021	8.2	2.1041	9.7	2.2721
6.8	1.9169	8.3	2.1163	9.8	2.2824
6.9	1.9315	8.4	2.1282	9.9	2.2925
7.0	1.9459	8.5	2.1401	10.0	2.3026
7.1	1.9601	8.6	2.1518		
7.2	1.9741	8.7	2.1633		
7.3	1.9879	8.8	2.1748		
7.4	2.0015	8.9	2.1861		

Table A.3.   Integrals*   (In all entries, *C* is an arbitary constant.)

(1) $$\int x^r \, dx = \frac{x^{r+1}}{r+1} + C \ (r \neq -1)$$

(2) $$\int \cos x \, dx = \sin x + C$$

(3) $$\int \sin x \, dx = -\cos x + C$$

(4) $$\int \sec^2 x \, dx = \tan x + C$$

(5) $$\int \csc^2 x \, dx = -\cot x + C$$

(6) $$\int \sec x \tan x \, dx = \sec x + C$$

(7) $$\int \csc x \cot x \, dx = -\csc x + C$$

(8) $$\int e^x \, dx = e^x + C$$

(9) $$\int \frac{dx}{x} = \log|x| + C$$

(10) $$\int a^x \, dx = \frac{a^x}{\log a} + C$$

(11) $$\int \log|x| \, dx = x(\log|x| - 1) + C$$

(12) $\int \dfrac{}{\sqrt{a^2 - x^2}} \quad \left(\dfrac{}{a}\right)$

(13) $\int \dfrac{dx}{a^2 + x^2} = \left(\dfrac{1}{a}\right) \arctan\left(\dfrac{x}{a}\right) + C$

(14) $\int \dfrac{dx}{\sqrt{x^2 + a^2}} = \log\left(\dfrac{x + \sqrt{x^2 + a^2}}{a}\right) + C$

(15) $\int \dfrac{dx}{\sqrt{x^2 - a^2}} = \log\left(\dfrac{x + \sqrt{x^2 - a^2}}{a}\right) + C$

(16) $\int \dfrac{dx}{a^2 - x^2} = \dfrac{1}{2a} \log\left(\dfrac{a + x}{a - x}\right)(x^2 < a^2) + C$

(17) $\int \dfrac{dx}{x^2 - a^3} = -\dfrac{1}{2a} \log\left(\dfrac{x + a}{x - a}\right)(x^2 > a^2) + C$

(18) $\int \dfrac{dx}{x\sqrt{a^2 - x^2}} = -\dfrac{1}{a} \log\left(\dfrac{a + a\sqrt{a^2 - x^2}}{x}\right)(0 < x < a) + C$

(19) $\int \dfrac{dx}{x\sqrt{a^2 + x^2}} = -\dfrac{1}{a} \log\left(\dfrac{a + \sqrt{a^2 + x^2}}{|x|}\right) + C$

(20) $\int \dfrac{x\,dx}{ax + b} = \dfrac{x}{a} - \dfrac{b}{a^2} \log|ax + b| + C$

(21) $\int \dfrac{x\,dx}{(ax + b)^2} = \dfrac{b}{a^2(ax + b)} + \dfrac{1}{a^2} \log|ax + b| + C$

(22) $\int \dfrac{dx}{x(ax + b)} = \dfrac{1}{b} \log\left|\dfrac{x}{ax + b}\right| + C$

(23) $\int \dfrac{dx}{x(ax + b^2)} = \dfrac{1}{b(ax + b)} + \dfrac{1}{b^2} \log\left|\dfrac{x}{ax + b}\right| + C$

(24) $\int \sqrt{a^2 - x^2}\,dx = \dfrac{x}{2} \sqrt{a^2 - x^2} + \dfrac{a^2}{2} \arcsin\left(\dfrac{x}{a}\right) + C$

(25) $\int \sqrt{x^2 + a^2}\,dx = \dfrac{x}{2} \sqrt{x^2 + a^2} \pm \dfrac{a^2}{2} \log|x + \sqrt{x^2 + a^2}| + C$

(26) $\int x^2\sqrt{a^2 - x^2}\,dx = -\dfrac{1}{4} x(a^2 - x^2)^{3/2} + \dfrac{a^2 x}{8} \sqrt{a^2 - x^2} + \dfrac{a^4}{8} \arcsin\left(\dfrac{x}{a}\right) + C$

(27) $\int \sec x\,dx = \log|\sec x + \tan x| + C$

(28) $\int \csc x\,dx = \log|\csc x - \cot x| + C$

(29) $\int e^{ax} \sin bx\,dx = \dfrac{e^{ax}(a \sin bx - b \cos bx)}{a^2 + b^2} + C$

(30) $\displaystyle\int e^{ax}\cos bx\,dx = \frac{e^{ax}(b\sin bx + a\cos bx)}{a^2 + b^2} + C$

(31) $\displaystyle\int x^n\log x\,dx = x^{n+1}\left[\frac{\log x}{n+1} - \frac{1}{(n+1)^2}\right] + C$

(32) $\displaystyle\int \sin^n x\,dx = \frac{-\sin^{n-1} x\cos x}{n} + \frac{n-1}{n}\int \sin^{n-2} x\,dx + C$

(33) $\displaystyle\int \cos^n x\,dx = \frac{\cos^{n-1} x\sin x}{n} + \frac{n-1}{n}\int \cos^{n-2} x\,dx + C$

(34) $\displaystyle\int x^n e^{ax}\,dx = \frac{x^n e^{ax}}{a} - \frac{n}{a}\int x^{n-1}e^{ax}\,dx + C$

(35) $\displaystyle\int \frac{dx}{ax^2 + bx + c} = \frac{2}{\sqrt{4ac - b^2}}\arctan\left(\frac{2ax + b}{\sqrt{4ac - b^2}}\right) + C\ (b^2 < 4ac)$

(36) $\displaystyle\int \frac{x\,dx}{ax^2 + bx + c} = \frac{1}{2a}\log|ax^2 + bx + c|$

$\qquad\qquad - \frac{b}{a\sqrt{4ac - b^2}}\arctan\left(\frac{2ax + b}{\sqrt{4ac - b^2}}\right) + C\ (b^2 < 4ac)$

Solutions to Selected Exercises

Chapter 1

Section 1.1

1. (a) Property 5
   (b) Property 5
   (c) False, $-x < -y$ by Property 6
   (d) Property 6, Property 5
   (e) False, e.g., $x = 1, y = -1$
   (f) False, e.g., $x = 1, y = -1$
   (g) Property 6

3. (a) $-\frac{1}{2} \le x \le \frac{1}{2}$
   (b) $-8 \le x \le 2$
   (c) $x > 6$ or $x < -2$
   (d) $x \le 2$
   (e) $\frac{3}{2} \le x \le \frac{5}{2}$
   (f) $-5 \le x \le 2$

6. 5 commercials

7. 16

8. 75

10. Exercise 3.
   (a) $[-\frac{1}{2}, \frac{1}{2}]$
   (c) $(-\infty, -2) \cup (6, \infty)$
   (e) $[\frac{3}{2}, \frac{5}{2}]$
   Exercise 4.
   (a) $(-\infty, -3] \cup [0, \infty)$
   (c) $[-3, 0]$
   (e) $[-\frac{1}{2}, 0]$
   (f) $(-\infty, -\frac{3}{2}] \cup [1, \infty)$
   (h) $[-1, 0]$

12. (a) Let $a = 1$, $b = 2$, $c = -3$, $d = -2$. Then $1 < 2$ and $-3 < -2$. But $-3 \not< -4$.

Section 1.2

2. (a) $\sqrt{52} = 2\sqrt{13}$
   (c) $\sqrt{410}$
   (e) $\sqrt{2}$

4. The cost of direct shipping is $5(20\sqrt{5}) = 100\sqrt{5}$ dollars.
   The cost of shipping via Indianapolis is $3(60) = \$180.00$.

5. 54 miles

6. (a) $d[(0, 0), (x, y)] - \sqrt{(x - 0)^2 + (y - 0)^2} = R$. Then $R^2 = x^2 + y^2$.
   (b) $x^2 + y^2 = 5^2$ or $x^2 + y^2 = 25$. Points $(2, 5)$, $(-1, 4)$ do not lie on the circle. Points $(3, -4)$, $(-5, 0)$, and $(0, -5)$ lie on the circle.

Section 1.3

1. (a) 7
   (c) $-\frac{37}{15}$
   (e) 0
   (g) -2

2. (a) $y = 7x - 8$
   (c) $y = -\frac{37}{15}x + \frac{43}{15}$
   (e) $y = 6$
   (g) $v = -?r + ?$

3. (a) $y = 5x + 12$
   (c) $y = -\frac{1}{2}x + \frac{3}{2}$
   (e) $y = 7$
   (g) $y = 7x - 77\frac{1}{3}$

5. Value in 1969 is $y = 80,000(1 - \frac{9}{13})$, $y = 24,615$.

7. Line through $(1, 2)$ with slope $-\frac{1}{5}$ is: $y = -\frac{1}{5}x + 2\frac{1}{5}$.

9. (a) $-\frac{1}{3}$
   (c) -2
   (e) $\frac{9}{4}$
   (g) $\frac{9}{4}$

10. (a) $(\frac{7}{6}, \frac{1}{6})$
    (c) No intersection, lines are parallel.
    (e) $(\frac{6}{11}, -\frac{5}{11})$

11. Equation of the line through $(-4, -5)$ with slope -4 is $y = -4x - 21$.

13. The line perpendicular to L is $y = -\frac{1}{2}x + 2\frac{1}{2}$. The point of intersection is $(\frac{11}{5}, \frac{7}{5})$. The distance from L to $(1, 2)$ is given by $d[(1, 2), (\frac{11}{5}, \frac{7}{5})] = \sqrt{(-6/5)^2 + (3/5)^2} = 3\sqrt{5}/5$.

Section 1.4

1. (a) Not a function.
   (c) (i) Function, domain is set of mothers and range is the set of eldest children.
       (ii) Not a function.
   (e) Function, the domain is the set of all people. Range is the set of letters of the alphabet.

2. (a) Domain of f is the set of real numbers.
   (b) (i) $f(2) = 3$
       (iii) $f(a + 2) = \dfrac{a + 3}{a + 1}$, $a \neq -1$
       (iv) $f(a^2) = \dfrac{a^2 + 1}{a^2 - 1}$, for $a \neq \pm 1$

3. (a) $g(x) = (1 - x)/(1 - x)$. This is not $h(x) = 1$ since there is a point missing.

5. (a)

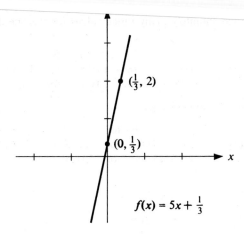

$\left(\frac{1}{3}, 2\right)$

$\left(0, \frac{1}{3}\right)$

x

$f(x) = 5x + \frac{1}{3}$

FIGURE 1.4–5 (a)

(c)

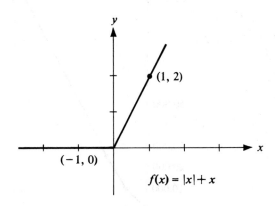

y

$(1, 2)$

x

$(-1, 0)$

$f(x) = |x| + x$

FIGURE 1.4–5 (c)

(e)

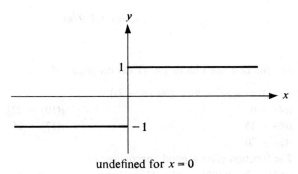

y

1

x

-1

undefined for $x = 0$

FIGURE 1.4–5 (e)

7. No, in a circle each value of the domain has two corresponding range values. No, a vertical line has infinitely many range values for the one domain value.

9. (a)

$$f(n) = \begin{cases} n & \text{for} & 1 \le n \le 10,000 \\ 10,000 + 1.5(n - 10,000) & \text{for} & 10,000 < n \le 20,000 \\ 25,000 + 2\,(n - 20,000) & \text{for} & 20,000 < n \end{cases}$$

This may be simplified as

$$f(n) = \begin{cases} n & \text{for} & 1 \le n \le 10,000 \\ 1.5n - 5,000 & \text{for} & 10,000 < n \le 20,000 \\ 2n - 15,000 & \text{for} & 20,000 < n \end{cases}$$

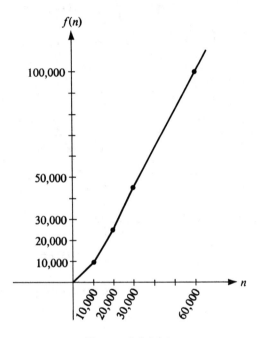

FIGURE 1.4–9 (a)

11. (a) For solution see Figure 1.4–11 (a) on page 375.

(b) For given $s(p) = 30 - [60/(p - 2)]$,

$s(4) = 0$ $\qquad\qquad\qquad$ $s(10) = 22\frac{1}{2}$

$s(6) = 15$ $\qquad\qquad\qquad$ $s(17) = 26$

$s(8) = 20$

The function gives a good approximation.

(c) $s(62) = 29$, $s(122) = 29\frac{1}{2}$. Note that the supply "levels off." As the price increases, the supply increases very little.

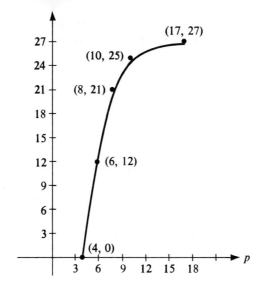

FIGURE 1.4–11 (a)

Section 1.5

1. (a) (i)

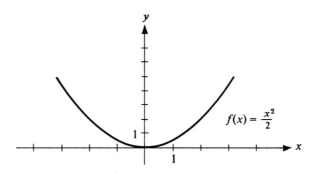

$f(x) = \dfrac{x^2}{2}$

FIGURE 1.5–1 (a) (i)

(a) (iv)

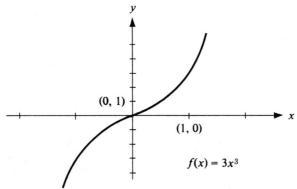

$(0, 1)$

$(1, 0)$

$f(x) = 3x^3$

FIGURE 1.5–1 (a) (iv)

(b)   (i)   even   (iii)   odd   (v)   odd

3.   The equation $S(t)$ does not describe the given problem. For $t > \frac{5}{2}$, the function is decreasing. It should be increasing. For $t = 10$, given any I, $S(t) = 0$. This is not meaningful.

5.   The receipts will be the greatest when there are 15 empty seats.

7.   (a) 2^5 (e) 2^{10} (h) 5^1

    (c) $5^{-1} = \frac{1}{5}$ (f) $2^2 \cdot 3^2$ (j) $= (2)^3 = 8$

9.   (a)   After one year the value of the investment is \$11,000. After 5 years the value of the investment is approximately \$16,000. After 25 years its value is approximately \$10,000(10.7608) = \$107,608.

    (c) $A = \$5000(1 + 0.04/4)^{4x}$ for x number of years
        For $x = 1$, $A = \$5000(1.01)^4 \cong \$5000(1.04) = \$5200$
        For $x = 3$, $A = \$5000(1.01)^{12} \cong \$5000(1.12) = \$5600$
        For $x = 20$, $A = \$5000(1.01)^{80} \cong \$5000(2.13) = \$10,650$

10.   (a)   2   (c)   5   (e)   4   (g)   2

11.   (a) $(3^3)^3 = 3^9 = 19,683.$ $3^{(3^3)} = 3^{27} = 3^9 \cdot 3^{18} = 19,683 \cdot 3^{18}$ which is larger than 19,683.

    (b) $2^{(2^{22})}$

13.   (i) $f(x + h) - f(x) = a^{x+h} - a^x = a^x \cdot a^h - a^x = a^x(a^h - 1)$

    (iii) $\dfrac{f(x + h)}{f(h)} = \dfrac{a^{x+h}}{a^h} = a^{(x+h)h} = a^{x+0} = a^x = f(x)$

15.   (a) $f(x + h) = f(x) = \log_a(x + h) - \log_a x = \log_a(x + h)/x$ by Property 3 $= \log_a(1 + h/x)$

Section 1.6

1.   (a) $[f + g](x) = 5x + 1$. The domain is all real numbers.

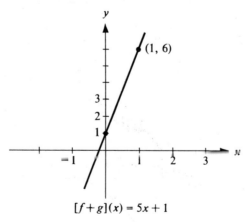

$[f+g](x) = 5x + 1$

FIGURE 1.6-1 (a)

    (c) $[f \circ g](x) = 4x + 4$. The domain is all real numbers. (See Figure 1.6-5 (b) on page 377.)

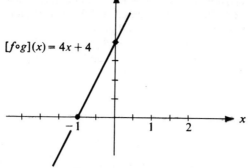

$[f \circ g](x) = 4x + 4$

FIGURE 1.6–1 (c)

3. (a) Yes. $[f + g](x) = f(x) + g(x) = g(x) + f(x) = [g + f](x)$.
    (b) Yes. $[f \cdot g](x) = f(x) \cdot g(x) = g(x) \cdot f(x) = [g \cdot f](x)$.
    (c) No (cf. Exercise 2).

4. $[g \circ f](x) = 1 + 3^x$
    No. $1 + 3^x \neq 3^{1+x} = 3 \cdot 3^x$

5. (a) $[g \circ f](x) = e^{\log x}$

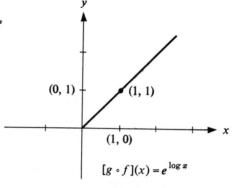

$[g \circ f](x) = e^{\log x}$

FIGURE 1.6–5 (a)

(b) $[f \circ g](x) = \log(e^x)$

$[f \circ g](x) = \log(e^x)$

FIGURE 1.6–5 (b)

6. (a) $h(x) = g[f(x)]$ where $f(x) = x^2 + 7$ and $g(x) = \log x$
    (c) $h(x) = g[f(x)]$ where $f(x) = x + e^{-x} + 1$ and $g(x) = \log x$
    (e) $h(x) = g[f(x)]$ where $f(x) = [\log (x^2 + 1)]/(x^4 + 1)$ and $g(x) = x^{1/3}$
    (g) $h(x) = g[f(x)]$ where $f(x) = 3^x$ and $g(x) = 2^x$
    (i) $h(x) = g[f(x)]$ where $f(x) = \log x$ and $g(x) = \log x$
    (k) $h(x) = g[f(x)]$ where $f(x) = \sqrt{1 + x}$ and $g(x) = e^x$

7. (a)

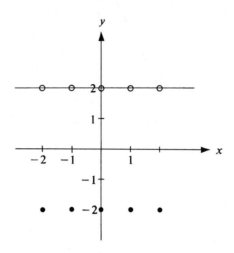

FIGURE 1.6–7 (a)

    (b) $f(2) = -2, f(\tfrac{1}{2}) = 2$

9. (a) By moving the graph of f up 7 units.
    (b) Yes. Leave the graph of f fixed and move the coordinate axis over 3 spaces to the right. $h(x) = [f \circ g]$, where $g(x) = x - 3$.

11. (a) $g(f(x)) = |\,|x| - 1\,|$, this has three "corners."

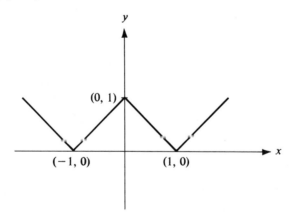

FIGURE 1.6–11 (a)

The function $f(x) = |\,|\,|x| - 1\,| - 1\,|$ has five "corners."

1. (a) $f(x) = 2 \sin x$

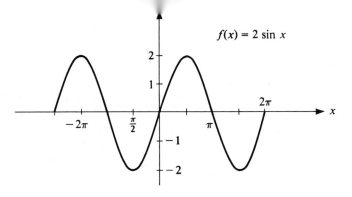

FIGURE 1.7–1 (a)

(c) $f(x) = |\sin x|$

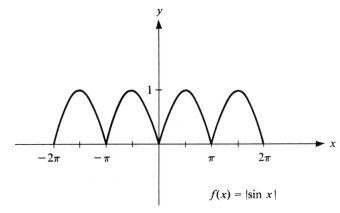

FIGURE 1.7–1 (c)

(e) $f(x) = 3 \cos x$

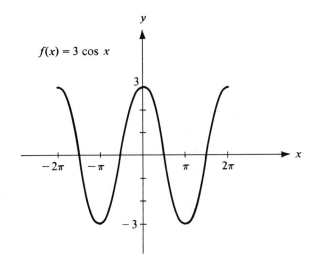

FIGURE 1.7–1 (e)

(g) $f(x) = |\cos x|$

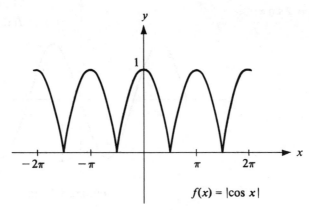

$$f(x) = |\cos x|$$

FIGURE 1.7-1 (g)

3. (a) $x = [(4n + 1)\pi]/4;$ $n = 0, 1, 2, \ldots, -1, -2, \ldots$
    (b) $x = [(2n + 1)\pi]/2;$ $n = 0, 1, 2, \ldots, -1, -2, \ldots$
    (c) $x = n\pi;$ $n = 0, 1, 2, \ldots, -1, -2, \ldots$

5. $g[f(x)] = 1 + \cos x, [f \circ g] = \cos (1 + x)$

7. $g[f(x)] = \sin x^2$
    $[f \circ g](x) = (\sin x)^2$

Section 1.8

1. (a) (i) $-7 \le x \le 1$
          (ii) $x > 1 - \sqrt{2}$ or $x < -1 - \sqrt{2}$
          (iii) $3 \le x \le 7$
    (b) $0 < x < 1$ or $x < -2$

2. (a) $a = -3$ and $b = 2$, or $a = 3, b = -2$
    (c) $c = 0$ (f) $b = 0$
    (d) $a + b + c = 0$ (g) $3a = -2b$

3. (a) Intersection is $(1, 0)$.
    (b) The line through $(1, 0)$ with slope -4 is $y = -4x + 4$.

5. If fuel consumption is to be no more than 10 gallons then A will have to make all deliveries.

7. (a) $[f + g](x) = x^2 + \sqrt{1 - x^2}$, domain is $-1 \le x \le 1$.
    (b) $[f \cdot g](x) = x^2 \sqrt{1 - x^2}$, domain is $-1 \le x \le 1$.

9. (a) Time from $(0, a)$ to $(x, 0) = \sqrt{x^2 + a^2}/s$. The time from $(x, 0)$ to $(b, 0) = (b - x)/r$.
         $T = \sqrt{x^2 + a^2}/s + (b - x)/r$ for $0 < x < b$.

    (b) $T = \sqrt{x^2 + \frac{9}{16}}/2 + (1 - x)/6$

x	0	$\frac{1}{4}$	$\frac{1}{2}$	$\frac{3}{4}$	1
T	0.54	0.52	0.54	0.57	0.61

11. The domain is all real $x \ge 0$, and the range is $0 < y \le 1$ for $y = 2^{-x}$.

13.
$$f \circ f = \frac{[(x + 3)/(2x - 1)] + 3}{2[(x + 3)/(2x - 1)] - 1} = \frac{7x}{7} = x$$

Section 2.2

1.

	$\lim\limits_{x\to a^-} f(x)$	$\lim\limits_{x\to a^+} f(x)$	$\lim\limits_{x\to a} f(x)$
a	yes	yes	no
c	yes	yes	yes
e	yes	yes	no
g	no	yes	no

2.

	$\lim\limits_{x\to a^-} f(x)$	$\lim\limits_{x\to a^+} f(x)$	$\lim\limits_{x\to a} f(x)$
a	yes	yes	no
c	yes	yes	yes
e	yes	yes	yes
g	yes	yes	no

3. (a)

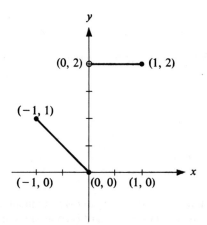

$(0, 2)$ ○——● $(1, 2)$

$(-1, 1)$

$(-1, 0)$ $(0, 0)$ $(1, 0)$

FIGURE 2.2–3 (a)

(b) $\lim_{x\to 0^-} f(x) = 0$, $\lim_{x\to 0^+} f(x) = 2$
(c) $\lim_{x\to 0} f(x)$ does not exist since $\lim_{x\to 0^-} f(x) \neq \lim_{x\to 0^+} f(x)$
(d) $\lim_{x\to -1^+} f(x) = 1$ and $\lim_{x\to 1^-} f(x) = 2$

Section 2.3

1. $\lim_{x\to -1} (x + 2) = 1$

3. $\lim\limits_{x\to 1} \left(\dfrac{2x - 2x^2}{x - 1} \right) = \lim\limits_{x\to 1} \left(\dfrac{-2x(x - 1)}{x - 1} \right) = -2$

5. (a) Yes   (c) No
   (b) No   (d) Yes

7. $\lim_{x\to 0} f(x) = 0$.

9. (a) $\dfrac{f(x) - f(2)}{x - 2} = \dfrac{x^2 - 4}{x - 2} = \dfrac{(x + 2)(x - 2)}{(x - 2)}$

   (b) $\lim\limits_{x\to 2} \dfrac{f(x) - f(2)}{x - 2} = \lim\limits_{x\to 2} (x + 2) = 4$

Section 2.4

1. (a) -1; Properties 1,2,4,5
   (b) $26\frac{2}{3}$; Properties 3,4,6
   (c) -15; Properties 1,2,3,4,5
   (d) $1\frac{3}{4}$; Properties 3,4,6
   (e) 36; Properties 1,2,3,4
   (f) 9; Properties 1,3,4

   (g) 3; Properties 1,2,4,6
   (h) 2; Properties 1,3,4,6
   (i) 0; Properties 1,2,4,6
   (j) 1; Properties 1,2,3,4,6
   (k) 2; Properties 1,2,3,4,6
   (l) 0; Properties 1,2,3,4,5,6

2. (a) $\lim\limits_{x \to 2} \dfrac{x^2 - x - 6}{x - 3} = 4$

   (b) $\lim\limits_{x \to 3} \dfrac{x^2 - x - 6}{x - 3} = 5$

   (c) $\lim\limits_{x \to 1} \dfrac{x^2 + 2x - 3}{x^2 + x - 2} = \dfrac{4}{3}$

   (d) $\lim\limits_{x \to -2} \dfrac{3(x^2 - 4)}{5x(x + 2)} = \dfrac{6}{5}$

3. (a) $\lim\limits_{x \to 2} \dfrac{f(x) - f(2)}{x - 2} = -\dfrac{1}{4}$

   (b) $\lim\limits_{x \to x_0} \dfrac{f(x) - f(x_0)}{x - x_0} = \dfrac{-2}{x_0^3}$

5.* $\lim\limits_{x \to a} \dfrac{x^n - a^n}{x - a} = \lim\limits_{x \to a} \dfrac{(x - a)}{(x - a)} (x^{n-1} + x^{n-2}a + x^{n-3}a^2 + \cdots + a^{n-1})$

   $\qquad = \lim\limits_{x \to a} (x^{n-1} + x^{n-2}a + \cdots + a^{n-1}) = na^{n-1}$

6.* Yes, 0.

7.* (a) $\lim\limits_{x \to 0} \dfrac{ax + b}{cx + d} = \dfrac{b}{d}$

   (d) $\lim\limits_{x \to \infty} \dfrac{ax^2 + bx + c}{dx^2 + ex + f} = \lim\limits_{x \to \infty} \dfrac{a + b/x + c/x^2}{d + e/x + f/x^2} = \dfrac{a}{d}$

Section 2.5

1. (b) $f(\frac{1}{2}) = \lim_{x \to (1/2)^-} f(x) = \lim_{x \to (1/2)^+} f(x) = 1$, so f is continuous at $x = \frac{1}{2}$. $f(=\frac{1}{2}) = \lim_{x \to -(1/2)^+} f(x) = \lim_{x \to -(1/2)^-} f(x) = -\frac{1}{2}$, so f is continuous at $x = -\frac{1}{2}$.

   (c) $f(0) = 1 \neq \lim_{x \to 0} f(x)$, so f is not continuous at 0. If we redefine f at 0 so that $f(0) = 0$, then f would be continuous at $x = 0$.

4. No, $\lim_{x \to 0} f(x) = \lim_{x \to 0} (1/x)$ does not exist.

5. $f(x) = e^{3x}$ is continuous.
   $f(x) = e^{-3x}$ is continuous.

6. $f(x) = \log (x - 1)$ is continuous on $(1, \infty)$. If $f(x) = \log (x - x_0)$, f is continuous for all $x > x_0$.

7. (a) As $x \to 50$, $c(x) = 0.50x$. $\operatorname{Lim}_{x \to 50} 0.50x = 25$. As $x \to 100^-$, $c(x) = 0.50x$. $\operatorname{Lim}_{x \to 100^-} 0.50x = 50$. As $x \to 100^+$, $c(x) = 0.45x$. $\operatorname{Lim}_{x \to 100^+} 45x = 45$. $\operatorname{Lim}_{x \to 100} c(x)$ does not exist. As $x \to 500^-$, $c(x) = 0.45x$. $\operatorname{Lim}_{x \to 500^-} 0.45x = 225$. As $x \to 500^+$, $c(x) = 0.42x$. $\operatorname{Lim}_{x \to 500^+} 42x = 210$. $\operatorname{Lim}_{x \to 500} c(x)$ does not exist.

   (b) $c(x)$ is not continuous at $x = 100, 500$.

9. (a) For solution see Figure 2.5–9 (a) on page 383.

   The graph is not continuous at $x = -1$ since $f(-1)$ is not defined, or at $x = 1$ since $\lim_{x \to 1} f(x)$ does not exist.

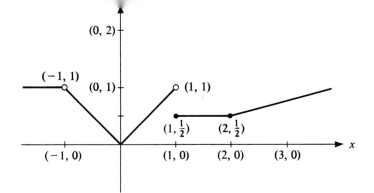

FIGURE 2.5-9 (a)

Section 2.6

1.

	$\lim\limits_{x \to a^-} f(x)$	$\lim\limits_{x \to a^+} f(x)$	$\lim\limits_{x \to a} f(x)$	continuity
2.6.1	yes	yes	no	no
2.6.3	yes	yes	yes	no
2.6.5	yes	no	no	no
2.6.7	yes	no	no	no

2. No, $\lim_{x \to 3} f(x) \neq f(3)$.

3. The function is continuous.

5. (a)

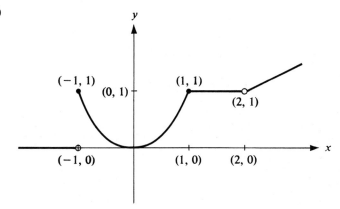

FIGURE 2.6-5 (a)

(b) $f(x)$ is not continuous at $x = -1$ since $\lim_{x \to -1} f(x)$ does not exist.

7. (a) $\frac{3}{2}$ (c) 3 (e) $\frac{1}{2}$

8. (a) $\lim\limits_{x \to 2} \dfrac{f(x) - f(2)}{x - 2} = \lim\limits_{x \to 2} \dfrac{[1/(x + 1)] - 1/3}{x - 2} = \lim\limits_{x \to 2} \dfrac{-(x - 2)}{3(x + 1)(x - 2)} = -\dfrac{1}{9}$

(b) $\lim\limits_{x \to x_0} \dfrac{f(x) - f(x_0)}{x - x_0} = \lim\limits_{x \to x_0} \dfrac{[1/(x+1)] - [1/(x_0+1)]}{x - x_0} = \lim\limits_{x \to x_0} \dfrac{-(x - x_0)}{(x+1)(x_0+1)(x - x_0)}$

$$= \lim\limits_{x \to x_0} \dfrac{-1}{(x+1)(x_0+1)} = -\dfrac{1}{(x_0+1)^2}.$$

9. (a) f is not continuous at $x = 2$.
   (b) f is continuous.

Chapter 3

Section 3.1

1. (a) $[f(x) - f(x_0)]/(x - x_0) = a$ is the slope of f when $f(x) = ax + b$ and so is independent of x and x_0.
   (b) $f(x) = ax + b$

3. $f(x) = x^2 - 1, f'(x) = 2x, f'(-1) = -2.$ ∴ The tangent line is $y = -2x - 2.$

4. (a) $f'(x) = 0$ for all x (d) $f'(2) = 4$
   (b) $f'(x) = 2$ for all x (e) $f'(1) = -2$
   (c) $f'(t) = 1$ for all t (f) $f'(1) = \frac{1}{2}$

5. (a)

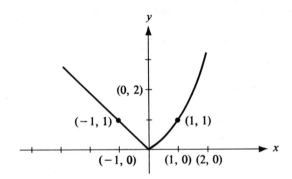

FIGURE 3.1–5 (a)

(b) Yes, since $\lim_{x \to 0} f(x)$ exists and equals 0.
   (c) No, since $\lim_{x \to 0} [f(x) - 0]/(x - 0) = \lim_{x \to 0} [f(x)]/x$ does not exist since $\lim_{x \to 0^+} [f(x)]/x = 0$ and $\lim_{x \to 0^-} [f(x)]/x = -1$.

7. $C(50) = 2501.$ $A(50) = 50.$ $C'(50) = 100$

Section 3.2

1. (b) (i) $f(x)$ is not differentiable at $x = 1$ since it is not continuous.
       (ii) $f'(1)$ does not exist.

3. $f(x)$ is not differentiable at $x = 1$ since
   $$\lim\limits_{x \to 1^+} \dfrac{f(x) - f(1)}{x - 1} = \lim\limits_{x \to 1^-} \dfrac{x - 1 - 0}{x - 1} = 1 \neq \lim\limits_{x \to 1^-} \dfrac{f(x) - f(1)}{x - 1} = \lim\limits_{x \to 1^-} \dfrac{-x + 1}{x - 1} = -1.$$

5. $f(x)$ is not differentiable at $x = 1, 2, 3, 4, 5, 6, 7$ since the function is not continuous at these points.

7. (b) f is not continuous at $x = 3$ since $\lim_{x \to 3} f(x) \neq f(3)$.
   (c) No, since $\lim\limits_{x \to 3^+} [f(x) - f(3)]/(x - 3) \neq \lim\limits_{x \to 3^-} [f(x) - f(3)]/(x - 3)$.

1. (a) $f'(x) = 3x^2, f'(1) = 3, f'(-3) = 27$
   (b) $f'(2) = \frac{1}{6}(2)^{-5/6}, f'(\frac{1}{4}) = \frac{1}{6}(\frac{1}{4})^{-5/6}$
   (c) $f'(x) = e^x, f'(2) = e^2, f'(0) = 1$
   (d) $f'(x) = 1/x, f'(\frac{1}{4}) = 4, f'(1) = 1$
   (e) $f'(x) = 4x^3, f'(1) = 4, f'(-1) = -4$
   (f) $f'(x) = \frac{3}{2}x^{1/2}, f'(4) = 3, f'(9) = \frac{9}{2}$
   (g) $f'(x) = \frac{4}{3}x^{1/3}, f'(8) = \frac{8}{3}, f'(27) = 4$

3. $y = (1/e)x$

5. (a) No, $\lim_{x \to 0} [f(x)]/x$ does not exist.

   (b) $f'(1) = 2$ since $\lim_{x \to 1} \dfrac{f(x) - 1}{x - 1} = 2, f'(-1) = 1, f'(2) = 4.$

   (c) $f'(x) = \begin{cases} 1 & \text{for } x < 0 \\ 2x & \text{for } x > 0 \end{cases}$

7. Let $x = $ length of side, $A(x) = x^2, P(x) = 4x, A'(x) = 2x$ which is $\frac{1}{2}P(x)$.

9. $A(x) = [s(100) - s(0)]/(100 - 0) = [2(100)^3]/100 = 20{,}000$ average costs. Marginal costs $s'(100) = 6(100)^2 = 60{,}000.$

11. $f(x) = x^n.$ $f'(x_0) = \lim_{x \to x_0} \dfrac{f(x) - f(x_0)}{x - x_0} = \lim_{x \to x_0} \dfrac{x^n - x_0^n}{x - x_0} = nx^{n-1}$ as shown in the text.

Section 3.4

1. (a) $6x^2 + 1/x$
   (b) $4e^x - 10x^{-11}$
   (c) $1 + 2x + 3x^2$
   (d) $4/x - 60x$
   (e) $-e^x - x^{-2} + 1/x$

   (f) $e^x - 1/x + 3x^2 - 4x^3$
   (g) $x^{-1/2} + 4x^3 + e^x$
   (h) $-\frac{15}{2}x^{-5/2} - 1/x$
   (i) $4x^3 + 6x^2 - 2x$
   (j) $-\dfrac{1}{x^2} + \dfrac{4}{3}x^{1/3} + 15x^4 + e^x$

3. $2y = 9x - 3$

5. $f'(x) = 12x^3 - 12x = 12x(x^2 - 1)$
   $f'(x) = 0$ for $x = 0, 1, -1$

7. (a) $243 + \dfrac{1}{3\sqrt[3]{3}}$
   (c) $8 + 2e$
   (e) $10 - 5e$

Section 3.5

1. (a) $e^x(x^2 + 2x)$
   (b) $1/x - x^9 - 10x^9 \log x$
   (c) $\dfrac{3x^2 - x^3}{e^x}$
   (d) $\dfrac{xe^x - 3e^x}{x^4}$
   (e) $\dfrac{1 - 5 \log x}{x^6}$
   (f) $5e^x + 5xe^x \log x + 5e^x \log x$

   (g) $20\dfrac{\log x}{x}$
   (h) $2e^{2x}$
   (i) $\dfrac{-1}{x^2\left(1 + \dfrac{1}{x}\right)^2\left(1 + \dfrac{1}{1 + 1/x}\right)^2} = \dfrac{-1}{(2x + 1)^2}$
   (j) $e^x\left(\dfrac{1}{x} + \log x\right) + \dfrac{25x^2 + 30x + 5}{2\sqrt{x}}$
   (k) $\dfrac{e^x}{(1 + e^x)^2}$
   (l) $\dfrac{x(2 + x)}{(1 + x)^2}$

(m) $\dfrac{e^x[(1/x) - \log x] + 1/x}{(1 + e^x)^2}$

3. $C'(x) = 3x^2 - x + 7;$ $C'(10) = 297$

5. (a) $R(x) = xF(x) = x[(1/x) + \tfrac{1}{2}x + 0.01x^2] = 1 + \tfrac{1}{2}x^2 + 0.01x^3$
   $R(10) = 1 + \tfrac{1}{2}(100) + \tfrac{1}{100}(1000) = 61$

   (b) $R'(x) = x + 0.03x^2,\ R'(10) = 10 + 0.03(100) = 13$

   (c) $T(x) = xF(x) - c(x) + 1 + \tfrac{1}{2}x^2 + 0.01x^3 + 0.30x + 0.001x^2$
   $T(10) = 1 + 50 + 10 - 3 - \tfrac{1}{10} = 57\tfrac{9}{10}$
   $T'(x) = x + 0.03x^2 - 0.3 - 0.002x,\ T'(10) = 12\tfrac{17}{25}$

   (d) Marginal revenue is $x + 0.03x^2$. Marginal cost is $0.30 + 0.002x$. They are equal when

$$x = \frac{-449 + \sqrt{258001}}{30} \cong \frac{-449 + 508}{30}, \qquad x \cong 2.$$

7. (a) $\tfrac{7}{2}$

   (b) -16

   (c) $1055\tfrac{4}{5}$

   (d) $7e + \dfrac{1}{e} + \dfrac{5}{2}$

   (e) $\dfrac{e^3 + 2e^2 + 2e - 2}{(1 + e)^2}$

Section 3.6

1. (a) $\dfrac{4x + 1}{2x^2 + x + 1}$

   (c) $\dfrac{3x^2 - 4x + 2}{(x + 1)^4}$

   (e) $\dfrac{e^x + 1}{e^x + x}$

   (g) $-4(x + \log x)^{-5}\cdot(1 + 1/x)$

   (i) $\dfrac{xe^x + e^x}{xe^x} = 1 + \dfrac{1}{x}$

   (k) $\dfrac{2(\log x)}{x} + 3x^2 e^{x^3}$

   (m) $\dfrac{4(x + 1)}{(x^2 + 2x + 1)}$

   (o) $e^{[(\log x/x)^{10}]} \cdot 10\left(\dfrac{\log x}{x}\right)^9 \cdot \dfrac{1 - \log x}{x^2}$

   (q) $\dfrac{1}{x \cdot \log x \cdot \log(\log x)}$

3. $x = 0,\ x = \dfrac{-1 \pm 7i}{2}$

5. $f'(x) = \dfrac{1}{e^x} \cdot e^x = 1 = \dfrac{d}{dx}(x)$

7. Slope is $-20\sqrt{11}/3267$. Given point is $(10, 2\sqrt{11}/33)$. Equation of the tangent is $y - (2\sqrt{11}/33) = [-(20\sqrt{11})/3267](x - 10)$.

Section 3.7

1. (a) $6x$

   (b) $(x^2 - 1)^{-1/2} - x^2(x^2 - 1)^{-3/2}$

   (c) $1/x$

   (d) $e^x \log x + \dfrac{e^x}{x} - \dfrac{e^x}{x^2}$

   (e) $64x^2(x + 1)e^{4x^2} + 8(2x + 1)e^{4x^2} + 8xe^{4x^2}$

   (f) $-4(x - 1)^{-3}$

   (g) $\dfrac{-(1 + \log x)}{x^2 \log^2 x}$

(i) $\dfrac{-2x(x^3 + \frac{4}{3}x^2 - x + \frac{2}{3})}{3(x^2 - 1)^{7/3}}$

(j) $\dfrac{(6x - 3x^4) + e^x[2 + 4x + x^2 - 4x^3 - 2x^4 + x^5 - 2x^2e^x]}{(x^3 + x^2e^x + 1)^2}$

(k) 2

(l) $18(x^3 - 9x^2 + 1)[4x^4 - 48x^3 + 135x^2 + x - 3]$

3. $y' = 2(1 - x)^{-2}$, $y'' = 4(1 - x)^{-3}$, $y''' = 12(1 - x)^{-4}$, $y'''(\frac{1}{2}) = 192$

5. $\dfrac{-[(f(x))^2 f''(x) + 2f(x)(f'(x))^2]}{(f(x))^4}$

7. $(g \circ f)(x) = \dfrac{1 - x^2}{2}$, $(g \circ f)'(x) = -x$, $(g \circ f)''(x) = -1$. $(f \circ g)(x) = \sqrt{1 - x^4}/4$.

Section 3.8

1. (a) $\lim (x^5 - 32)/(x - 2) = \lim (x^4 + 2x^3 + 4x^2 + 8x + 16) = 80$, or note this is simply the derivative of $f(x) = x^5$ evaluated at $x = 2$. $f' = 5x^4$ and $f'(2) = 5 \cdot 16 = 80$.

   (b) $f'(x) = \dfrac{-1}{(x + 2)^2}$, $f'(2) = \dfrac{-1}{4^2} = -\dfrac{1}{16}$

2. (a) $\dfrac{x^3 - 8x}{(x^2 - 4)^{3/2}}$

   (c) $4x + 3x^2e^x + x^3e^x$

   (e) $\dfrac{5(2x + 2)}{x^2 + 2x + 1}$

   (g) $8/x$

   (i) $2xe^{-x^2} - 2x^3e^{-x^2} - e^{-x} \log x + \dfrac{e^{-x}}{x}$

3. Slope is $\frac{1}{2}$ and tangent goes through $(1, 0)$. The equation is $2y = x - 1$.

5. $y' = x^2 + x + 1$. Find points where $y' = 0$. If $x^2 + x + 1 = 0$, then
$$x = \dfrac{-1 \pm \sqrt{1 - 4}}{2}.$$
These are not real numbers. Hence there are no points on the curve that have tangents parallel to the x axis.

7. $R(x) = x(5 - x/20)^2$. Marginal revenue is $R'(x) = (5 - x/20)^2 - (x/10)(5 - x/20)$. $(5 - x/20)^2 - (x/10)(5 - x/20) = 0$ when $x = 100$. The corresponding fare is 0.

9. $\dfrac{dy}{dx} = f'(x^2) \cdot 2x = 2x\sqrt{3x^4 - 1}$

11. $d'(p) = -45 = 2p$ and $d'(25) = 5$

13. $f(x) = (x - 5)^{-4/5}\left(\dfrac{1}{x} + 2\right)^{1/6}$, $f'(x) = \dfrac{25 - 29x - 48x^2}{30(x - 5)^{9/5}[(1/x) + 2]^{5/6}}$

15. $y'(x) = \dfrac{2 - 2x^2}{(1 + x^2)^2}$; $y'(x) = 0$ when $x = \pm 1$. $y''(x) = \dfrac{4x^3 - 12x}{(1 + x^2)^3}$; $y''(x) = 0$ when $x = 0$, $\pm\sqrt{3}$.

17. $\dfrac{d}{dx} F(ax + b) = F'(ax + b) \cdot a$

Using the chain rule, this equals $aG(ax + b)$ since $F'(x) = G(x)$.

Chapter 4

Section 4.1

1. (a) local and absolute max at $x = 0$
   (b) local max at $x = -1, 1$; max at $x = -1, 1$; local min at $x = 0$; min at $x = 4$
   (c) local max at $x = 1, 2, 3, \ldots$ and at each $x \leq 0$ other than integer values. Local min at $x = -1, -2, -3, \ldots$ and at each $x \geq 0$ other than integer values.
   (d) local max at $x = -1, 1, 2$; local min at $x = 0$; min at $x = 0$
   (e) local max at $x = -1, \frac{1}{4}$; max at $x = \frac{1}{4}$; local min at $x = -2, 1$; min at $x = 1$

2. (a) max at $x = 10$, min at $x = 0$ (g) max at $x = 1$, min at $x = -1$
   (c) max at $x = -3$, min at $x = 5$ (i) max at $x = 1$, min at $x = 4$
   (e) max at $x = 4$, min at $x = 1$ (k) max at $x = 1$, min at $x = 0$

3. Maximum area when $x = 8$, $y = 8$.

4. Maximum product occurs when $x = 10$, $y = 10$.

Section 4.2

1. $x = 50\sqrt{2}$, $h = 25\sqrt{2}$, $A = 1250$

3. $3'' \times 6''$ 6. $62{,}500$ 9. $V = \dfrac{500\pi\sqrt{3}}{9}$

4. $x = 3\sqrt{2}$ in.   7. 3456 10. $d = \sqrt{34}$

5. $x = \dfrac{8}{3}$ in.   8. $\dfrac{500\pi\sqrt{6}}{9}$ 11. $A - 1$

Section 4.3

1. (a) $f' > 0$ for $-2 < x < 1$; $f' < 0$ for $1 < x < 3\frac{1}{2}$. $f'' > 0$ for $-1 < x < 0 \cup 2 < x < 3\frac{1}{2}$; $f'' < 0$ for $-2 < x < -1 \cup 0 < x < 2$.
   (c) $f' > 0$ for $-1 < x < \frac{1}{2} \cup 1 < x < 2$; $f' < 0$ for $\frac{1}{2} < x < 1 \cup 2 < x < 4$. $f'' < 0$ for $1 < x < 3$.

2. (a) Increasing for $x > \frac{5}{2}$; decreasing for $x < \frac{5}{2}$; concave up for all x.
   (d) Increasing for $x > 1 \cup x < -1$; decreasing for $-1 < x < 1$. Concave up for $0 < x$; concave down for $x < 0$. See Figure 4.3–2 (d) on page 389.
   (f) Increasing for $x < 0$; decreasing for $x > 0$; concave down for all x.
   (g) Increasing for $x > 1$; decreasing for $0 < x < 1$; concave up for all x.
   (j) Increasing if $x > 0$; decreasing if $x < 0$; concave down on domain.

3. $b \geq \frac{3}{2}$

5. No, f is not differentiable on $(-1, 1)$. There is no derivative at $x = 0$.

Section 4.4

1. (a) Increasing for $x > \frac{5}{2}$; decreasing for $x < \frac{5}{2}$; concave up for all x.
   (c) Increasing for $x > 6 \cup x < -1$; decreasing for $-1 < x < 6$. Concave up for $x > \frac{5}{2}$; concave down for $x < \frac{5}{2}$.
   (e) Increasing for $x > 2 \cup x < -\frac{4}{3}$; decreasing for $-\frac{4}{3} < x < 2$. Concave up for $x > \frac{1}{3}$; concave down for $x < \frac{1}{3}$.
   (g) Increasing for $0 < x < \frac{2}{3}$; decreasing for $x < 0 \cup x > \frac{2}{3}$. Concave up for $x < \frac{1}{3}$; concave down for $x > \frac{1}{3}$.
   (i) Increasing for $x > 1$; decreasing for $x < 1$; concave up for all x.

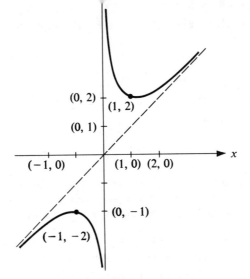

$$(0, 2)$$
$$(1, 2)$$
$$(0, 1)$$
$$(-1, 0)$$
$$(1, 0) \quad (2, 0)$$
$$x$$
$$(0, -1)$$
$$(-1, -2)$$

FIGURE 4.3–2 (d)

(j)   Increasing for $x > \sqrt[3]{4}/2$; decreasing for $x < \sqrt[3]{4}/2$. Concave up for $x < -1 \cup$ $x > 0$; concave down for $-1 < x < 0$.

2. $x = 1/e$ gives minimum A. Increasing for $x > 1/e$. Concave up for $1 \le x \le 4$.

3.   The function is concave downward for all x.

Section 4.5

1.   (a)   local min at $x = 3$
    (b)   local max at $x = -4$, local min at $x = 3$
    (c)   no max or min
    (e)   local max at $x = 0$
    (g)   local min at $x = \sqrt[3]{2}$

3.   min $d = \sqrt{15/8}$

4.   min P at $x = \sqrt{2A}/2$, $y = \sqrt{2A}$

5. $a = 4, b = 6, d = -5, c = 0$

8.   Increasing for $0 < x < e$. Decreasing for $x > e$. Concave up for $x > e^{3/2}$. Concave down for $x < e^{3/2}$. As $x \to \infty$, $f(x) \to 0$.

9.   (a)   maximum at $x = 3$, minimum at $x = 0$
    (c)   no max, no min   (g)   min at $x = \frac{1}{2}$, no max
    (e)   min at $x = 0$, max at $x = 3$

Section 4.6

1. $x = 10$ gives maximum total profit.

2.   max area $= 2500/\pi$, min area $= 2500/(\pi + 4)$ (total wire used for circle)

3.   max $A = \dfrac{2}{\sqrt{2}} e^{-1/2}$

4. max $A = \dfrac{200}{(4 + \pi)^2} [400 + \pi]$

5. Minimize area. Hence $y = 4\sqrt{3}$, $x = (10\sqrt{3})/3$ gives a minimum area, hence minimum cost of the page.

7. $x = 400$

9. min $f \cong 1300$ when $x = 20$

10. The garden is 50×100.

12. $s = 30$ gives max profit

14. $A = \frac{4}{9}\sqrt{3}$

Section 4.7

1. (a) Increasing for $x < -5 \cup x > 1$; decreasing for $-5 < x < 1$. Concave up for $x > -2$; concave down for $x < -2$. Max at $x = -5$; min at $x = 1$.
    (c) Increasing for $x < 0$; decreasing for $x > 0$. Concave up for all x. Max at $x = 0$; min at $x = \pm 1$.
    (e) Increasing for $x < 0$; decreasing for $x > 0$. Concave up for $x > 3/\sqrt{2} \cup x < -3/\sqrt{2}$; concave down for $-3/\sqrt{2} < x < 3/\sqrt{2}$. Max at $x = 0$.

2. Max occurs at $x = \sqrt{2}$, max total revenue is $4\sqrt{2}$.

3. $T(x) = \dfrac{8x}{4 + x^2}$, max $T(x) = 2$

6. The inflection point is at $x = 11.67$. This is the point of diminishing returns.

7. (a) $y' = 0$ and $y'' = 2$. By the second derivative test (b) we know y has a local min at x_0.
    (b) $y' = 0$ and $y'' = 0$. By the second derivative test (c), we have no information.
    (c) $y' = 0$ and $y'' = -2$. By the second derivative test (a), x_0 is a local max.

9. Length = width = height = $\sqrt[3]{12}$

10. $A = 1250$

11. $x = 2$

12. (a) $x = 6$ (b) \$9 per item   (c) \$34 a week

Chapter 5

Section 5.1

1. (a) No   (c) No
    (b) $F(x) = -x^{-1} + C$ (d) No

2. (a) $\frac{2}{5}x^{5/2} + C$ (e) $\frac{2}{3}x^{3/2} - \frac{2}{5}x^{5/2} + C$ (h) $(x^2/2) + C$
    (c) $(t^3/3) + (t^2/2) + C$ (f) $(x^2/2) + C$ (j) $(3x^2/2) - 5e^x + C$

3. (a) $h' = -x(1 - x^2)^{-1/2}$ (b) $\sqrt{1 - x^2} + C$

5. (a) $\frac{1}{2}x^2 - (1/x) + C$ (d) $\frac{2}{3}t^{3/2} + 2t^{1/2} + C$
    (c) $(x^3/3) - \frac{2}{3}x^{3/2} + C$ (f) $(at^{9001}/9001) + C$

Section 5.2

1. $\dfrac{x^3}{3} - \dfrac{3x^2}{2} + 9x + C$

5. $-\dfrac{3}{2}x^{2/3} + \dfrac{6}{5}x^{5/6} - \dfrac{9}{8}x^{8/9} + C$

7. $-2e^x + e^x + C$

9. $\log x - \dfrac{3}{x} + \dfrac{3}{4}x^{4/3} - 3x^{1/3} + C$

11. $\dfrac{x^5}{5} - \dfrac{4x^3}{3} + C$

13. (a) $f(x) = x^2 - \frac{3}{2}x^4$ \qquad\qquad (c) $f(x) = 2x^4 - x^2 - 20$
    (b) $x^3 - \frac{7}{2}x^2 + 2x + 3 = f(x)$

15. $f'(x) = \dfrac{2}{3}x^{3/2} + 2x^{1/2} - \dfrac{2}{3}$, $f(x) = \dfrac{4}{15}x^{5/2} + \dfrac{4}{3}x^{3/2} - \dfrac{2}{3}x + C$

Section 5.3

1. (a) $dy = (3x^2 + 2e^{2x})\,dx$ \qquad (g) $dy = 4x^3\,dx$

   (c) $dy = 2xe^{(x^2+1)}\,dx$ \qquad\qquad (i) $dy = \dfrac{dx}{x}$

   (e) $du = \dfrac{x\,dx}{\sqrt{x^2+1}}$

3. $\dfrac{3}{8}x^{8/3} + \dfrac{6}{7}x^{7/6} - \dfrac{1}{x} + C$ \qquad\qquad 17. $2e^{\sqrt{x}} + C$

5. $\frac{3}{4}(x^2 + 2x + 3)^{2/3} + C$ \qquad\qquad 19. $-\sqrt{1-x^2}\log\dfrac{\sqrt{1-x^2}}{e} + C$

7. $\frac{2}{5}u^{5/2} - \frac{2}{3}u^{3/2} + C$ \qquad\qquad 21. $\frac{1}{3}e^{x^3} + (x^2 + 1)^{3/2} + C$

9. $\frac{2}{3}u^{3/2} + 18u^{1/2} + C$ \qquad\qquad 23. $f(x) = \frac{1}{3}e^{x^3+3x} + \frac{2}{3}$

11. $\dfrac{(\log x)^2}{2} + C$ \qquad\qquad 25. $f(x) = 3x - 5$

13. $\log(\log x) + C$ \qquad\qquad 27. $f(x) = -\frac{2}{63}(x^2 - 1)^2 + \frac{128}{63}$

15. $(\log x)^2 + C$

Section 5.4

1. (a) $\frac{28}{3}$ \qquad (c) 1 \qquad\qquad (e) $2(e - e^{-3/2})$
   (b) $\frac{4}{3}$ \qquad (d) $\frac{10}{9}$ \qquad\qquad (g) $\log 2 + \frac{1}{2}$

3. (a) $\frac{19}{9}$ \qquad (e) $\frac{1}{6}$ \qquad\qquad (h) 50
   (c) $\frac{4}{9}(\sqrt[4]{27} - \sqrt[4]{8})$ \quad (f) $4 - \frac{1}{4}[(1/e) + 1]^4$ \quad (j) $\frac{1}{4}[(\log 5)^2 - (\log 2)^2]$

5. Yes, since $F(x) = -x^2/2$ on $[-1, 0)$; $F(x) = x^2/2$ on $(0, 1]$; and when $x = 0$, $F(x) = 0$.
   $\int_{-1}^{1} f(x)\,dx = 1$.

Section 5.5

1. (a) $x^2\left(\dfrac{\log x}{2} - \dfrac{1}{4}\right) + C$ \qquad (d) $x^3\left(\dfrac{\log x}{3} - \dfrac{1}{9}\right) + C$

   (b) $(1 + x)^{3/2}\left(\dfrac{2}{7}x^2 = \dfrac{8x}{35} + \dfrac{16}{105}\right) + C$ \quad (e) $\dfrac{2x^{3/2}}{3}\left(\log x - \dfrac{2}{3}\right) + C$

   (c) $e^x(x^3 - 3x^2 + 6x - 6) + C$ \qquad\qquad (f) $x\log x - x + C$

2. (c) $\dfrac{2048}{105}$

(d) $\dfrac{3\sqrt{2} + \log(\sqrt{2} - 1)}{8}$

5. Let $u = x$, $dv = g'(x)\,dx$ and $du = x$, $v = g(x)$.

7. $\displaystyle\int_a^b \dfrac{dx}{x} = \log\dfrac{b}{a}$

Section 5.6

1. $\frac{32}{3}$

5. $\frac{64}{3}$

9. $\frac{4}{3}$

3. $e - 1$

7. $\frac{5}{6} - e^{-1/3}$

11. $b = -2$ and $b = 4$

13. (a) $\frac{1}{2}\pi a^2$

(b) $\int_0^a \sqrt{a^2 - x^2}\,dx = \frac{1}{4}\pi a^2$

15. (a) $F'(x) = \sqrt{1 - x^2}$

(b) Use chain rule; let $f(x) \int_0^x \sqrt{1 - t^2}\,dt$ and $g(x) = x^2$, then $F(t) = f(g(x))$. $F'(t) = f'(g(x)) \cdot g'(x) \cdot F'(x) = \sqrt{1 - x^4} \cdot 2x$.

Section 5.7

1. Yes, about 4 years.

3. $C_s = \frac{555}{8}$

5. $P_s = 12 - 8e$

7. average value $= \dfrac{1}{t_2 - t_1}$

$\displaystyle\int_{t_1}^{t_2} f'(t)\,dt = \dfrac{f(t_2) - f(t_1)}{t_2 - t_1}$

9. $a\sqrt{3} - \frac{1}{3}a$

11. approximately 6 years

13. approximately $1213

Section 5.8

1. (a) $\frac{5}{3}x^3 - x^{-9} - 15x^{-2/3} + C$

(k) $2 \log 2 - \frac{3}{4}$

(c) $\dfrac{1}{10e^9}(e^{20} - 1)$

(m) $-\dfrac{(3 \log x + 1)}{9x^3} + C$

(e) $-(1 - x^2)^{1/2} + C$

(o) $2e^4(2e^2 - 1)$

(g) $\dfrac{(4x^2 - 9)^6}{48} + C$

(q) $\sqrt{11} - \sqrt{2}$

(i) $\frac{2}{3}x(x - 1)^{3/2} - \frac{4}{15}(x - 3)^{5/2} + C$

3. $\frac{4}{15}$

5. $\frac{53}{3}$

7. $s = 5x + 10 \log x + \dfrac{x^2}{2} \log x - \frac{1}{4}x^2$. $s(1) = 4\frac{3}{4}$, $s(5) = 18\frac{3}{4}$. The change in sales is $s(5) - s(1) = 16$.

9. $\int_{-1}^1 (x^3 - x)\,dx = 0$

11. $330

13. 6.3%

Section 5.10

1. $\frac{16}{3}a^3$

3. $\sqrt{3}/2$

5. (a) $\pi/2$

(d) $\pi(\frac{43}{2}\sqrt{2} - \frac{115}{4})e^{2\sqrt{2}} + (\pi/4)e^2$

(b) $\frac{111}{4}\pi$

(e) $\pi(2 \log^2 2 - 4 \log 2 + 2)$

(c) $(\pi/3)(2\sqrt{2} - 1)$

(f) $(\pi/123)(6^{41} - 5^{41})$

9. $\frac{56}{13}\pi$

Chapter 6

Section 6.1

1. (a) $\dfrac{-y}{x}$

   (b) $\dfrac{-6xy - 4y^2}{3x^2 + 8xy}$

   (c) $-ye^x$

   (d) $-\dfrac{1}{x \log x}$

   (e) $\dfrac{y(-y^3 - 5x^5 \log 2y)}{x(x^5 + 3y^3 \log x)}$

   (f) $\dfrac{-y(y + 3x)}{x(x + 3y)}$

   (g) $\dfrac{-2y}{x}$

   (h) $-\dfrac{9x}{4y}$

3. $A'(2AB + C^2) + B'(2BC + A^2) + C'(2CA + B^2) = 0.$

5. $u'(x)e^{u(x)} + [v'(x)]/[v(x)]$

7. $u'[2v^2 + 1/(u - 2w) + e^{w^2 + u}] + v'[4uv] + w'[2/(2w - u) + 2we^{w^2 + u}]$

9. $(y + a) = -\frac{1}{4}(x - 2a)$

Section 6.2

1. (a)

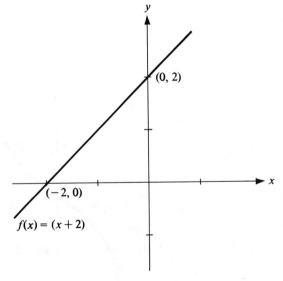

FIGURE 6.2-1 (a) (i)

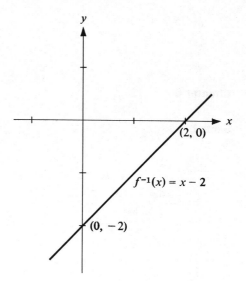

FIGURE 6.2–1 (a) (ii)

(c)

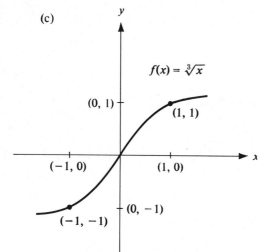

FIGURE 6.2–1 (c) (i)

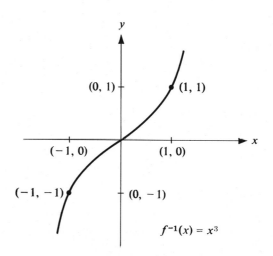

FIGURE 6.2–1 (c) (ii)

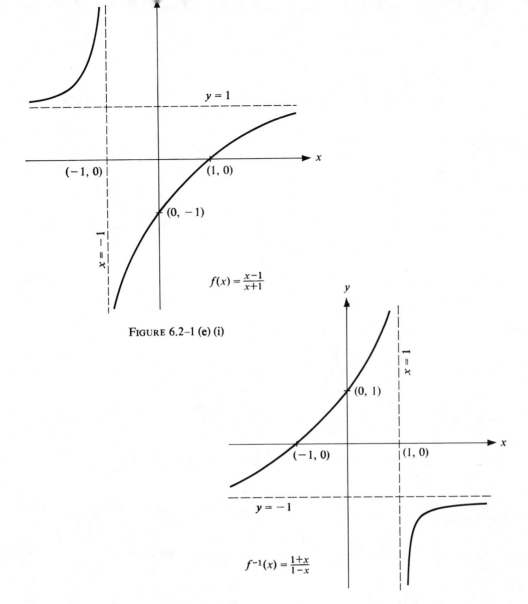

$$f(x) = \frac{x-1}{x+1}$$

FIGURE 6.2–1 (e) (i)

$$f^{-1}(x) = \frac{1+x}{1-x}$$

FIGURE 6.2–1 (e) (ii)

3. $f'(x) = -x/\sqrt{1-x^2}$. $f' < 0$ for all $x > 0$. By argument in Exercise 2, f^{-1} exists and vertical tangent at $x = 1$ is in its domain.

5. $\dfrac{dh}{dx} = \dfrac{dg^{-1}(x)}{dx} = \dfrac{1}{g'(g^{-1}(x))} = \dfrac{1}{4(h^3(x))} = \dfrac{1}{4(x - 13)^{3/4}}$

$h(x) = \sqrt[4]{x - 13}$

$h'(14) = \frac{1}{4}$

7. $f'(x) = 5x^4 + 3x^2 > 0$; hence f^{-1} exists. $x = (f^{-1})^5 + (f^{-1})^3 + 1$; $(f^{-1})' = 1/[5(f^{-1})^4 + 3(f^{-1})^2]$.
    $f^{-1}(1) = 0$; $f^{-1}(-1) = -1$; $f^{-1}(41) = 2$. $(f^{-1})'(1)$ undefined; $(f^{-1})'(-1) = \frac{1}{8}$; $(f^{-1})'(41)$
    $= \frac{1}{92}$.

9. (a) $h'' = -\dfrac{h'(x) \cdot g''(h(x))}{[g'(h(x))]^2} = -\dfrac{g''(h(x))}{[g'(h(x))]^3}$

    (b) $h(x) = \dfrac{1 + x}{1 - x}$, $h''(x) = \dfrac{4}{(1 - x)^3}$

11. $f'(x) = a$, if $a \neq 0$. As in Exercise 2, f' is never 0 and hence $f(x)$ is $1 - 1$. $f'(x)$ exists by
    Exercise 2 arguments. $f^{-1}(x) = (x - b)/a$, which has a straight line graph.

Section 6.3

1. (a) $12 - 2e^4$
    (b) Yes, when $0 = 3t^2 - 2e^{2t}$.

3. (a) 104   (c) -8 (e) $3\sqrt{2}/2$
    (b) $24e^{36} - 1$ (d) $\frac{22}{9}$ (f) $-\frac{5}{2}$

5. (a) 4800   (c) 360,000   (e) -4800

7. At $h = 60$, $\dfrac{dh}{dt} = \dfrac{3}{25\pi}$. At $h = 30$, $\dfrac{dh}{dt} = \dfrac{12}{25\pi}$.

9. $\dfrac{dA}{dt} = \dfrac{4}{5}(225\pi)^{1/3}$

11. $s(t) = -3t^2 + 25t$, $s(\frac{25}{6}) = \frac{625}{6}$.

13. $a(t) = 12$, $s(t) = 6t^2 + C_1 t + C_2$
    $s(t) = 6t^2 + 40t - 76$

Section 6.4

1. (a) 1   (d) 0
    (b) 1   (e) undefined
    (c) 0

3. 0

5. 1

1. (a) $-5 \sin 5x$

Section 6.5

(b) $-\dfrac{1}{2\sqrt{1 - x}} \cdot \cos \sqrt{1 - x}$

(c) $-\dfrac{\cos (1 - x)}{2\sqrt{\sin (1 - x)}}$

(d) $3 \cos (3xe^{\sin 3x})$

(e) $-2 \tan 2x$

(f) $\dfrac{3(x + 5)^2}{\cos^2 (x + 5)^3}$

(g) $\dfrac{e^x}{\cos^2 e^x}$

(h) $\dfrac{2}{(1 - x)^2 \sin\left(\dfrac{1 + x}{1 - x}\right) \cos\left(\dfrac{1 + x}{1 - x}\right)}$

(i) $-\dfrac{(\sin x)[\cos (\cos x)]}{\sin (\cos x)}$

(j) $\dfrac{e^x(\sin x - 2 \cos x)}{(\sin x)^3}$

(k) $\dfrac{1}{x \cos^2 (\log x)}$

(l) $(\cos [\sin (\sin x)])(\cos (\sin x))(\cos x)$

(m) $\dfrac{e^{\tan x} 2(1 + \sin x) - \cos^3 x}{2 \cos^2 [x(1 + \sin x)^{3/2}]}$

3. $\dfrac{d}{dx}(\cos x) - \dfrac{d}{dx}\left(\sin\left(x + \dfrac{\pi}{2}\right)\right) \cdots$

7. Let $y = \text{arc cos } x$, then $\cos y = x$.

$$-\sin y\,\dfrac{dy}{dx} = 1 \quad \text{or} \quad \dfrac{dy}{dx} = -\dfrac{1}{\sqrt{1 - x^2}}$$

Since $\sin^2 y + \cos^2 y = 1$, $\sin y = \sqrt{1 - x^2}$.

8. (a) $\dfrac{2x}{\sqrt{2x - x^2}}$ (c) $\dfrac{1}{2\sqrt{x}(1 + x)}$ (f) $\dfrac{2x\cos x^2}{1 + (\sin x^2)^2}$

10. $\dfrac{d\theta}{dt} = \dfrac{5}{148}$

11. $\dfrac{dx}{dt} = 800\,\pi$ ft/min

Section 6.6

1. (a) $\dfrac{dy}{dx} = \dfrac{2\sqrt{y}(1 - 2y)}{4x\sqrt{y} - 2\sqrt{y} + 1}$ (e) $\dfrac{dy}{dx} = \dfrac{y - 3x^2}{3y^2 - x}$

    (b) $\dfrac{dy}{dx} = \dfrac{x - xy^2}{x^2 y - y}$ (f) $\dfrac{dy}{dx} = -\dfrac{y}{x^2}$

    (c) $\dfrac{dy}{dx} = \dfrac{5x^4 - 4(x + y)^3 - 4(x - y)^3}{4(x + y)^3 - 4(x - y)^3 - 5y^4}$ (g) $\dfrac{dy}{dx} = -\dfrac{ye^y}{x[ye^y \log xy + e^y - 2y^2]}$

    (d) $\dfrac{dy}{dx} = \dfrac{y(3x^2 - 1 - \log xy - 2xe^{x^2 y})}{x(xye^{x^2 y} + 1)}$

3. $r' = \frac{3}{16}$

5. $dx/dt = 60$ mi/hr. Find dr/dt when $x = -5$.

$$12^2 + x^2 = r^2, \quad x\,\dfrac{dx}{dt} = r\,\dfrac{dr}{dt}$$

$$-5(60) = 13\,\dfrac{dr}{dt}; \quad \dfrac{dr}{dt} \doteq -23\tfrac{1}{13}\ \text{mi/hr}$$

7. $s(t) = 12t - \frac{1}{2}\cdot 9t^2$

    $0 = 12t - 9t$

    $s(\frac{4}{3}) = 12\cdot\frac{4}{3} - \frac{9}{2}(\frac{4}{3})^2 = 8$

8. (a) $\frac{1}{4}$ (c) ∞ (f) $1/e$

9. (a) Yes   (b) Yes   (c) No   (d) No

11. (a) $v(t) = -pt + 88$, $s(t) = -\dfrac{p}{2}t^2 + 88t$. $p \cong 38.72 \cong 39$.

    (b) $s(t) = 25$ ft

Chapter 7

Section 7.1

1. (a) $\frac{1}{5}\sin 5x + C$ (d) $e^{\tan x} + C$

    (b) $-\frac{1}{2}\cos x^2 + C$ (e) $-\cos x \log |\cos x| + \cos x + C$

    (c) $-\frac{1}{2}\log \cos 2x + C$ (f) $-\frac{1}{2}x\cos 2x + \frac{1}{4}\sin 2x + C$

3. $(\pi/2) - 1$

5. $\pi^2/4$

7. $\frac{3}{2}\sqrt{3}$

Section 7.2

1. (a) $x\sqrt{x^2 + \frac{3}{4}} + \frac{3}{4}\log|x + \sqrt{x^2 + \frac{3}{4}}| + C$

   (b) $-\dfrac{1}{3}\log\dfrac{3 + \sqrt{9 - 2x^2}}{x\sqrt{2}} + C$

   (c) $\left(\text{Let } u^2 = \dfrac{x - 1}{x + 2}\right),\ \sqrt{(x - 1)(x + 2)} - \dfrac{3}{2}\log\dfrac{\sqrt{x - 1} + \sqrt{x + 2}}{\sqrt{x - 1} - \sqrt{x + 2}} + C$

   (d) $e^{8x}\left(\dfrac{x^3}{8} - \dfrac{3x^2}{64} + \dfrac{3x}{256} - \dfrac{3}{2048}\right) + C$

   (e) Let $u = e^x$, then let $u = \sqrt{5}/3\tan\theta$.
   $(\sqrt{15}/150)\arctan[(\sqrt{3/5})\,e^x] + (\sqrt{15}/3300)\sin 2[\arctan(\sqrt{3/5})\,e^x] + C$

   (f) $\dfrac{x^4 e^{2x}}{2} - x^3 e^{2x} + \dfrac{3x^2 e^{2x}}{2} - \dfrac{3xe^{2x}}{2} + \dfrac{3e^{2x}}{4} + C$

   (g) $\displaystyle\int \dfrac{x\,dx}{\sqrt{x^2 - 9}} - \int \dfrac{3\,dx}{\sqrt{x^2 - 9}} = (x^2 - 9)^{1/2} - 3\log\left(\dfrac{x + \sqrt{x^2 - 9}}{3}\right) + C$

   (h) $-\dfrac{1}{3}\log\left(\dfrac{3 + \sqrt{9 - x^2}}{x}\right) + C$

   (i) $\dfrac{x}{5} - \dfrac{4}{25}\log|5x + 4| + C$

   (j) $-\dfrac{1}{4}x(25 - x^2)^{3/2} + \dfrac{25x}{8}\sqrt{25 - x^2} + \dfrac{625}{8}\arcsin\left(\dfrac{x}{5}\right) + C$

3. (a) $[1/(2\sqrt{2})][x\sqrt{x^2 + 3}/2 - \frac{3}{2}\log(x + \sqrt{x^2 + 3/2})] + C$

   (b) $\frac{1}{2}\log\frac{5}{3}$

   (c) $\frac{3}{16}\arcsin x^2 - \frac{3}{16}x^2\sqrt{1 - x^4} - \frac{1}{8}x^6\sqrt{1 - x^4} + C$

5. $(e^2/4) - \frac{3}{4}$

Section 7.3

1. (a)

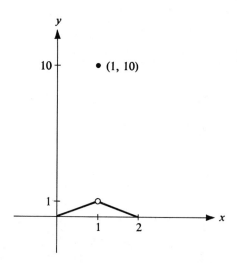

FIGURE 7.3–1 (a)

3. (a)

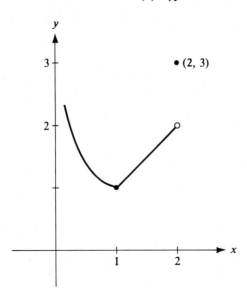

FIGURE 7.3–3 (a)

4. (a) $\frac{6}{5}$ (c) diverges   (e) diverges

5. (a) 1   (c) $\sqrt{-3}$
   (b) 4   (d) divergent

6. (a) diverges   (c) -1 (e) diverges

7. (a) for $n < 1, 1/(1 - n)$ (b) for $n > 1, 1/(n - 1)$

9. $A = \lim_{N \to \infty} \int_1^N (1/x)\, dx = \lim_{N \to \infty} (\log N - \log 1) = \infty$, diverges.

$V = \lim_{N \to \infty} \int_1^N \pi(1/x^2)\, dx = \pi.$

Section 7.4

1. 1.1949

3. $\pi/4$

5. $A(x) = 22.33$ sq ft
   $V = 117{,}902.4$ cu ft

Section 7.5

1. (a) $\dfrac{1}{6} \log \left(\dfrac{3 + x}{3 - x} \right)$

   (b) $\dfrac{1}{2} \log \left(\dfrac{x + \sqrt{x^2 + 9/4}}{3/2} \right) + C$

   (c) $\sqrt{x^2 + 2x - 3} + \log \left(\dfrac{x + \sqrt{x^2 + 2x - 3}}{2} \right) + C$

(d) $\sqrt{x^2 + 9} + 2 \log \left(\dfrac{x + \sqrt{x^2 + 9}}{3} \right) + C$

(e) $\dfrac{\sqrt{3x}}{2} \sqrt{x^2 + 5/3} + \dfrac{5\sqrt{3}}{6} \log |x + \sqrt{x^2 + 5/3}| + C$

(f) Let $u = e^x + 1$, then $u - \frac{1}{2} = w$; $-\log \dfrac{e^x + 1}{e^x}$.

(g) Let $u = 2x^2 + 4x + 1$; $(2x^2 + 4x + 1)[\log (2x^2 + 4x + 1) - 1] + C$

(h) $\dfrac{1}{6(3x + 6)} + \dfrac{1}{36} \log \left| \dfrac{x}{3x + 6} \right| + C$

(i) $-\dfrac{1}{4} x(25 - x^2)^{3/2} + \dfrac{25x}{8} \sqrt{25 - x^2} + \dfrac{625}{8} \arcsin \dfrac{x}{5} + C$

(j) $\log \left(\dfrac{e^x + \sqrt{e^{2x} + 9}}{3} \right) + C$

2. (a) No   (c) $\frac{1}{3}$ (e) No

3. $A = \int_0^\infty xe^{-x^2}\, dx = \frac{1}{2}$

5. This is $F'(2)$ where $F'(x) = e^{-x^2}$. $\therefore F'(2) = e^{-4}$.

7. (a) 1.114 approximately
   (b) 1.108 approximately

Chapter 8

Section 8.1

1. $y(x) = (x^4/4) + C_1 x + C_2$, then $y'(x) = x^3 + C_1$, $y''(x) = 3x^2$.

3. $dy/dx = C_1 e^x + C_2(e^x + xe^x)$, and $d^2y/dx^2 = C_1 e^x + C_2 e^x + C_2(e^x + xe^x)$.
   $(d^2y/dx^2) - [2 (dy/dx)] + y = e^x(C_1 + 2C_2 + C_2 x) - 2e^x(C_1 + C_2 + C_2 x) + e^x(C_1 + C_2) = e^x(C_1 + 2C_2 + C_2 x - 2C_1 - 2C_2 - 2C_2 x + C_1 + C_2 x) = e^x(0) = 0$.

5. (a) $y = (x^5/5) - x^3 + 1$
   (b) $y = (x^6/6) + e^x + 2$
   (c) $y = -\frac{1}{2}e^{-x^2} + 1$
   (d) $y = x^2\{[(\log x)/2] - \frac{1}{4}\} - (x^2/2) + x - \frac{1}{4} = \frac{1}{2}x^2(\log x - \frac{3}{2}) + x - \frac{1}{4}$

7. (a) $y' = \dfrac{3x^2}{2} + x + C_1$ and $y = \dfrac{x^3}{2} + \dfrac{x^2}{2} + C_1 x + C_2$

   (b) $f(x) = \dfrac{x^3}{2} + \dfrac{x^2}{2} + 2x + 1$

   (c) $f(x) = \dfrac{x^3}{2} + \dfrac{x^2}{2} + \dfrac{1}{2}x$

Section 8.2

1. (a) $y = Ce^{x^3/3}$
   (b) $y\, dy = x\, dx$, then $y^2/2 = (x^2/2) + C_1$ or $y^2 = x^2 + C$, $y = \sqrt{x^2 + C}$ is a set of solutions. $y = -\sqrt{x^2 + C}$ is also a set of solutions.
   (c) $(y - y^2)\, dy = (x + x^2)\, dx$, then $(y^2/2) - (y^3/3) = (x^2/2) - (x^3/3) + C_1$, $3y^2 - 2y^3 = 3x^2 + 2x^3 + C$
   (d) $e^y\, dy = e^x/(1 + e^x)\, dx$, $e^y = \log (1 + e^x) + C$, $y = \log [\log (1 + e^x) + C]$
   (e) $y = 4 + Ce^{x^1/4}$

$$(A+P)^B$$

Section 8.3

1. (a) $y = e^{-x} + Ce^x$
   (c) $y = (x^3/2) + 3x^2 - (2x \log x) + Cx$
   (e) $y = \frac{1}{2}x^3 + Cx^3 \cdot e^{x^{1/2}}$
   (f) $y = \frac{1}{2} \log x + C/\log x$
   (g) $y = \frac{1}{2}x^2 - \frac{1}{2} + Ce^{-x^2}$

2. (a) $y = \frac{1}{2}x^2 - \frac{1}{2} + 2e^{1-x^2}$
   (c) $y = 2 + e^{1+(1/4)e - 1/4 e^{2x}}$
   (e) $y = \frac{1}{2}x^2 e^{-x^2} + e^{-x^2}$

3. $K = (b/a - c/a^2)(1 - e^{-ax}) + (C/a) x$

5. (a) $Y = y^2$, then $dY/dx = 2y \, (dy/dx)$. Then $2xy(dy/dx) + y^2 = x$ becomes $x \, (dY/dx) + Y = x$.
   (b) The general solution is $y = \pm \sqrt{x/2 + C/x}$.

Section 8.4

1. $N(t) = N_0 e^{Kt} = 2000 e^{Kt}$, $5000 = 2000 e^{K \cdot 1}$, $2.5 = e^K$. $N(t) = 2000(2.5)^t$,
   $N(5) = 2000(2.5)^5 = 195,672.5$.

3. $N(10) = 300$, $K = 0.06$, find $N(0)$. $N(10) = 300 = N_0 e^{.06 \cdot 10} = N_0 e^{0.6} = N_0(1.8221)$
   $N_0 = \dfrac{300}{1.8221} = \$164.65$.

5. $dy/dx = -a$, $y(x) = -ax + C$. $y(12) = 82$, then $82 = -12a + C$. $y(0) = 100$, then $100 = C$ and $a = \frac{3}{2}$. Then $y(x) = -\frac{3}{2}x + 100$. At 2 years, $x = 24$. $y(24) = -\frac{3}{2} \cdot 24 + 100 = 64$. After x months $y(x) = -\frac{3}{2}x + 100$. $y = 0$ when $x = 66\frac{2}{3}$, then 66 months is the maximum length of service of the stewardesses.

7. $N(t) = N_0 e^{Kt}$ where $N(t)$ is the amount in t years. $N_0 = \$2000$ and $K = 0.07$. Then $N(t) = 2000 e^{0.07t}$. In 25 years $N(25) = 2000 e^{0.07 \cdot 25} = (2000)(5.7618) = \$11,523.60$. Find t when $N(t) = \$20,000$. $20,000 = 2000 e^{0.07t}$, $10 = e^{0.07t}$, then $\log 10 = 0.07t$, and $2.3026 = 0.17t$, $t = 32.89$. In 33 years the amount will be $20,000.

9. $dN/dt = -KN$, $N = N_0 e^{-Kt}$, $\frac{1}{2}N_0 = N_0 e^{-Kt}$, $\frac{1}{2} = e^{-K \cdot 5600}$. $-\log 2 = -K \cdot 5600$, $K = \log 2/5600$, $N = N_0 e^{-t(\log 2/5600)} = N_0 e^{-t(0.000124)}$. $N(500) = N_0 e^{-500 \cdot 0.000124} = N_0 e^{-0.062} = N_0 \cdot 0.93$. 93% remains and 7% disintegrates. $N(50,000) = N_0 e^{-5000 \cdot 0.000124} = N_0 e^{-0.062} = N_0(0.5381)$. 54% remains.

11. $dN = -1.5 \cdot dM/M$. $N = -1.5 \log M + C$, then $100 = -1.5 \log 10,000 + C$. $C = 100 + 1.5 \log 10,000$. $N = -1.5 \log M + 100 + 1.5 \log 10,000$

14. Eighty-nine machines added at the end of the year.

Section 8.5

1. (a) $y = x^4/12 + x^5/20 + 3e^x + C_1 x + C_2$
   (b) $y = e^{-x^3/3} \int e^{x^3/3} \, dx$
   (c) $y = x \log x + 2x^{3/2} + xC$
   (d) $y = \frac{3}{2}x^4 + x^2 C$
   (e) $y = e^{\sqrt{2 \log xC}}$
   (f) $y = \dfrac{x^3 - 3x + C}{3(x-1)^4}$
   (g) $y = \dfrac{(Cx)^3 + 2(x+1)^3}{(x+1)^3 - (Cx)^3}$
   (h) $y = \frac{2}{3}ax - 1/x + C/\sqrt{x}$

3. The integrating factor is $y = -(x+1)(C - e^{-x^2})/2x$.

5. $dN/dt = -KN^2(t)$, $dN/[N^2(t)] = -K\,dt$ or $[N^{-1}(t)]/-1 = -Kt - C$, $-1/[N(t)] = -Kt - C$ or $N(t) = 1/(Kt + C)$. When $t = 0$, $N_0 = 1$. $1 = 1/C$, $C = 1$. When $t = 1$, $N(t) = \frac{1}{2}$, $\frac{1}{2} = 1/(K + C)$, $K = 1$. The solution is $N(t) = 1/(1 + t)$.

7. (a) Present value is $606.50.
    (b) It takes 13.9 years to double its value.

9. $N = 1/(1 + Ke^{-at})$

Chapter 9

Section 9.1

1. (a)

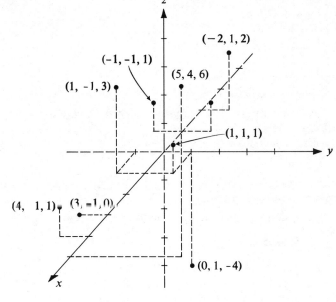

FIGURE 9.1–1 (a)

3. (a)

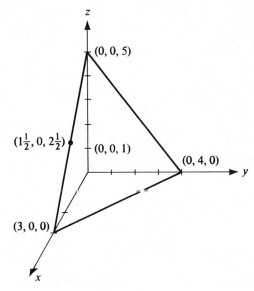

FIGURE 9.1–3 (a)

Section 9.2

1. No

2. The equation of a torus is: $z^2 + (\sqrt{x^2 + y^2} - c)^2 = a^2$.

5. (a) 3   (c) 27   (e) 27
   (b) 4   (d) 4   (f) 6

6. (a)

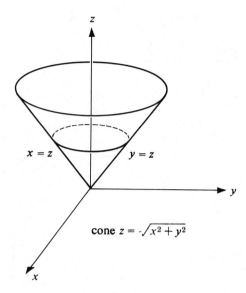

cone $z = \sqrt{x^2 + y^2}$

FIGURE 9.2–6 (a)

(c)

$z = (x^2 + y^2)^{1/4}$

FIGURE 9.2–6 (c)

(d)

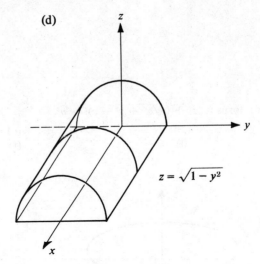

$z = \sqrt{1 - y^2}$

FIGURE 9.2–6 (d)

(f)

(j)

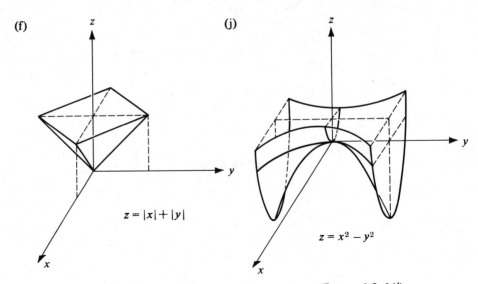

$z = |x| + |y|$

$z = x^2 - y^2$

FIGURE 9.2–6 (f)

FIGURE 9.2–6 (j)

Section 9.3

2. No

3. f is not continuous at $(0, 0)$.

4. $\lim_{x \to 0} f(x, 0) = 0$
    $\lim_{x \to 0} f(x, x) = \frac{1}{2}$
    We conclude that the limit doesn't exist.

1. $\dfrac{\partial z}{\partial x} = 1 + 2xy + 9x^2y. \quad \dfrac{\partial z}{\partial y} = x^2 + 6x^3y$

2. (a) $\dfrac{\partial z}{\partial x} = 2ye^{2xy}; \dfrac{\partial z}{\partial y} = 2xe^{2xy}$

   (b) $\dfrac{\partial z}{\partial x} = \log y; \dfrac{\partial z}{\partial y} = \dfrac{x}{y}$

   (c) $\dfrac{\partial z}{\partial x} = \dfrac{4}{y}; \dfrac{\partial z}{\partial y} = -\dfrac{4x}{y^2} + 2y$

   (e) $\dfrac{\partial z}{\partial x} = \dfrac{2yx}{x^2 + y}; \dfrac{\partial z}{\partial y} = \log(x^2 + y) + \dfrac{y}{x^2 + y}$

3. $\dfrac{\partial f}{\partial x} = \dfrac{1}{\sqrt{y^2 + 1}}.$ At $(1, 2),$ $\dfrac{\partial f}{\partial x} = \dfrac{1}{\sqrt{5}}.$

4. $\dfrac{\partial z}{\partial x}\Big|_{(x_0, y_0)} = A, \quad \dfrac{\partial z}{\partial x}\Big|_{(x_0, y_0)} = B$

5. $\dfrac{\partial^2 z}{\partial x^2} = 2y^5 + e^{x+y}; \dfrac{\partial^2 z}{\partial x^2}\Big|_{(1, 0)} = e. \quad \dfrac{\partial^2 z}{\partial y^2} = 20x^2y^3 + e^{x+y}; \dfrac{\partial^2 z}{\partial y^2}\Big|_{(-1, 1)} = 21$

7. $\dfrac{\partial^2 f}{\partial x^2} + \dfrac{\partial^2 f}{\partial y^2} = e^x \cos y - e^x \cos y = 0$

9. $\dfrac{\partial^2 f}{\partial x \, \partial y} = 2x, \dfrac{\partial^2 f}{\partial x^2} = 2y. \quad \therefore x\dfrac{\partial^2 f}{\partial x^2} - y\dfrac{\partial^2 f}{\partial x \, \partial y} = 2xy - 2xy = 0.$

10. $\dfrac{\partial z}{\partial x} = 2, \dfrac{\partial z}{\partial y} = 0.04y. \quad \dfrac{\partial z}{\partial x} = 2$ means that for every dollar the customer is willing to put down on the house the banker will loan \$2. $\dfrac{\partial z}{\partial y} = 0.04y$ means that for every dollar in the customer's salary the banker will loan an additional four cents.

12. (a) $\dfrac{\partial f}{\partial x} = -2; \dfrac{\partial f}{\partial y} = 1; \dfrac{\partial g}{\partial x} = 1; \dfrac{\partial g}{\partial y} = -1.$ f and g are competitive.

   (b) $\dfrac{\partial f}{\partial x} = -2; \dfrac{\partial f}{\partial y} = 1; \dfrac{\partial g}{\partial x} = -2; \dfrac{\partial g}{\partial y} = -3.$ f and g are neither competitive nor complementary.

13. (a) $\dfrac{\partial z}{\partial u} = 1, \dfrac{\partial z}{\partial v} = 1$ at $u = 1, v = 1$

   (b) $\dfrac{\partial z}{\partial u} = 0, \dfrac{\partial z}{\partial v} = 13$ at $u = 1$ and $v = \frac{1}{2}$

   (c) $\dfrac{\partial z}{\partial u} = \dfrac{\partial z}{\partial v} = 1$ at $u = 1 = v$

Section 9.5

1. (a) $\dfrac{\partial f}{\partial x} = 3x^2 + 1.$ $3x^2 + 1$ not 0. Hence $\dfrac{\partial f}{\partial x} \neq 0.$

   (c) $\dfrac{\partial f}{\partial x} = 2x + 4y + 1.$ $\dfrac{\partial f}{\partial y} = 2y + 4x.$ Solution $(\frac{1}{6}, -\frac{1}{3}).$

(e) $\dfrac{\partial f}{\partial x} = 2x + 24y$. $\dfrac{\partial f}{\partial y} = 16y + 24x$. Solution $(0, 0)$.

(g) $\dfrac{\partial f}{\partial x} = 3x^2 - 3y^2$. $\dfrac{\partial f}{\partial y} = -6xy + 3y^2$. Solution $(0, 0)$.

2. $\dfrac{\partial f}{\partial x} = \dfrac{\partial f}{\partial y} = 0$ at the origin; but f does not have a relative maximum or minimum there.

3. (a) $\dfrac{\partial f}{\partial x} = -2xe^{-(x^2+y^2)}$. $\dfrac{\partial f}{\partial y} = -2ye^{-(x^2+y^2)}$. Solution $(0, 0)$.

This point is an absolute maximum.

(c) $\dfrac{\partial f}{\partial x} = e^{-xy}(1 - xy - y^2)$. $\dfrac{\partial f}{\partial y} = e^{-xy}(1 - x^2 - xy)$. Solution

$$\left(\frac{\sqrt{2}}{2}, \frac{\sqrt{2}}{2}\right)\left(-\frac{\sqrt{2}}{2}, -\frac{\sqrt{2}}{2}\right).$$

These are both saddle points.

6. Saddle point

Section 9.6

1. The maximum occurs at $(\tfrac{1}{2}, \tfrac{1}{2})$. The minimum occurs at $(0, 0)$, $(1, 0)$, $(0, 1)$, $(1, 1)$.

(b) $\dfrac{\partial f}{\partial x} = \dfrac{\partial f}{\partial y} = 0$ at $(3, -2)$. $f(3, -2) = -22$ is minimum value.

(c) $\dfrac{\partial f}{\partial x} = \dfrac{\partial f}{\partial y} = 0$ at $(\tfrac{23}{17}, -\tfrac{13}{17})$. This is a saddle point.

(d) $f(0, 0)$ is a maximum on the interval. The minimum then occurs on the boundary.

(e) Maximum at $(3, 2)$, minimum at $(0, 5)$.

2. f has a maximum at $f(\sqrt{2}/2, 0)$.

3. There is no maximum, there is no minimum.

4. Minimum cost for $x = 2\sqrt[3]{10}$, $y = 3\sqrt[3]{10}$, $z = \sqrt[3]{10}$.

5. $x = 2\sqrt[3]{150}$, $y = 2\sqrt[3]{150}$, $z = \sqrt[3]{150}$

6. $y = 2$ and $x = \log 4000 - 1 \approx 7.3$

7. $x = 23$, $y = 41$, $u = 36$ gives minimum fuel costs.

Section 9.7

1. Maximum occurs at $x = \sqrt{3}/3$, $y = \sqrt{3}/3$, $z = \sqrt{3}/3$.

2. Maximum volume when $x = \tfrac{4}{3}\sqrt{15}$, $y = \tfrac{2}{3}\sqrt{15}$, $z = \tfrac{10}{3}\sqrt{15}$.

3. Minimum occurs at $x = 2\sqrt[3]{150} = y$, $z = \sqrt[3]{150}$.

5. Maximum profit when $x = 14$, $y = 4$, and $z = 5$.

Section 9.8

1. The regression line has the equation $y = x$.

2. Regression line has the equation $15y + 21x + 8 = 0$.

3. Regression line has the equation $47y = -57x + 102$.

Section 9.9

1. (a) $\dfrac{\partial z}{\partial x} = 6x^2y + ye^{zy}; \dfrac{\partial z}{\partial y} = 2x^3 + xe^{zy}$ (g) $\dfrac{\partial z}{\partial x} = 2y^2; \dfrac{\partial z}{\partial y} = 4xy$

   (d) $\dfrac{\partial z}{\partial x} = \dfrac{y^2}{(x+y)^2}; \dfrac{\partial z}{\partial x} = \dfrac{x^2}{(x+y)^2}$ (i) $\dfrac{\partial z}{\partial x} = e^z; \dfrac{\partial z}{\partial y} = \dfrac{1}{y}$

2. The absolute maximum is 1 at $(0, 0)$ and the absolute minimum occurs at points on the boundary $x^2 + y^2 = 1$. It is 0.

3. $z(\tfrac{1}{2}, \tfrac{2}{3}) = \tfrac{43}{108}$ is absolute maximum. The absolute minimum is at $(0, 0), (0, 1), (1, 0), (1, 1)$.

4. Absolute minimum at $(0, 0)$ is 1. Absolute maximum at $(0, 2)$ is 4.

5. Absolute minimum occurs at $(0, 0)$ and is 0. The absolute maximum is 2 and occurs at $(1, 1), (-1, -1), (1, -1)$, and $(-1, 1)$.

6. Minimum occurs at $(\tfrac{11}{7}, -\tfrac{1}{7}, \tfrac{12}{7})$, and is $\tfrac{4}{7}\sqrt{14}$.

7. An absolute maximum does occur when $x = 30, y = 70$.

8. The maximum volume is $20\sqrt{5}$.

9. Maximum occurs when $x = 10, y = 8$, and $z = 15$.

10. The equation of the regression line is $123y = 325x - 1070$.

Index

E F G H I J   9 8 7 6 5 4 3